Management of Tourism Ecosystem Services in a Post Pandemic Context

Tourism and ecosystem services are interdependent and face unique challenges. This book explores the challenges faced by destinations regarding the management and restoration of their ecosystem services.

Responding to the effects of the COVID-19 pandemic, this book offers unique management solutions based on best practices from Europe, America, Asia, Africa, Indonesia and island destinations. The management techniques and strategies proposed are adaptive in nature, and they are meant to protect and sustain natural and cultural ecosystem services utilized by the tourism industry.

Drawing from a rich collection of international case studies, the book adopts a user-friendly pedagogic approach, while seeking to be an essential future reference to scholars, researchers, academics and industry practitioners, destination management organizations and restoration agencies.

Vanessaa G.B. Gowreesunkar, PhD, is an Associate Professor at Anant National University, India.

Shem Wambugu Maingi, PhD, is a Lecturer within the Department of Hospitality and Tourism Management, Kenyatta University, Kenya.

Felix Lamech Mogambi Ming'ate, PhD, is a Senior Lecturer in the Department of Environmental Studies and Community Development, Kenyatta University, Kenya.

Routledge Insights in Tourism Series

This series provides a forum for cutting edge insights into the latest developments in tourism research. It offers high quality monographs and edited collections that develop tourism analysis at both theoretical and empirical levels.

Millennials, Spirituality and Tourism
Edited by Sandeep Kumar Walia and Aruditya Jasrotia

Tourism, Safety and COVID-19
Security, Digitization and Tourist Behaviour
Salvatore Monaco

COVID-19 and the Tourism Industry
Sustainability, Resilience and New Directions
Edited by Anukrati Sharma, Azizul Hassan and Priyakrushna Mohanty

Management of Tourism Ecosystem Services in a Post Pandemic Context
Global Perspectives
Edited by Vanessaa G.B. Gowreesunkar, Shem Wambugu Maingi, and Felix Lamech Mogambi Ming'ate

For more information about this series, please visit: www.routledge.com/Routledge-Insights-in-Tourism-Series/book-series/RITS

Management of Tourism Ecosystem Services in a Post Pandemic Context

Global Perspectives

Edited by
Vanessaa G.B. Gowreesunkar,
Shem Wambugu Maingi, and
Felix Lamech Mogambi Ming'ate

Routledge
Taylor & Francis Group
LONDON AND NEW YORK

First published 2023
by Routledge
4 Park Square, Milton Park, Abingdon, Oxon OX14 4RN

and by Routledge
605 Third Avenue, New York, NY 10158

Routledge is an imprint of the Taylor & Francis Group, an informa business

British Library Cataloguing-in-Publication Data
A catalogue record for this book is available from the British Library

Library of Congress Cataloging-in-Publication Data
A catalog record has been requested for this book

ISBN: 978-1-032-24808-0 (hbk)
ISBN: 978-1-032-24809-7 (pbk)
ISBN: 978-1-003-28021-7 (ebk)

DOI: 10.4324/b23145

Typeset in Times New Roman
by SPi Technologies India Pvt Ltd (Straive)

Contents

Figures

Tables

Foreword

Hugues Seraphin
The University of Winchester

Despite the volume and variety of research (books, book chapters, journal articles, etc) published every year in tourism and related fields of research, blind spots remain in many areas, such as the achievement of sustainability (Burrai, Buda & Stanford, 2019; Visser, 2015); how to make the industry more ethical (Lovelock & Lovelock, 2013); and the range and role of stakeholders involved in the industry (Sun, Rodriguez, Wu & Chuang, 2013; Todd et al., 2017). Like Samuel Johnston in Rasselas Prince of Abyssinia (Smith, 1996), it appears that so far, research is providing conclusions in which nothing is concluded. To unlock changes, a certain level of transparency in the industry is needed, alongside a greater engagement of stakeholders (Visser, 2015). The thorough investigation of the tourism ecosystem provided in that book is going towards unlocking the changes needed for future fitness of the industry. The investigation of the tourism ecosystem is to some extent an investigation of the tourism stakeholders and the interaction among them. Indeed, it is believed that one of the main reasons why some issues related to the industry, such as sustainability, has still not been achieved is due to a lack of clear, articulated, meaningful, and genuine collaboration among the partners (Burrai et al., 2019; Font & Sallows, 2002; McTiernan, Musgrave & Cooper, 2021; Todd, Leask & Ensor, 2017), which is stopping radical innovations, in other terms, innovations which are moving away from status-quo (incremental innovations), and disrupting established practices in the industry (Brooker & Joppe, 2014). This non-effective collaboration among stakeholders which is stopping innovation is shedding light on a pivotal construct which is not really explored in tourism and related industries, namely "trust." Indeed, trust is presented by McTiernan et al. (2021) as a transformative and regulatory tool which enables all stakeholders to work effectively together, towards the achievement of a common goal. The construct of trust in tourism is also calling the emerging and popular concept of transformative management research. Until the tourism industry operates radical changes in the way it operates, mistrust, conflicts, immobility, and unethical practices will remain predominant and will prevent positive changes(Hudson & Hudson, 2017; Lovelock & Lovelock, 2013; McTiernan et al., 2021; Séraphin, Sheeran & Pilato, 2018).

References

Brooker, E., & Joppe, M. (2014). Developing a tourism innovation typology: Leveraging liminal insights. *Journal of Travel Research*, 53(4), 500–508.

Burrai, E., Buda, D.M., & Stanford, D. (2019). Rethinking the ideology of responsible tourism. *Journal of Sustainable Tourism*, 27(7), 992–1007.

Font, X., & Sallows, M. (2002). Setting global sustainability standards: The Sustainable Tourism Stewardship Council. *Tourism Recreation Research*, 27(1), 21–31.

Hudson, S., & Hudson, L. (2017). *Marketing for Tourism, Hospitality and Events*. London: Sage.

Lovelock, B., & Lovelock, M. (2013). *The Ethics of Tourism*. London: Routledge.

McTiernan, C., Musgrave, J., & Cooper, C. (2021). Conceptualising trust as a mediator of pro-environment tacti knowledge transfer in small and medium sized tourism enterprises. *Journal of Sustainable Tourism*. DOI: 10.1080/09666958.2021.1942479.

Séraphin, H., Sheeran, P., & Pilato, M. (2018). Over-tourism and the fall of Venice as a destination. *Journal of Destination Marketing & Management*, 9, 374–376.

Smith, D.H. (1996). Patterns in Samuel Johnson's Rasselas. *Studies in English Literature, 1500-1900*, 36(3), 623–639.

Sun, Y.Y., Rodriguez, A., Wu, J. H., & Chuang, S.T. (2013). Why hotel rooms were not full during a hallmark sporting event: The 2009 World Games experience. *Tourism Management*, 36, 469–479.

Todd, L., Leask, A., & Ensor, J. (2017). Understanding primary stakeholders' multiple roles in hallmark event. *Tourism Management*, 59, 494–509.

Visser, W. (2015). *Sustainable Frontiers. Unlocking Change through Business, Leadership and Innovation*. Sheffield: Greenleaf Publishing.

Preface

Tourism and ecosystem services (ES) are two sides of the same coin, as the former depends on the latter. An enormous part of tourism activities globally are produced and consumed using natural and cultural ecosystem services. As a result, destinations imperatively need the support of ES in order to run their tourism businesses. On another note, it is well documented that tourism is the driving force behind many changes in society; tourism provides several benefits but at the same time, the industry makes extensive use (if not overuse) of the natural ecosystem. This is reflected in mass tourism and development-driven land-use conversion, which have had irreparable damage to the physical environment and community livelihoods. Likewise, tourism, as co-created experience with locals, inevitably involves the consumption of cultural resources (traditional food, local products and services, indigenous experiences, among others). Expressed in these terms, tourism's dependence on both natural and cultural ES is obvious. Hence, there is a need to understand how ES are converted into tourism resources and how they are managed to achieve socio-economic sustainability. It is crucial to examine the interplay between tourism and ES, not only as natural and cultural ES but also as an economic component contributing to the benefit of the society. Drawing from a rich collection of worldwide case studies and research notes, the book *Management of Tourism Ecosystem Services Post Pandemic: A Global Perspective* enlightens readers on issues and challenges faced by destinations regarding the management and restoration of their ecosystem services. At a time when the world is still striving to develop references to cope with the pandemic, the book proposes unique management solutions based on best practices from Europe, America, Asia, Africa, Indonesia and island destinations. The management techniques and strategies proposed are adaptive in nature, and they are meant to protect and sustain natural and cultural ecosystem services utilized by the tourism industry. The book adopts a user-friendly pedagogic approach, while seeking to be an essential future reference to scholars, researchers, academics and industry practitioners, destination management organizations and restoration agencies.

Prof. Vanessaa G.B. Gowreesunkar
Dr. Shem Wambugu Maingi
Dr. Felix Lamech Mogambi Ming'ate

Acknowledgements

The book *Management of Tourism Ecosystem Services in a Post Pandemic Context: Global Perspectives* would not have seen the light of day without the support and effort of some special people. We wish to offer our gratitude to the contributors, the reviewers, the Routledge team and all those who indirectly accompanied us during this wonderful journey. We owe a depth of gratitude to the TRINET academic community, which responded to our call for chapters overwhelmingly and with great interest. We gratefully acknowledge the endorsements of the following eminent personalities of tourism:

- Prof. Anna Farmaki, Cyprus University of Technology
- Prof. Marina Novelli, University of Brighton
- Prof. Dogan Gursoy, Washington State University
- Prof. John S. Akama, Kisii University, Kenya
- Prof Robin Nunkoo, University of Mauritius

It is a matter of pride to have the blessing of the above professors, who are models and references to the tourism community worldwide. We also wish to acknowledge the Anant National University (India) and the Kenyatta University (Kenya) for supporting us throughout this exceptional journal. Last but not the least, we wish to acknowledge the unstinting support of our families, and the trust of our contributors. In adversity, we found opportunity and altogether as a terrific team, we are glad to offer readers a memorable gift produced during the reign of COVID-19, Delta and Omicron.

Vanessa, Shem, Felix
Book Editors

Editors

Vanessaa G.B. Gowreesunkar, PhD, is an Associate Professor at Anant National University, India; Deputy Chair for the Tourism, Infrastructure and Energy cluster of the African Union Economic and Social Council; and National Coordinator for Women Advancement for Economic and Leadership Empowerment (WAELE), Mauritius.

Shem Wambugu Maingi, PhD, is a Lecturer within the Department of Hospitality and Tourism Management, Kenyatta University. He is an African researcher on sustainable tourism development in Africa and currently has published a book on global tourism destination management post pandemics.

Felix Lamech Mogambi Ming'ate, PhD, is a Senior Lecturer in the Department of Environmental Studies and Community Development, Kenyatta University. He is an African researcher specialized in natural resources co-management and sustainable livelihoods research in African context.

Contributors

Jim Ayorekire, PhD, is a Senior Lecturer, researcher and consultant on tourism policy and planning. He is a recent Fulbright scholar with strong interest in promoting regional and international collaboration in tourism research and capacity development.

Kindieneh Awoke is a tourism and hospitality educationist from Ethiopia. He has done Master's in Tourism and Heritage Management from the University of Gondar and Bachelor of Arts in Tourism from Jimma University, Ethiopia. He is associated with the Department of Tourism Management, Assosa University, since 2017. His primary research interests are sustainable tourism, tourism events, indigenous tourism, tourism marketing and heritage management.

Janis Bikse is a PhD candidate and Lecturer at Vidzeme University of Applied Sciences, Valmicra and an Assistant Professor at Maritime Academy, Latvia. He is a member of several boards and organizations, such as Systems Terminal, Systems Logistics, and Systems Recycling and VEF Board. His area of research includes IT, tourism and transport, geo-tourism and spring tourism. He is the corresponding author of this article.

Jacek Borzyszkowski, PhD, is an Associate Professor at the WSB University in Gdańsk. He is the author of nearly 100 scientific publications on contemporary tourism, mainly in tourism policy, tourism economy, and tourism organization and management (his major focus is the operation of destination management organizations in European countries). He is also a member of the Polish Tourism Organization Council and vice-president of the board of the Municipal Tourism Organization in Sianów, Poland. Jacek Borzyszkowski is the coordinator of the "management and quality science" discipline in the Scientific Federation of WSB-DSW Universities. He is the author of numerous expert opinions, and planning and strategic documents on tourism function commissioned by tourist organizations and local self-governments.

Lidia Caballero-Galeote is a PhD candidate in the Faculty of Tourism and Economy at the University of Málaga. Among her recent papers indexed in the *Journal Citation Reports* are "Stakeholders' Perception on the Impacts of Tourism on Mass Destinations: The Case of Seville" (2021) and "Qualitative Impact Analysis of International Tourists and Residents' Perceptions of Málaga-Costa Del Sol Airport" (2020). Her fields of research are air transport, tourism, new technologies and sustainable development.

Crispin Dale, PhD, is a Senior Lecturer in the Wolverhampton Business School at the University of Wolverhampton. Crispin lectures in tourism management and marketing and has published widely in books and peer-reviewed journals. His research interests focus on destination management and strategy.

Iveta Druva-Druvaskalne, PhD, is a Lecturer and Head of Spatial Research Laboratory of the Institute of Social, Economic and Humanities Research of Vidzeme University of Applied Sciences, Latvia.

Lance Dubos, PhD, received his JD from Tulane University and practiced law for nine years in the United States. He is currently an adjunct lecturer at James Cook University Singapore and Embry-Riddle Aeronautical University Asia in business law.

Matilde Encabo, PhD, Magister Scientiae, National Researcher Category I. She is a Retired Associate Professor. She has expertise in recreation and tourism in natural environments from a biodiversity conservation perspective, with tools of recreation ecology. Recreation and Tourism in Conservation Group – Faculty of Tourism – National University of Comahue. Buenos Aires 1400, Neuquén Capital, Argentina. Her academic and research topics include management and planning of tourism and recreation in natural areas, particularly in protected natural areas.

Mohammad Osman Gani works as an Assistant Professor in the Department of Marketing at Bangladesh University of Professionals (BUP). He has completed his MSc in Development Science from the Graduate School for International Development and Cooperation, Hiroshima University. He holds a BBA and an MBA from the Department of Marketing, University of Dhaka. His research interests are tourism and hospitality management, consumer behaviour, anthropology, entrepreneurship, and international business. He has published a couple of articles in different peer-reviewed journals.

Subhajit Das, PhD, is currently working as an Assistant Professor at the Geography department of Presidency University, Kolkata, India. He did his PhD in Geography from the University of Kalyani, West Bengal, India, with specialization in Geography of Tourism. Some of the courses in the UG and PG levels that he teaches in his department are introduced by him as the exclusive modules on tourism geography. His major areas of research are related to the issues of tourism in ruralscape. Presently, he is also focusing on the several issues pertaining to the walking trails at different kinds of tourists' destinations and the resultant touristscapes.

Anisur R. Faroque, PhD (International Business and Entrepreneurship), is the post-doctoral researcher (Assistant Professor) at the School of Business and Management at LUT University, Finland. His research interests are in the marketing, international business and entrepreneurship domains and include opportunity recognition, market orientation, entrepreneurial orientation, internationalization, networks and cognitive heuristics and biases in decision making. His research has been published in journals such as *International Business Review, Journal of Marketing Theory and Practice, Journal of Business & Industrial Marketing, Asia Pacific Journal of Marketing & Logistics, International Journal of Entrepreneurship and Small Business, International Review of Entrepreneurship, Anatolia* and *International Journal of Emerging Markets*, among others.

Mahender Reddy Gavinolla, PhD, is a Guest Lecturer at Vidzeme University of Applied Sciences, Valmiera, Latvia, and an Assistant Professor at the National Institute of Tourism and Hospitality Management. His research and publications area includes responsible and sustainable tourism, and he published several papers in Sage, Routledge and Emerald.

Afera Gebremedihn is associated with the Department of Tourism and Hotel Management, Debark University, since 2018. Apart from teaching and research, Mr Afera is a professional and certified tour guide of Ethiopia. He primarily organizes and escorts tour in and around Simien Mountain National Park. He has been working as a tour guide since 2012. His research interests concern ecotourism, wildlife tourism, mountain tourism, sustainable tourism, tour guiding and destination development.

Francis Gitagia, PhD is a Lecturer in the Department of Accounting and Finance, Kenyatta University. He holds a PhD in Finance from Kenyatta University. His area of interest is general corporate finance.

Raúl González, PhD, is Principal Researcher of the National Scientific and Technical Research Council of Argentina (CONICET) and Professor in the School of Marine Sciences of the Comahue National University. He is a member of the Advisory Commission on Biodiversity and Sustainability (CAByS) of the Argentine Ministry of Science and Technology. His research topics are focused on the assessment of the abundance and spatial distribution and patterns of marine mammals, and the impact of fisheries and tourism on whales and dolphins in Patagonia.

Vikas Gupta holds a PhD in Hospitality. He has rich and extensive experience of teaching for more than 13 years in both India and abroad with renowned names such as Café Coffee Day, Fiji National University, Amity University and various central and state IHMs in India. He is presently working with Amity University, Noida, Uttar Pradesh, as an Assistant Professor in the field of hospitality. He has widely published in national and international journals (including Emerald SCOPUS indexed journals) such as *International Journal of Contemporary Hospitality Management, British Food Journal, Journal of Culinary Science & Technology, Tourism Review, Worldwide Hospitality and Tourism Themes, Journal of Wine Research* and *International Journal of Tourism Cities.*

Zeynep Gulen Hashmi, PhD, is an award-winning scholar-practitioner with more than 26 years of experience in the hospitality industry. She is a holder of the Socrates "Best Manager of the Year 2016" award from the United Kingdom. Her doctoral specialization lies in tourism and hotel sustainability strategies, change management and action research. Her publications are on systemic collaborative challenges of our time, such as collaboration for tourism and sustainability, societal and corporate well-being and sustainable luxury management. Currently, she is teaching tourism destination management and sustainability in hospitality courses at NUST University, Islamabad.

Unathi Sonwabile Henama, PhD, is a Lecturer in Tourism Marketing at the Department of Tourism Management at the Tshwane University of Technology. He has a PhD from Mid Sweden University and a Master's from the University of the Free State. He has written numerous articles and presented research papers at both local and international conferences. He is the leading tourism commentator in South Africa, and his views are highly sought after by TV, print and online news outlets.

Michael Johansson, is a PhD holder and a researcher in Environmental Strategy at the Department of Service Management and Service Studies, Lund University, Sweden. His research interest mainly focuses on sustainable urban and regional planning. The research highlights the value of ecosystem services, shared mobility, political sustainable strategies and urban development. Most of his research is in cooperation with municipalities and organizations.

Bitok Kipkosgei, PhD, has seven years of teaching and postgraduate supervision. He has made contributions to various refereed journals in tourism and is now on his initial journey in contribution to book chapters. In addition, he is a reviewer for *Asian Journal of Education and Social Studies* and has also participated in the review of the national tourism policy in Kenya. His research interests include sustainable tourism, ecotourism, policy and planning, and human resource development in tourism.

Fredrick Nyongesa Kassilly, PhD, is a Biodiversity Fellow of the United Nations University, a North-South Dialogue Scholar and a Professor of Conservation Biology. He holds post-doctorate training in Biodiversity Conservation from the United Nations University, Tokyo (2003); an MSc degree in Applied Leadership and Management from the Management University of Africa, (2015); a doctorate degree (Dr Nat.Techn) in

Conservation Biology (2001) and a post-doctorate training certificate in Conservation Hunting (2008) from the University of Natural Resources and Life Sciences, Vienna; and an MSc degree in Ethology (1993) and a BSc degree in Range Management (1988), from the University of Nairobi.

Munaza Kazmi holds a MPhil in Management Sciences (Bahria University Islamabad, 2020). She is a travel writer, an author and a co-author of scientific contributions in national and international publications. Her main areas of research include tourism and quality management.

Isaac Kimunio, PhD, is a Lecturer in the Department of Applied Economics, Kenyatta University. He holds a PhD in Economics, specializing in monetary economics in African context.

Aman Kumar, PhD and UGC NET qualified, is an Assistant Professor working at Chandigarh college of Hospitality, Landran, Mohali (Pb.), and has 17 years of teaching experience. He has presented 13 research papers at conferences, which have also been published. He has attended many workshops and FDPs and has received certificates. Aman Kumar is a life-time member of ITHC (Indian Tourism and Hospitality Congress). He has received the best teacher award in the year 2011, the award of excellence in the year 2019 and the award of appreciation in the year 2019 in CGC.

Agita Livina is a Professor in the Faculty of Social Sciences at the University of Applied Sciences (VUAS), and Lead Researcher, Director of the Institute of Social, Economic and Humanities Research of VUAS, Latvia. She is a chairperson of UNESCO chair on Biosphere and Man in the university as well as chair of the University Eco Council. She is an expert of Latvian Council of Science in the fields of economics and management. Her research interests are tourism planning, including heritage and protected areas, and talent attraction for small and medium urban areas.

Adrian Lubowiecki-Vikuk, PhD, has been teaching, researching and consulting the leisure of various communities for over ten years. His scientific interests include consumer behaviour in the areas of sport, tourism and health. An important research subject he has focused on is medical tourism management.

Apleni Lwazi teaches tourism at the University of Zululand, KwaZulu Natal, in the Republic of South Africa. He holds a master's degree in Management Sciences from the University of Science and Technology Beijing (China) obtained in 2015. In his Master's journey, he focused on "The Implications of Carbon Tax in South Africa's Tourism Industry." Lwazi is the recipient of the Chinese Scholarship Council (2013–2015) scholarship. As part of his career growth and development, he is currently doing his PhD (specializing in Local Economic Development) at the University of Johannesburg, having registered for the programme in June 2020. His research interests include sustainable development, tourism development and tourism management.

Mirosław Marczak, PhD, is an author, with over 80 scientific publications. His main areas of scientific interest are management and marketing in tourism, especially implemented at the national level (by national tourism organizations) and the local level (tourism management implemented by the units of territorial self-government); functioning of DMOs (Destination Management Organizations) at the national and regional levels, first of all, national tourism organizations (NTOs) as the main entities responsible for promotion of a particular country as an attractive tourism destination; tourist destination brand management; creating the image of a tourist destination; promotion campaigns implemented by NTOs; organizational structures of DMOs; systems of tourism organization and promotion operated in the world; branding in the activities of national tourism organizations; and management of brand tourism products in the destination area.

Bantu Mbeko Msengi, PhD, teaches entrepreneurship at the Durban University of Technology in the province of KwaZulu Natal in South Africa. He previously taught tourism development and policy development at the Cape Peninsula University of Technology. He holds a Master's in Business Leadership from the University of South Africa (UNISA) Business School of Leadership. The focus of the research for his dissertation was on corporate governance. He is currently enrolled for a DBA (specializing in corporate governance) at the University of KwaZulu Natal. His research interests include corporate governance, sustainable development, local economic development and tourism development.

Anda Mežgaile is a research assistant and PhD student at Vidzeme University of Applied Sciences. She is Deputy Head of UNESCO chair on Biosphere and Man in the university. Her particular research interests are biosphere reserve management issues and tourism development. She has a rich international experience as she had represented NVBR reserve in 2019 MAB Youth Forum in China and UN Youth Climate Summit 2019, United States, and continues working in EuroMAB Youth network.

Kidanu Melese is a Lecturer at the Department of Tourism Management, Assosa University, Ethiopia.

Madiseng Messiah Phori, PhD, is a tourism lecturer within the Department of Tourism Management at Tshwane University of Technology, South Africa. He holds a Master of Management Sciences, specializing in Hospitality and Tourism from Durban University of Technology in KwaZulu Natal, South Africa. He is currently in the final year completing doctoral study in Tourism from Tshwane University of Technology. His research focus includes community-based tourism, sustainable tourism development, heritage and cultural tourism, rural tourism, as well as safety and security in tourism.

Francis Mugizi, PhD, is a Lecturer and researcher in tourism development and community-based tourism. He has strong networks with tourism actors and local community groups in Uganda's protected areas involved in community-based tourism enterprises.

Joseph K. Muriithi, PhD, is a Senior Lecturer based at the Department of Environmental Studies and Community Development, Kenyatta University, Kenya. His area of teaching and research specialization focuses on community conservation issues, with a special interest in role of tourism in conservation and development. He has taught, researched and published on ecotourism practices and certification, natural resource governance, and urban greening in Kenya. His current research pursuits are in the area of green entrepreneurship with a special focus on urban tree growing and trade and its role in enhancing sustainable livelihoods among urban poor.

Au Yong Hui Nee holds a PhD from USM, Master in Economics from University of Tsukuba and Bachelor in Resource Economics from UPM. She is an Associate Professor at the Faculty of Business and Finance, Universiti Tunku Abdul Rahman. She has two decades of professional experience managing supply chain and compliance risk management at Fortune 500 multinational corporations and a state agency. She is a chief editor of World Scientific Publishing's Belt and Road Initiative book. She has received over USD 150,000 of research grants. She is a recipient of Monbukagakusho Scholarship.

Jan Henrik Nilsson, PhD, is a reader in human geography and associate professor in economic geography at the Department of Service Management and Service Studies, Lund University, Sweden. His research interest lies mainly in the field of tourism geography, in recent years mainly with an urban focus. His publications cover areas like the economic geography of tourism, urban tourism and destination development, sustainable urban tourism, political geography, aviation policy, and tourism and hotel history. Most of his research takes place in different parts of the Baltic Sea Area, such as Sweden, Denmark, Estonia and Germany.

Joseph Obua, PhD, has extensive experience in research, supervision and capacity building in ecotourism and environmental forestry. His research interests include analysis of local community participation in conservation and sustainable tourism.

Ade Oriade, Research Assistant, is the Deputy Head of UNESCO chair Biosphere and Man. He is a PhD student at the Vidzeme University of Applied Sciences, Latvia.

Michael Pompeia, PhD, is presently the Head of Business Support Services at SME Mauritius Ltd, the National Agency for the Promotion of Entrepreneurship in Mauritius. Formerly, he was Lecturer at the Faculty of Management, Curtin Mauritius. Dr Pompeia has proven track record in marketing strategies for various well-known brands and MSMEs within the industry. Dr Pompeia's resume showcases his experience across well-known organizations and in different sectors of activities in Mauritius. He is passionate about unlocking future growth potential and has experience in both developing and developed markets, mainly in the Indian Ocean.

Elena Cruz-Ruiz, PhD, is Professor of the Department of Economics and Business Administration in the Marketing and Market Research Area at the University of Malaga. The main lines of research are related to tourism marketing, where the contributions are related to wine tourism, gastronomy and cruise tourism. She has published books, book chapters and articles indexed in the *Journal Citation Reports* related to the territory brand, consumer perception, wine routes and sustainable development in the field of tourism, among which stands out *Análisis de las tipologías de cruceristas: Imagen, satisfacción y lealtad a la ciudad de Málaga* (2015) published by the Costa del Sol Tourist Board and in collaboration with other authors; *El progreso del turismo de cruceros en España: de las élites a las masas* (2016), in the cruising field. Some of the most recent publications are *Sustainable Tourism and Residents' Perception towards the Brand: The Case of Malaga* (Spain) (2019) and *Transmission of Place Branding Values through Experiential Events: Wine BC Case Study* (2021). Likewise, her work as a reviewer in top-level indexed journals actively stands out.

Neil Robinson, PhD, lectures at Salford Business School teaching at both the undergraduate and postgraduate levels. Neil has research interests in dark tourism and destination management and has published widely in tourism journals.

Prakash Chandra Rout, PhD, is a tourism professional with seven years of experience. Presently, he works as an Assistant Professor at Assosa University, Ethiopia. Prior to his current assignment, he was associated with the International Centre for Integrated Mountain Development (ICIMOD), Nepal, G.B. Pant National Institute of Himalayan Environment, India and H.N.B. Garhwal (A Central) University, India. He has extensively worked in the Himalayan landscape for more than five years. His research interests are community-based ecotourism, value chain approaches, mountain tourism, tourism innovation, indigenous tourism and neuromarketing.

Hiran Roy, PhD, is a Lecturer of International School of Hospitality, Sports, and Tourism Management at Fairleigh Dickinson University, Vancouver, Canada. He holds a PhD in Management from the University of Canterbury, Christchurch, New Zealand. His research interests include sustainable tourism, sustainable local food systems, local food marketing, city branding through food, hospitality luxury branding and sustainability. Hiran's work has been published in a variety of book chapters such as Routledge, Elsevier, and Emerald and high impact prestigious leading academic journals.

Guadalupe Sarti is a graduate in Tourism and a Doctoral fellow in Geography at the National Scientific and Technical Research Council (CONICET) and the Félix de Azara Natural History Foundation. Her areas of research interest is whale watching management and its impacts on whales, on visitors and on natural protected spaces.

Abdelhak Senadjki, PhD, is an Associate Professor of Economics at the Faculty of Business and Finance, Universiti Tunku Abdul Rahman, Malaysia. He obtained his PhD and Master's of Economic Management from Universiti Sains Malaysia and a bachelor's degree of Economics from the University of Algiers. He was awarded the USM fellowship. His research interests include energy economics and economic development. He has been involved in various research grants. He has published widely in various local and international refereed journals, ISI, Scopus, chapters in books and research papers. He is a reviewer for a number of refereed journals.

I Nengah Subadra obtained PhD in Tourism at the University of Lincoln, United Kingdom. Currently, he is the Dean of the Faculty of Tourism, University of Triatma Mulya, Bali-Indonesia. His research interests include cultural tourism, heritage tourism, community-based tourism and disaster mitigation in tourism.

Alelign Takele works as a Lecturer at the Department of Tourism Management, Assosa University, Ethiopia. He holds both master's and bachelor's degrees in Tourism and Heritage Management from the University of Gondar. His research interests include indigenous tourism, sociological tourism, heritage management and tourism marketing. Mr Alelign is an avid reader and philanthropist who too devotes his time for community service and voluntary work.

Yip Chee Yin, PhD, is an Assistant Professor of Economics at the Faculty of Business and Finance, Universiti Tunku Abdul Rahman, Malaysia. He obtained his PhD in Applied Statistics from Universiti Sains Malaysia, Master in Statistics from Universiti Sains Malaysia, and Bachelor of Science in Physics from Universiti Malaya and Bachelor of Science in Mathematics from the University of London. His research interest is economics. He has been involved in various research grants. He has published widely in various local and international refereed journals, Web of Science, Scopus and research papers. He is a reviewer for a number of refereed journals.

Abbreviations and Acronyms

AGIC	Arizona Geographic Information Council
ALRIS	Arizona Land Resources Information System
ASALs	Arid and Semi-arid Lands
AWR	Ajai Wildlife Reserve
BR	Biosphere Reserves
BSAMPA	Bahia de San Antonio Marine Protected Area
BTB	Bangladesh Tourism Board
CBC	Community-based Conservation
CBE	Community-based Enterprise
CBNRM	Community-based Natural Resource Management
CBT	Community-based Tourism
CES	Cultural Ecosystem Services
CICES	Common International Classification of Ecosystem Services
CSO	Central Statistical Office
CSV	Creating Shared Value
CVM	Contingent Valuation Method
DMO	Destination Management Organization
ES	Ecosystems Services
FAO	Food and Agriculture Organization of the United Nations
GCF	Global Conservation Fund
GDP	Gross Domestic Product
GoK	Government of Kenya
GSTC	Global Sustainable Tourism Council
GWP	Gross World Product
HTD	Health Tourism Destination
IBA	Important Bird Areas
ICB	Instituto de Conservacion de Ballenas
IPCC	Intergovernmental Panel on Climate Change
ISH	Islamabad Serena Hotel
IUCN	International Union for Conservation of Nature
KFS	Kenya Forest Services
KII	Key Informant Interviews
KWS	Kenya Wildlife Services
LAC	Limits of Acceptable Change

MAB	Man and Biosphere Programme
MEA	Millennium Ecosystems Assessment
MFCA	Murchison Falls Conservation Areas
MFNP	Murchison Falls National Park
MICE	Meetings Incentives Conventions and Exhibitions
MMWCA	Maasai Mara Wildlife Conservancies Association
MPA	Marine Protected Area
MSME	Micro, Small and Medium Enterprises
NACOSTI	National Commission for Science, Technology and Innovation
NAVTTC	National Vocational and Technical Training Commission
NGO	Non-Governmental Organization
NPO	National Productivity Organization
NRT	Northern Rangelands Trust
NVBR	North Vidzeme Biosphere Reserve
OECD	Organization for Economic Co-operation and Development
OECS	Organization of Eastern Caribbean States
PES	Payment for Ecosystems Services
PPP	Public Private Partnerships
PTO	Polish Tourism Organization
RaTiC	Recreation and Tourism in Conservation
RTO	Regional Tourism Organization
SADC	Southern African Development Community
SDGS	Sustainable Development Goals
SEED	Social and Environmental Education Development
SES	Spring Ecosystems Services
SIDS	Small Island Development States
SIT	Special Interest Tourism
SJHTS	Subak Jatiluwih Heritage Tourism Site
SMNP	Simien Mountains National Park
SRW	Southern Right Whale
SSI	Semi-Structured Interviews
TBL	Triple Bottom Line
UK	United Kingdom
UN	United Nations
UNCTAD	United Nations Conference on Trade and Development
UNDP	United Nations Development Programme
UNEP	United Nations Environmental Programme
UNESCO	United Nations Educational, Scientific and Cultural Organization
VFR	Visiting Friends and Tourism
WBPCB	West Bengal Pollution Control Board
WCS	Wildlife Conservation Society
WFH	Working from home
WHO	World Health Organization
WHS	World Heritage Sites
WTC	Willingness To-accept Compensation

WTO	World Tourism Organization
WTP	Willingness To Pay
WTTC	World Travel and Tourism Council
WW	Whale watching
WWF	World Wildlife Fund
ZIMCODD	Zimbabwe Coalition on Debt and Development

Part I

Introduction

1 Introduction

Status of the Tourism Ecosystem Services: Marking More Realities

Vanessaa G.B. Gowreesunkar, Shem Wambugu Maingi, and Felix Lamech Mogambi Ming'ate

As the world continues to face the cold blows of the COVID-19 pandemic, new realities seem to be marking the tourism ecosystem services, and in the most unexpected and unpredictable manner. Two years into the pandemic, countries are still struggling with the fifth and sixth waves with all types of travel restrictions so much so that the global tourism industry looks like going into a third year of uncertainty. As the different mutations of the Coronavirus continue travelling around the world, the unavailability of vaccines in poor economies and the protests of the "antivaxxers" in significant parts of industrialized countries continues to affect national economies, businesses, health services and social life. Destinations that were just finding their footing after nearly two years of devastation are facing new uncertainties; for instance, the collective travel bans by some countries seem to be the new norm while others are marketing themselves as COVID-free destinations. Likewise, unaligned restrictions in travel requirements are also destabilising tourism businesses. For example, in Austria, face masks are not mandatory in essential shops and in public transport whereas in Vienna, masks remain compulsory in public transport and in pharmacies (BBC News, 2021). The pandemic followed different trajectories globally, with double and multiple waves in India and Spain respectively as well as inconclusive vaccination effects in France and Germany. Likewise, the prolonged downturn in the travel and tourism sector has also made islands that rely heavily on foreign tourism very concerned about their finances. For instance, the island of Mauritius closed its border to India overnight and to many more countries, thus creating more cancellations; travellers who were in the middle of their trip were urged to meet new travel requirements while others simply travelled back to their respective countries. The COVID-19 mutations are affecting tourism businesses in such an unpredictable manner that it becomes challenging to make assumptions about travel behaviour, let alone the tourism ecosystem services. The words of Tony Wheeler, co-founder of Lonely Planet, illustrate the situation:

DOI: 10.4324/b23145-2

> In the travel game, it's tough even to understand what's going on in the present. Some countries (Australia) won't let people out, other countries (America) won't let people in, even when they're coming from a place with a better virus story. Or you can leave (the UK) and go somewhere else (the list changes daily) only to find (typically at 4 a.m.) all sorts of restrictions on your return.
>
> (BBC News, 2021)

As travel restarts in some parts of the world, a large number of questions are being raised about its impacts on the tourism ecosystem services in the near and long term (Business Standard, 2021). For instance, the ongoing pandemic context is causing further issues such as inequality (the collective ban on South Africa), issues on freedom of travel (closure of borders by a number of countries), including additional impacts on the natural and cultural ecosystem services. While on the one hand, tourism is being praised for giving incentives to preserve heritage and cultural sites (see İsmail Yağcı and Avni Can Yağcı, 2021; Gowreesunkar et al., 2021), on the other hand, due to lockdown and travel restrictions, the cultural ecosystem services are still being affected (Wahome and Gathungu, 2021; Varriale et al., 2021). According to Al-Said (2020), world heritage sites in Southeast Asia saw a decline in visitors of up to 99% in April 2020. These sites represented vital sources of employment, not only to the local population but also to cultural organizations, institutions, associations, archaeologists, and artisans. Likewise, in Afghanistan, restoration work on the Topdara Stupa (dating since early AD) was put to a halt due to the pandemic (Al-Said, 2020). Similarly, an extensive rehabilitation project on the Tomb of Askia in Mali was put on hold. The tomb, a unique pyramidal structure that contains two mosques and a cemetery, is representative of the monumental mud-building traditions of the West African Sahel (Al-Said, 2020).

Regarding the natural ecosystem services, favourable remarks have been shared by several researchers. For instance, tourist places (like Venice, Amsterdam and Malta) suffering from overtourism, were found to be restored to normalcy due to the absence of visitors (Seraphin et al., 2018; Maingi, 2019; Gowreesunkar and Reddy, 2020). Likewise, the pandemic situation significantly improved air quality in different cities across the world, reduced GHGs emissions, lessened water pollution and noise, and reduced pressure on several tourist destinations (Rume and Ul Islam, 2020; Gowreesunkar et al., 2021). The other side of the coin shows that natural ecosystem services have also been affected in various ways: For instance, negative environmental consequences such as increase in medical waste; haphazard use and disposal of disinfectants, masks, and gloves; and burden of untreated wastes were observed (Rume and Ul Islam, 2020; Lama and Rai, 2021). As poverty increases, the subsistence strategies of local communities that depend directly on surrounding resources for their survival lead to increased illicit exploitation of natural resources, such as higher consumption of firewood, food, traditional medicine ingredients, and self-employment materials. According to Rajao, et al. (2020), 2% of agricultural estates are responsible for 62% of all potentially illegal deforestation in Latin America. The impact of the ongoing pandemic on tourism has cut the sector's contribution to GDP and employment, and, in some cases, hampered the maintenance of national parks (ECLAC, 2020). The natural heritage in Latin America and the Caribbean is very important for tourism, which has declined sharply as a result of lockdown and prevention measures and border closures. A comparison of international tourist arrival statistics between May 2019 and May 2020 shows a 51-percentage-point drop in Chile and a 73-percentage-point decline in Mexico (Rajao, 2020).

Based on the above-documented evidence, a mixed picture of the status of the natural and cultural ecosystem services is captured; on the one hand, resources (food, safe water, biodiversity, fire, and electricity) are vital for controlling the crisis; on the other, they are affected by the consequences of the crisis (the use of fuels, wood, and minerals). According to the United Nations World Tourism Organisation (UNWTO), it is still unclear how the pandemic will further affect the industry, and another year of loss of USD 910 billion to USD 1.2 trillion in international tourism business with 100–120 million direct tourism jobs at risk globally is forecasted (UNWTO, 2021). The World Travel & Tourism Council 2021 Economic Impact Research shows that in 2020, the global travel and tourism sector lost almost USD 4.5 trillion dollars with over 62 million jobs lost. Its GDP contribution fell to USD 4.7 trillion in 2020 from nearly USD 9.2 trillion the previous year, a decrease of 49% (WTTC Report, 2021). The sector's contribution to the global economy decreased from 10.4% to just 5.5% last year, leisure spending decreased by 49.4% while business spending by 61%. According to the CEO of the WTTC, Gloria Guevara, the pandemic crisis was 18 times bigger than the global financial crisis of 2008, and it has resulted in global challenges, economic and healthcare crises, and posed spillover impacts on the tourism ecosystem worldwide (WTTC Report, 2021). As a result, it is imperative that the entire tourism industry continuously innovate and share its best practices in order to successfully manage and restore the global tourism ecosystem services.

Against this background, "Management of Tourism Ecosystem Services in a Post Pandemic Context: Global Perspectives" proposes a synthesis of views and case studies that foster understanding of the natural and cultural tourism ecosystem services in an ongoing pandemic context. Published at a time when the world is still striving to develop resources and references to cope with the continued caprices of COVID-19, Delta, and Omicron, the book positions itself as a just-in-time strategy for the global tourism industry. In the absence of a guide on destination management post-pandemic context, the textbook serves as a workable option; readers are interested in understanding the impacts and implications of the pandemic on ecosystem services and practitioners are interested to have hand-on solutions that can be adapted for management and restoration of their ecosystem services. Drawing together from 23 chapters written by 46 authors with varied backgrounds and interdisciplinary interests, the book proposes a selection of illustrative global case studies from 16 countries, namely India, Kenya, Poland, Argentina, Lamu island, Mauritius island, Spain, Bali, Uganda, Latvia, Nigeria, Sweden, Pakistan, Sri Lanka, China, and South Africa. The book centres efforts on providing some practical insights and solutions regarding the management and restoration of ecosystem services serving the global tourism industry. The diversity of case studies proposed from different countries adds richness to the overall textbook. They are seen as useful references in destination management, strategy formulation, and resource restoration.

The book is conceptualized around five key thematic areas, namely ecosystem services in tourism, environmental conservation in tourism, cultural ecosystems services in tourism, restoration of ecosystems services and sustainable tourism development, and global trends in ecosystems services. The aim is to provide readers with an enriching insight into emerging trends of ES following the impacts of the post-COVID-19 pandemic on the global tourism industry.

Part I of the book, which is dedicated to ecosystem services in tourism, offers a glimpse of the different forms of ecosystem services serving the global tourism industry. The management and marketing challenges facing a number of European and African destinations are also unveiled. Post-pandemic solutions and challenges to destination ecosystem services are illustrated through case studies from Europe and Africa. The broad themes covered in this section relate to branding and marketing of ecosystem services, Ecosystem services as a marketing tool for tourism destinations, stakeholders' role, partnership, and ethical consumption of ecosystem services in the tourism industry.

Part II of the book addresses ecosystem services from an environmental conservation lens. Case studies on Kenya, Argentina, and India are helpful to demonstrate challenges faced by destinations in the restoration of natural ecosystems serving the tourism industry. This section also provides a synthesized understanding of the central role that tourism plays in the realization of biodiversity conservation, ecosystems restoration, and community development goals. Broad themes covered in this section are conservation of protected areas ecosystems services, implications of whale watching, as a natural tourism ecosystem service, ecological integrity and responsible tourism, and rural tourism.

Part III of the book is dedicated to management and restoration of cultural ecosystems services in tourism. This section proposes readership a rich variety of case studies from Lamu Island, Uganda, Spain, Bali, India, and Latvia. The chapters proposed therein seek to provide improved understanding of the impacts of cultural ecosystem services on livelihoods and challenges related to the use of cultural ecosystem services in tourism. Themes covered are role of residents in management and restoration of cultural ecosystem, impact of gender on tourism ecosystem services, community-based conservation for sustainable ecosystem services, cultural ecosystem services and heritage tourism, cultural ecosystem services in urban national parks, and restoration of spring ecosystem services.

Part IV of the book proposes an overview on sustainable development goals and ecosystem services. Case studies on India, Sweden, Nigeria, Asia, China, Sri Lanka, and Spain offer a rich insight into how sustainable development goals are achieved through management and restoration of natural and cultural ecosystem services. The chapters also shed light on stakeholder's involvement in the management and restoration of ecosystem services for different tourism products The themes covered are the sustainable livelihood approach and the lives of women, urban sustainable tourism, tourism ecosystem services potential to achieve sustainable goals, sustainable tourism development, restoration and sustainable tourism development of an eco-tourist destination, and stakeholders' role in sustainability of ecosystem services.

Part V addresses the global trends in ecosystem services. This section mainly focuses on the impact of COVID-19 on the management and restoration of ecosystem services. Case studies from Kolkata, Malaysia, South Africa, China, India, Sri Lanka, and Ethiopia provide substance to support the ongoing pandemic and emerging future trends. The tourism communities depending on ecosystem services have been experiencing serious livelihood challenges due to lock-down and the absence of economic activities. The topics proposed in the overall chapters address the impacts of the COVID-19 pandemic on ecosystem services.

Contributors to this volume have provided an in-depth coverage of each conceptual and practical topic so that each chapter can serve as a trusted source of reference and hence provide essential practical knowledge on management and restoration of tourism ecosystem services. Management solutions proposed therein may be adopted and/or rather adapted by destinations facing similar challenges, especially that the unprecedented impacts of the pandemic demonstrate signs of hatching more outbreaks, a point also echoed in the words of Vice President Kamala Harris: "At the same time that the world works to get through this pandemic, we also know we should prepare for the next" (USA Today, 2021). Likewise, Tom Friedman, one of the columnists from the *New York Times*, opined recently that the current generation will come to think of BC and AC as Before Corona and After Corona (Friedman, 2020).

References

Al-Said, N. (2020). The impact of COVID-19 on the protection of cultural heritage. IPI Global Observatory, 17 June. https://theglobalobservatory.org/

BBC News. (2021). Will our day ever be the same! https://www.bbc.com/worklife/article/20201109

Business Standard. (2021). COVID-19: Total travel bans collective punishment. 30 November. https://www.business-standard.com/article/current-affairs/covid-19

ECLAC (Economic Commission for Latin America and the Caribbean). (2020). "Addressing the growing impact of COVID-19 with a view to reactivation with equality: New projections." COVID-19 Special Report, No. 5, Santiago, July.

Friedman, T. L. (2020). Our new historical divide: B.C. and A.C. — the world before corona and the world after. The New York Times. https://www.nytimes.com/2020/03/17/opinion/coronavirus-trends.html

Gowreesunkar, V., Maingi, S., Roy, H. and Micera, R. (2021). *Tourism destination management in a post-pandemic context: Global issues and destination management solutions*, First Edition. Emerald Publishing Limited, UK.

Gowreesunkar, V. and Reddy, M. (2020). *Urbanism and overtourism: Impacts and implications for the city of Hyderabad*. Routledge Handbook of Tourism Cities, First Edition. Emerald Publishing, UK.

İsmail Yağcı, M. and Avni Can Yağcı, A. (2021). Effect of perceived risk on tourist behavioral intentions post COVID-19: The case of Turkey. In Gowreesunkar, V., Maingi, S., Roy, H. and Micera, R. (Eds.) *Tourism destination management in a post-pandemic context: Global issues and sestination management solutions*, First Edition. Emerald Publishing Limited, UK.

Lama, R. and Rai, A. (2021). Challenges in developing sustainable tourism post-COVID-19 pandemic. In Gowreesunkar, V., Maingi, S., Roy, H. and Micera, R. (Eds.) *Tourism sestination management in a post-pandemic context: Global issues and destination management solutions*, First Edition. Emerald Publishing Limited, UK.

Maingi, S.W. (2019). Sustainable tourism certification, local governance and management in dealing with overtourism in East Africa. *Worldwide Hospitality and Tourism Themes*, 11(5): 532–551. DOI:10.1108/WHATT-06-2019-0034

Rajao, R., Soares-Filho, B., Nunes, F., Börner, J., Machado, L., Assis, D., … and Figueira, D. (2020). The rotten apples of Brazil's agribusiness. *Science*, 369(6501), 246–248.

Rume, T. and Ul Islam, S. D. (2020). Environmental effects of COVID-19 pandemic and potential strategies of sustainability. *Heliyon*, 6(9), 1–8. https://doi.org/10.1016/j.heliyon.2020.e04965

Seraphin, H., Sheeran, P. and Pilato, M. (2018). Over-tourism and the fall of Venice as a destination. *Journal of Destination Marketing & Management*, 9, 374–376.

UNWTO (2021) Secretary General's Policy brief on Tourism and COVID-19: Tourism and COVID-19 unprecedented economic impacts. United Nations World Tourism Organization. Retrieved from: https://www.unwto.org/tourism-and-covid-19-unprecedented-economic-impacts

USA Today. (2021). Vice President Harris to kick off national vaccination tour in Greenville. 11 June. https://www.usatoday.com/story/news/politics/2021

Varriale, L., Volpe, T. and Noviello, V. (2021). Enhancing cultural heritage in times of the COVID-19 outbreak: A portrait of Campania region museum experiential strategies through ITCs. In Gowreesunkar, V., Maingi, S., Roy, H. and Micera, R (Eds.) *Tourism destination management in a post-pandemic context: Global issues and destination management solutions*, First Edition. Emerald Publishing Limited, UK.

Wahome, E. and Gathungu, J. (2021). Redefining sustainability in the conservation and promotion of the cultural heritage tourism product in Kenya. In Gowreesunkar, V., Maingi, S., Roy, H. and Micera, R. (Eds.) *Tourism destination management in a post-pandemic context: Global issues and destination management solutions*, First Edition. Emerald Publishing Limited, UK.

World Travel & Tourism Council. (2021). Opening of WTTC's Global Summit hears praise for private sector's united approach to building recovery. https://wttc.org/News-Article/Opening-of-WTTCs-Global-Summit-hears-praise-for-private-sectors-united-approach-to-building-recovery

Part II

Management of Natural Ecosystem Services

2 Role of conservancy-based tourism in the management of natural ecosystem services in the Maasai Mara, Kenya

Joseph K. Muriithi

2.1 Introduction

Protected areas like parks and nature reserves are some of the greatest biodiversity hotspots. They are the cornerstone of global preservation of biodiversity conservation efforts (Tabor et al., 2018). Hence, they are very important in ensuring sustainable provision of ecosystem services to humanity. Moreover, the creation of protected areas has been the main approach to conserving biodiversity such as wildlife and their habitats, and tourism has been the main use for protected areas (Adams, 2013). It has been established that tourism plays an important role in conservation financing (Stronza, Hunt and Fitzgerald, 2019). It also promotes development in adjacent community areas (Snyman and Bricker, 2019). Tourism is therefore regarded as one of the most compatible human activities in conservation of ecosystems (Job, Becken and Lane, 2017). In addition, Walpole, Goodwin and Ward (2001) argued that tourism activities provide the main economic justification for establishing protected and other conservation areas that enable ecosystems to thrive.

However, protected areas also face a myriad of external and internal challenges that threaten ecosystem services, upon which the tourism industry also depends (Hausmann et al., 2016). According to Okello and Kiringe (2004), in Kenya, protected areas and adjacent dispersal areas are spaces where the greatest threats that contribute to loss and degradation of biodiversity happen. Loss and degradation of biodiversity remain a major threat to the Kenya tourism industry, much of which is wildlife based (Western, Waithaka and Kamanga, 2015). These contradictions around the practice and role of tourism in conservation and development have evolved over time, leading to establishment of the wildlife conservancy model in areas beyond the conventional protected areas across the country. Tourism is central to the success of the wildlife conservancy model in Kenya and has been at the heart of mitigating the effects of biodiversity degradation and loss that compromises the provision of ecosystem services.

In Kenya, the Maasai Mara area provides the best case study to examine the role that the wildlife conservancy model plays in conservation and local development through tourism. As a fairly recent concept in tourism and conservation practice cycles in Kenya, very limited studies have been undertaken

DOI: 10.4324/b23145-4

to examine how the model is contributing towards mitigation against biodiversity loss, restoring ecosystems and enhancing the ability of the ecosystems to supply various ecosystems and goods and services. In light of these gaps, this paper describes how conservancy-based tourism in the Maasai Mara ecosystem contributes in supporting biodiversity conservation, ecosystem restoration and supply of ecosystem goods and services with a special focus on provisioning and cultural services.

2.2 Literature review

2.2.1 The wildlife conservancy concept in Kenya

According to the Kenyan Wildlife Management and Conservation Act-2013, which sets the legal framework for establishment of a conservation area, a conservancy is defined as "land set aside by an individual land owner, group of land owners, body corporate or a community for the purpose of wildlife conservation and other compatible land uses" such as tourism (GoK, 2013). In the context of the global biodiversity conservation agenda, the Kenyan wildlife conservancies can be placed under IUCN category V of protected and conservation areas also known as a "protected landscape" (Dudley, 2008). The objective of this category is to safeguard regions that have built a distinct character in regard to their ecological, biological, cultural, or scenic value (Pullin et al., 2013). In this sense, it becomes apparent that this category is unique in terms of allowing a wider range of conservation aims, as captured in the way wildlife conservancies function where wildlife, people and their livestock co-exist.

In Kenya, the establishment of the conservancy model has been based on scientific evidence According to Western, Russell and Cuthill (2009), Kenya's conventional protected areas (national parks and national reserves) cannot accommodate all the wildlife within their borders. The study revealed that 65% of all wildlife in Kenya live outside the formal parks and reserves in the human-dominated landscapes. Moreover, Western et al. (2020) also suggest that the old traditional thinking of exclusive nature conservation is no longer tenable, practical and sustainable. Therefore, there has been a need to expand spaces for wildlife conservation in the human-dominated landscapes adjacent to protected areas (Okello et al., 2014; Western, Waithaka and Kamanga, 2015). Consequently, new innovative approaches that extend conservation beyond conventional parks and reserves have recently emerged as wildlife conservancies across the Kenyan rangelands. The widespread acceptance of this model by local people is the basis of the 2013 amendment of the Wildlife Management and Conservation law to accommodate this reality. In this regard, wildlife conservancies have become an important part of the evolving conservation landscape that allows integration or co-existence of wildlife, local people and their livestock (Allan et al., 2017). In all, in wildlife conservancies as in conventional protected areas, tourism is the main economic activity that provides the incentive for the partnership between landowners and investors when establishing the parks.

2.2.2 The Maasai Mara ecosystem threats and the conservation-restoration-tourism nexus

The Maasai Mara area has been in the lead for establishment of wildlife conservancies. Wildlife conservancies have become a characteristic feature of the ecosystem that provides new habitats for wildlife thus allowing co-existence between the local community and their traditional livestock herding practices (Osano et al., 2013). However, even with the establishment of conservancies, conservation challenges that have faced the ecosystem over the years continue to persist, thus weakening the health of the ecosystem and its ability to provide ecosystem goods and services. Waithaka (2004) examined the ecological and anthropogenic challenges that have threatened the Maasai Mara ecosystem for a long time. Among the challenges and threats identified and which persist today are land tenure change, land subdivision, crop farming, habitat fragmentation and destruction, and blockage of wildlife migratory corridors. These are largely human-induced threats and are incompatible with the conservation and restoration of the Maasai Mara ecosystem.

The emergence of wildlife conservancies has underlined the importance of tourism in the landscape and how it has helped to mitigate these threats. Osano, de Leeuw and Said (2017) indicated that through tourism, wildlife is now regarded as a prime ecosystem service that provides important incentive for local landowners to allow their land to be secured as wildlife habitats through conservancy-based tourism in the Mara. This has also helped change local communities' attitudes towards wildlife conservation through the benefits they derive from the ecosystem through tourism. This is in agreement with other studies that show that tourism is an important component in biodiversity conservation, in that its benefits provide positive attitudes among local communities to conserve biodiversity in their ecosystem (Walpole and Goodwin, 2001). However, fences remain the greatest challenge facing the Maasai Mara wildlife conservancies (Løvschal et al., 2017; Weldemichel and Lein, 2019). Fences are erected following the acquisition of individual land titles and a desire to secure their own land after land subdivision process that has been going through in much of Kenyan rangelands. Moreover, the erection of fences creates barriers to free movement of wildlife, thus preventing them from accessing habitats in conservancies. This wildlife blockage is linked to loss of wildlife in the Mara (Veldhuis et al, 2019). Other current threats to biodiversity conservation and degradation of the ecosystem include settlement dwellings and even improperly tourism facilities that had no environmental and community benefits considerations and standards (Li et al., 2020). These threats impede the sustainable provision of ecosystem services, especially to the local community.

2.2.3 Conservancy-based tourism and payment of ecosystem services

Wildlife conservancies are important in securing and maintaining health ecosystems and consequently ensuring sustained provision of ecosystem services.

According to the Millennium Ecosystem Assessment Report (2005), ecosystem services are the benefits that people derive from the ecosystem, including provisioning, regulating, supporting and cultural services. Bullock et al. (2011) suggested that the concept of ecosystem services can be used to encourage conservation of ecosystems to stem the degradation and loss of biodiversity. Valuation of ecosystem services provides the basis for enlisting local people in biodiversity conservation by suggesting the economic worth or value of ecosystem services that can be paid if people contribute to their conservation. This way, farmers or herdsmen who are landowners are compensated for agreeing to undertake sustainable land use activities that contribute to sustained provision of ecosystem services. Underlying this economic compensation for supporting conservation is the concept of payment of ecosystem service (PES). PES is defined by Wunder (2015) as a "direct, voluntary and conditional transaction relating to a well-defined ecosystem services between buyers or service users and sellers or service providers". Therefore, according to Wunder, the idea of PES works by external beneficiaries of ecosystem services making direct, contractual or even conditional payments to landholders or users as an incentive to encourage them to adopt practices to secure conservation and restoration of ecosystems. Moreover, Ingram et al. (2014) suggest that the notion of payment of ecosystem services is useful in promoting sustainable management of ecosystem services and supporting local development. Accordingly, PES mechanism has become popular around the world in an attempt to stem the tide of deteriorating ecological situations of ecosystems in the world.

In the context of Maasai Mara conservancies, the notion of PES can be applied to ecotourism. Tourists pay to enjoy and appreciate the services in the conservancies established through lease of land from landowners for conservation of wildlife and their habitats. For leasing their land for conservation of wildlife, landowners are paid a guaranteed monthly lease fees and tourist night fees as a compensation for leasing their land for wildlife conservation (Mehta, 2021). Consequently, conservancies are designed as partnership agreements between the local Maasai landowners and private tourism operators. Therefore, the purpose of the partnership agreements is to secure space for wildlife in settled areas for tourists to see wildlife and the local landscape.

2.2.4 Protected area tourism in the post-pandemic context

Protected areas and wildlife conservancies as well have been significantly affected by the COVID-19 pandemic. Studies have been conducted that capture the impacts that protected areas have suffered during the COVID-19 period and have suggested pathways in which to handle future crises and uncertainties when they happen. Spenceley et al. (2021) examined how protected area tourism was disrupted by COVID-19 pandemic. The case studies revealed that the damage caused by the impacts of COVID-19 was huge ranging from loss of conservation finances, tourism businesses and people's livelihoods to disruption of supply of goods and services. As a strategy to face the post-pandemic reality, they suggest that a shift from the business as usual kind of protected area tourism to a future form of tourism that is sensitive to climate change and biodiversity loss concerns is inevitable. In other words, a new form of protected area tourism that integrates inclusive, equitable and sustainable development principles should be envisaged.

A study by Waithaka (2020) conducted at the height of the pandemic focusing on the situation of protected areas in Africa indicates that conservation work in and outside the protected areas was significantly hard hit by the pandemic. There has been destruction of ecosystems, biodiversity loss and unsustainable use of natural resources, deterioration of security in and outside of protected areas and loss of community livelihoods. As a way forward, the study suggested that there is a need for national and regional dialogues, as well as continent-wide networking on protected areas conservation strategies so as to reduce the impacts of COVID-19 and such kind kinds of uncertainties in the future.

In another study by Anand and Kim (2021) focusing on pandemic-induced changes in the economic activities in protected areas, the study suggests that across all protected areas in Africa spanning all the IUCN protected areas categories, there was significant reduction in tourism-related economic activities. The findings show that interventions put in place to control the pandemic had direct bearing on people's livelihoods and enterprise development and posed serious challenges to protected area conservation programmes. The study suggests a post-pandemic approach to protected area tourism that prioritizes recovery efforts, with governments putting more focus on rebuilding the economies to mitigate the biting effects of the pandemic, especially on local communities that rely on protected areas tourism for their livelihoods.

In Kenya, the government also commissioned a study focusing on the entire tourism sector (GoK, 2020). While the study guided the government on measures to take to mitigate the impact of pandemic on the tourism sector, the study is important in the sense it frames the post-pandemic context in light of past uncertainties and crises that have faced the country's tourism sector, such as terrorist attacks, climate-related phenomena like *el Nino* flooding, political violence and the global economic downturn in planning preparedness of future similar occurrences in Kenya.

2.3 Methodology

2.3.1 Study area

This study was conducted in five Maasai Mara conservancies (Olare Motorogi, Nashulai, Mara North, Naboisho and Enonkishu), which are located within the Maasai Mara ecosystem. The Maasai Mara ecosystem is a smaller portion of the wider Serengeti-Mara ecosystem that straddles the Kenya–Tanzania border and covers an area of 40,000 km^2. The Serengeti–Mara ecosystem comprises protected areas like Serengeti National Park on the Tanzanian side and the Maasai Mara National Reserve on the Kenyan side of the international

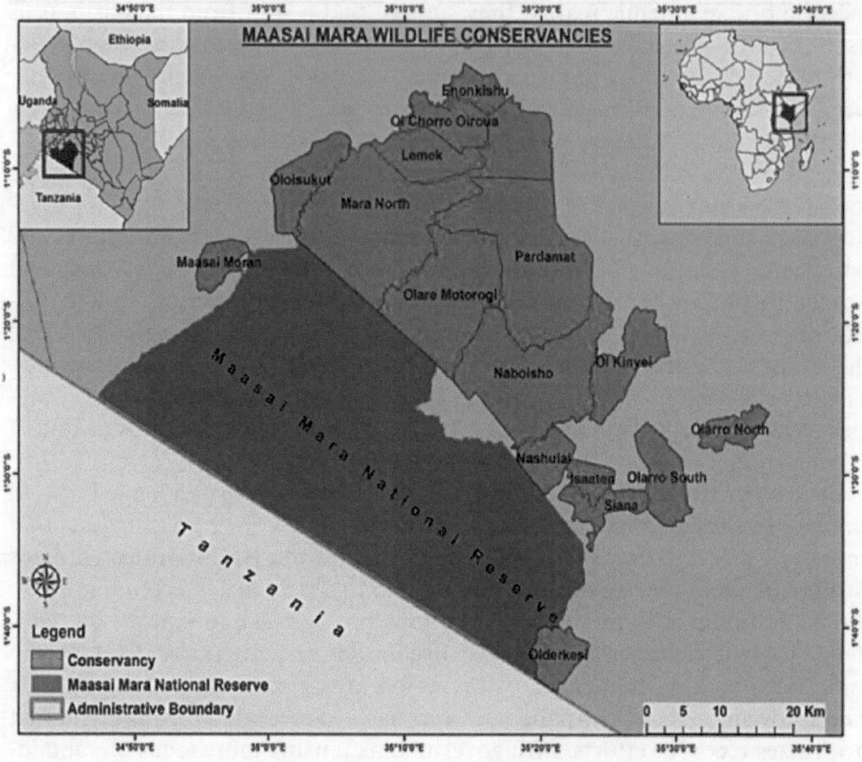

Figure 2.1 The Maasai Mara ecosystem showing the Maasai Mara National Reserve and the wildlife conservancies.

border. The Maasai Mara National Reserve, together with the surrounding conservancies, constitutes the Maasai Mara ecosystem. The Maasai Mara ecosystem covers an area of about 6,500 km^2, of which the Maasai Mara National Reserve is about 1,500 km^2 (Figure 2.1) (Thompson et al., 2009). The area under the wildlife conservancies is about equal to the size of Maasai Mara National Reserve (MMWCA, 2021). The Maasai Mara ecosystem is home to one cto Bedelian and Ogutu (2017), it has the highest wildlife density in East Africa and is therefore a very popular wildlife tourism attraction in the world (Bedelian and Ogutu, 2017). The annual wildlife migration to and from the Serengeti National Park in Tanzania, involving about 1.6 million wildebeest, 0.6 million zebra and 0.4 million gazelles, creates a great tourism spectacle to behold (Sinclair, 1995), making the ecosystem a most sought after tourist destination. The wildlife migration represents a unique and irreplaceable Kenyan heritage (Løvschal et al., 2017), thus making tourism and recreation one of the best ecosystem services to humanity.

2.3.2 Data collection and analysis

This study adopted a case study approach to examine the objectives of the study. Specifically, a multi-site, mixed-method, collective case study design was used. A collective case study involves studying multiple cases simultaneously or sequentially as a way of generating a broader appreciation of an issue (Crowe et al., 2011). In this respect, a single conservancy served as a single case that is located on a particular site. The multiple sites selected were the five more established conservancies of Olare Motorogi, Naboisho, Mara North, Nashulai and Enonkishu. The determination of the choice of the five conservancies was also based on key interviews with officials of the Maasai Mara Wildlife Conservancies Association (MMWCA). Data collection was done using multiple sources of data, including documentary evidence, desk review of numerous internet downloads from the MMWCA and the Kenya Wildlife Conservancies Association (KWCA) websites, websites of the sampled conservancies, and hardcopies of reports from the headquarters of the MMWCA in Aitong town located within the Mara North Conservancy. The website information was useful in corroborating other information collected using other methods after conservancies' officials referred the researcher to the websites. This data was supplemented with wide-ranging semi-structured interviews (SSI) with nine community members and six key informant interviews (KIIs) with conservancies' officials and the Maasai Mara Wildlife Conservancies Association officials. Data analysis was done by first categorizing data into themes and subthemes that had been pre-determined during the planning of the study. The categorization into themes enabled easy summaries of data, which were then used to do the write-up.

2.4 Results and discussion

Results of this study are presented and discussed in three sections: provisioning services, cultural services and community benefits. First, synthesized results on provisioning ecosystem services identified from the five case study conservancies are discussed. These include wildlife, livestock, beekeeping, water, fuelwood and livestock fodder and forage in the name of grass. A discussion of the second category of cultural ecosystems services revolve around the identified services, including recreation and tourism, aesthetics, cultural heritage, research and education and knowledge. The final section relates to community benefits, which are more direct benefits people get from the practice of ecotourism in the conservancies. These are treated here as indirect ecosystem benefits that arise out of the direct impacts of tourism activities in the conservancies.

2.4.1 Provisioning ecosystem goods and services

2.4.1.1 Maintenance of biodiversity in the conservancies

The results have revealed that wildlife in the conservancies is one of the most important ecosystem goods in the Maasai Mara conservancies. Tourism in the conservancies is largely dependent on the maintenance of wildlife in their numbers within the conservancies. Furthermore, through wildlife conservation, the practice of ecotourism has itself introduced many principles that have positively impacted on ecosystem conservation and restoration. Wildlife in the conservancies therefore influences the practice of ecotourism and consequently the wellbeing of the local Maasai people, who are the owners of the land on which wildlife thrives and tourism takes place. In the words of a key informant,

> … what brings the tourism partners and the Maasai land owner together is not land per se, but wildlife. Wildlife is profitable in nature and if wildlife is not there, they are not benefiting anyone. Not the Maasai person or the tourist partner.
>
> – KII3

Therefore, the establishment of the wildlife conservancies has been a unique and timely intervention. The conservancies have played an important role in maintaining and conserving biodiversity by serving as an important buffer zone between the Maasai Mara National Reserve and the local community. They provide that extra space to allow wildlife migration within the Maasai Mara in what is regarded as the Loita plains migration (Msoffe et al., 2019) as well as across the Maasai Mara–Serengeti ecosystem. Both the internal and external migrations act as big attraction for tourism, which impacts on lives of humanity from across all parts of the universe through tourism. Through calls to do away with the fences that were put up when land was subdivided, this has been a positive biodiversity maintenance and enhanced conservation value by way of unblocking wildlife dispersal areas (Western, Waithaka and Kamanga, 2015) and avoiding the severe fragmentation of the ecosystem and wildlife habitats (Løvschal et al., 2017). More specifically, the officially stated focus of the Mara North Conservancy is rehabilitation and restoration of the wildlife habitats that have been degraded for a long time and protecting wildlife from poaching and snaring. Additionally, in the Naboisho Conservancy, the key stated goals of the conservancy are the rejuvenation of the vital wildlife corridor of the Mara ecosystem that runs through the conservancy. These measures have played a key role in minimizing human-wildlife conflicts, and ultimately playing a key role in the normal functioning of the ecosystem.

2.4.2 Enhanced co-existence of wildlife and livestock in the conservancies

One of the most important opportunities offered by way of establishing wildlife conservancies is co-existence between wildlife and livestock. The logic of this opportunity is evidenced by studies and conservancy residents' opinions that have suggested that wildlife conservation and pastoralism are compatible with land uses (Weldemichel and Lein, 2019; Western et al., 2020). This was a key message mentioned in the Maasai Mara, as captured in the statement of a resident in Nashulai Conservancy:

> For many years before all this land was chopped into small pieces [plots], the Maasai cows used to graze freely with wildebeests and zebras. It is

only recently when we started these conservancies that I started seeing this kind of grazing.

– SSI 7

This balanced approach has enhanced opportunities for provision of ESs. Through the lease agreements between the tourism partners and landowners, conservancies have allowed mixed use of wildlife–livestock grazing areas. In Nashulai and the Mara North conservancies, the mixed-use planning has allowed wildlife and livestock to co-exist as the best conservation land-use model. Through use of new grassland management planning that entails zoning of grazing areas, it has become possible to engage in rotational grazing between wildlife and livestock. This has been part of the ecosystem restorative strategies adopted to fit the grazing plans for both wildlife and livestock. The implication of these kinds of plans is that livestock is now seen and accepted as part of biodiversity in the Mara ecosystem. As a result of such planning strategies, it has been established that in Enonkishu Conservancy, vegetation has "improved by 38% and livestock health has improved drastically" (Enonkishu Conservancy, 2021). However, challenges have emerged in the implementation of the grazing management interventions with disputes arising between tourism partners and landowners over access to grazing areas by community members (Weldemichel and Lein, 2019) and where grazing rules outlaw certain areas where livestock cannot graze (Bedelian and Ogutu, 2017).

Some conservancies have become focal point areas for sustainable rangeland management and ecosystem restoration. The Enonkishu Conservancy is a demonstration site for regenerative grazing to showcase how grazing regulation for livestock and wildlife can be undertaken. Hence, regenerative grazing planning demonstrates how the new sedentarized Maasai pastoral lifestyle has fitted into the new conservancy model where conventional free-range herding of the traditional Maasai is not possible. Moreover, the Enonkishu Conservancy has also enabled adoption of new Boran cattle breed that can fit with the conservancy or ranch model of land use, thereby ensuring diversified ecosystem goods and services. The new breed is less vulnerable to disease and drought and has a higher market value and is therefore adapted well to the ecosystem. This has broadened the range of ecosystem goods and services in the conservancy. However, it is also leading to loss of the traditional Maasai Zebu cattle breed, which is adapted to pastoralist nomadic free-ranging conditions. This has implications on the Maasai cultural heritage, which is a key tourism resource and an important ecosystem service.

2.4.2.1 Ecosystem goods and services and greening initiatives in the conservancies

Camps and lodges in the conservancies are at the forefront of promoting greening by tapping into specific ecosystem goods and services like water and solar energy. This is in recognition that camps and lodges in the conservancies have to observe the highest levels of environmental management as captured in ecotourism practices. This has provided the ideal setting for tourists as they enjoy nature in the conservancies. Across all the conservancies, lodges and camps, most of which are eco-certified have set high standards in green credentials. Through the practice of ecotourism, camps and eco-lodges have laid the groundwork to transition to a low carbon future across the Mara conservancies. For example, all eco-certified lodges and camps have adopted sustainable renewable energy, such as solar energy and biogas. They are also into sensitizing local people to abandon charcoal burning and instead encouraging the adoption of eco-friendly charcoal briquettes, as opposed to cutting down trees and shrubs for home cooking. Furthermore, the eco-lodges are at the forefront of adoption of electric vehicles for game drives within the conservancies. One of the recently introduced eco-lodges, the Emboo River Camp (Emboo River Camp, 2021), introduced an electric vehicle for use in game drives by guests, thus contributing to reduction in carbon emissions. Additionally, the eco-lodges are in the lead in reforestation campaigns through tree planting programs on their premises and across the conservancies. A common practice in the camps visited is to organize tree planning activities by guests as a way of offsetting carbon footprint of their travel. These greening initiatives are, however, not confined to the camps. Local people through their community groups have also come up with initiatives like waste collection and management in the shopping centres in the conservancies. In Nashulai Conservancy, residents devote specific times in the year to organize cleanup exercises in their conservancy. A resident of Sekenani Township captured it this way:

> Over the last three years everyone is mobilized to take part in environmental cleanup exercises. This happens every time there are heavy rains when River Sekenani floods carrying with it all sorts of rubbish as well as during special days such the world environmental day.
>
> – SSI2

Such exercises are important ecosystem restorative activities. With river Sekenani being a critical watershed on whose lives many people, wildlife and livestock depend, it provides essential ecosystem services as a component of the great Maasai Mara ecosystem.

2.4.3 Cultural ecosystem goods and services

2.4.3.1 Ecotourism as a cultural ecosystem service

In the Maasai Mara, biotic and abiotic cultural goods and services combine to make the conservancies the ideal ecotourism destination. Biotic goods and services like the wildlife, birds, shrubs and trees combine with abiotic goods and services like the scenic landscape, the Maasai Manyattas and the Maasai culture to present a complete package that is a source of recreation and ecotourism targeted by tourists in the nearly 60 tourist camps and lodges in the conservancies. This is because wildlife (representing the biotic goods and services) resonates very well with the cultural heritage of the Maasai people in the same conservation setting. Therefore, through ecotourism, wildlife conservation and the preservation of the Maasai culture have become mutually beneficial to each other. This has further enhanced co-existence between people and wildlife and increased people's appreciation of wildlife more than in the past. This has also increased the rate of opening up of more partnerships between landowners and tourism investors in establishing more conservancies from the current 16 conservancies. This composite picture is also used in marketing individual conservancies online on their websites. It is also the same picture that emerges in the eco-lodges, which are constructed using locally available material, and most reflect the local culture. Furthermore, in the operations of the lodges, all employees are local Maasai from tour guides, to cooks, to rangers and to scouts who monitor the wildlife and protect the tourists.

In terms of observing ecotourism principles of maintaining the ecological integrity of the conservancies therefore assuring the ability of the conservancies to provide ecosystem goods and services, there is strict enforcement of sustainability practices. For instance, there is strict enforcement of tourist number entrance limits to the conservancy, therefore reducing the vehicular crowds. Personal vehicles are also not allowed for game drives as a matter of minimizing impacts on the environment. These ecotourism values are also reflected in lease agreements between landowners and tourism investors, thereby ensuring parties agree to abide by their obligations. Tourism partners as managers of the conservancies are obligated to ensure that those values are upheld.

2.4.3.2 Research values in the conservancies

Research is an important aspect of cultural ecosystem services. As the host of the highest concentration of different wildlife species on earth, the Mara ecosystem offers great opportunities for research in different disciplines in the natural, social sciences and the humanities. The Maasai cultural heritage has been a great source of research interest to both national and international researchers. Hence, the ecosystem is a naturally occurring laboratory for scientific and human research recognized across the world. Moreover, the Mara ecosystem is a key circuit for field research excursions and activities even for many basic education institutions like high schools and also university and college students. The ecosystem has been an important research hub before the establishment of the conservancies and even after the establishment of the conservancies. A number of research programs have been hosted in the conservancies. Currently, three research programs are hosted within the conservancies. These are the Mara Predator Conservation Program management by the Kenya Wildlife Trust, the Mara Elephant Project and the Big Cats Research Project, all hosted in Olare Motorogi and Naboisho conservancies.

2.4.3.3 Education and knowledge values

Related to the research values above, the Maasai Mara conservancies are also sites for great education and knowledge values, useful not only in the local context but also in the national and global context. Maasai traditional ecological knowledge has been known to be useful in wildlife conservation. For instance, over the years before the sedentarization of their ways of life due to land subdivision, they were a migratory people, moving from one place to another to allow regeneration of pastures in places they had left. It is this pastoralist migratory knowledge for regeneration of pastures that is being combined with scientific concepts like planned zoning, where livestock is supposed to graze for some time before they are allowed elsewhere. Furthermore, the Maasai indigenous knowledge is used to set up core conservation zones, where livestock grazing is not supposed to take place. These are areas within conservancies which are habitats of wildlife. Just as in the traditional setup, the Maasai had certain places which were sensitive to grazing, where they could release their cattle to grazing. In the words of a key informant in Nashulai Conservancy:

> Today, the Maasai people are combining traditional grazing knowledge with scientific knowledge when implementing modern zonal grazing system that has been introduced by the conservancy. First of all, we have removed all the fences in this area and gone back to the way we have been rotating our cattle grazing like in communal land.
>
> – KII2

This is a powerful statement that shows how relevant indigenous knowledge and education is useful under the current changed conservancy arrangement. Bringing down the fences, which is a hindrance to free movement of wildlife, combined with rotation grazing, has helped increased wildlife presence in the conservancies while allowing livestock to also thrive.

2.4.4 Community benefits from conservancy-based ecotourism

Through ecotourism in the conservancies, the local community has witnessed a great deal of benefits from conservation measures that have ensured continued provision of ecosystem services. According to Kiss (2004), the rationale of introducing tourism in an area is premised on the principle that through tourism, the conservation of biodiversity will generate economic benefits to the local people. In the Maasai Mara conservancies, local community members derive a wide range of benefits spanning the economic, social and cultural spheres. Although they cannot be discussed in that general version, this section identifies and discusses specific range of conservation-tourism benefits that local people derive from the membership of their conservancies. The range of benefits to communities may vary from conservancy or eco-lodge or from one camp to another. However, some benefits may be the same across conservancies because most community needs addressed by tourism and conservation actors are the same across all the conservancies and attend to the community common good.

To start with, the most important form of benefit is the land lease fees. Land lease fees are monies that landowners get paid as a result of agreeing to lease their land for wildlife conservation. Land lease fees are guaranteed to all landowners who enter into land lease agreement with a tourism partners or investor. Lease fees are paid directly into a landowner's bank account on a monthly basis (Bedelian and Ogutu, 2017). This way the fees play a major role in providing livelihoods to landowner's households. Additionally, conservancy lease fees help landowners to diversify their livelihoods from the traditional pastoralism livelihood means of livestock keeping.

Sustainability of livelihoods is ensured by the fact that the leases are signed on long-term basis of between 15 and 25 years. Moreover, these leases are renewable. The longevity of the leases further creates certainty, especially to landowners who may wish to use the lease agreement to secure loans for various reasons. For instance, the Olare Motorogi Conservancy has entered into loan guarantee with a leading national bank to provide loan to landowners with the lease agreement serving as the collateral. In the words of a key informant in Talek township in Olare Motorogi Conservancy:

> One of the reasons why the leading commercial bank set base here is because of acknowledging that in this area whose key businesses revolves around wildlife, there are many customers they can do business with. Since the bank was set up here, many of our members (land owners) have merely used a copy of the leave agreement to secure loans. We have assured the bank through an MOU that members have monthly incomes in land lease fees to repay their loans.
>
> – KII1

The leases therefore enable landowners to make long-term financial decision and planning. Lease fee payments are dependent on the size of land per acre leased to the conservancy. The monthly lease fees vary from one conservancy to another. In Naboisho conservancy, the monthly lease is USD 85 with an annual 8% upward adjustment for inflation (Mehta, 2021). In addition, there is USD 80 per bed night in the conservancy, which is also paid directly to landowners. Besides, these payments go a long way to justify to community members that wildlife conservation pays, hence that wildlife is profitable by way of attracting tourist, who pays to watch them. The fees therefore serve as an incentive to other landowners to pool their lands for profitable conservation purposes.

Employment opportunities is the second set of benefits that community members derive from conservancies. Employment opportunities come mainly from conservation and tourism activities in the conservancies and in the eco-lodges. For instance, all community scouts and rangers employed in the conservancies are Maasai. Likewise, all workers in the camps and eco-lodges apart from the senior management are local Maasai people. Rangers and community scouts are employed to protect visitors when in the camps, enforcing controlled grazing regulations and wildlife surveillance. It is however noted that skill sets among the local people are inadequate, especially those working in the eco-lodges and camps. Two of the key areas of skill set inadequacy are in the area of computer literacy and foreign languages. Consequently, some eco-lodges sponsor their workers for training in relevant areas of work like computer studies, front office operations and foreign language for tour guides. Many camps also offer internship opportunities to many local students who are in their final year of training. In most cases, these students get absorbed in the camps upon completion of their internship. Most of the training institutions are in the Mara area to take care of the training needs of the local people. One of the best-known training schools and centres includes Koiyaki Guiding School, which is a local leader in certificate course training in tour guiding and operations and other relevant areas related to conservation and tourism in the Mara region.

Another important area of community benefit is in the social sector support and especially in the health and education. On health, support has been in projects focusing on setting up health facilities to bring services closer to the people, especially in regard to maternal diseases, HIV and Aids, TB, malaria, as well on waterborne diseases. To fight waterborne diseases, camps in Naboisho and Olare Motorogi conservancies have done hydrological surveys and sunk boreholes in key community centres to improve hygiene. This programme support is especially prominent in Olare Motorogi and Naboisho conservancies supported by BaseCamp Foundation, which is the corporate social responsibility arm of Basecamp Explorer's five camps in the Maasai Mara. In the education sector, many camps in the conservancies have supported education in the community. Most of these programmes relate to construction of classrooms and supporting in school fees payment through bursaries to children in extreme need. Conservancies themselves have also provided school fees payment by giving bursaries. A good example is Olare Motorogi Conservancy. A key informant put it this way:

> In Olare Motorogi, we have made sure that income that the conservancy gets goes towards supporting education. Two children per family are supported with Ksh 5000 (about 47USD) per term. We have realized this is the only way the conservancy can support the poorest among its members.
>
> – KII1

A final area of community benefit identified is support in building the capacity of community members, especially women with entrepreneurial skills. The capacity-building programmes once again are spearheaded by Basecamp Foundation in Naboisho and Olare Motorogi conservancies, focusing on empowerment of women to start income-generating activities such as handicraft making, livestock trade, beekeeping and shopkeeping. In terms of handicraft making, women are the sole suppliers of handcrafts and curio items in the Basecamp camps. They are also supported to market their curios and other handcraft like beads and necklaces overseas.

To sum up on the results, this paper has highlighted important implications that need to be emphasized pertaining to the role played by wildlife conservancies in the management and supply of ecosystem services. Having conducted the study at the height of the COVID-19 pandemic, the importance of coordinating how the Maasai Mara ecosystem is conserved and the degraded biodiversity restored around the idea of conservancy was observed. By implication, the paper has revealed that in Kenya, the organizational and governance arrangements of conservancies are devolved with effective structures at the micro, meso and macro levels. Hence, at the micro level, there is a local independent wildlife conservancy. Then at the meso, or regional, level there is a regional wildlife conservancy association such as the MMWCA in the case of the Maasai region that coordinates mainly conservation and related activities of the conservancies in a region. Then, at the macro level, the regional wildlife conservancies' associations join together to be members of the national level Kenya Wildlife Conservancies Association.

Therefore, during the COVID-19 period, these kinds of structures anchored around the wildlife conservancy model were very useful in mobilizing relief funds to ensure essential services and activities continued in the conservancies when there were no tourists visiting. For instance, the MMWCA was fundraised to ensure that the Mara Conservancies' basic operations continued during the period that travel restrictions were in place. On the other hand, at the macro level, the KWCA lobbied the government to secure funds to mitigate the consequences of escalation of environmental crimes like poaching by ensuring that community rangers across the country who had been laid off were reinstated and their salaries secured for a period of six months. These arrangements have implications to the post-pandemic contexts of conservancy-based tourism. Since Kenyan tourism in the past has been prone to uncertainties such as post-election violence cycles, acts of terrorism and climate-related phenomenon like *el Nino* and *la Nina*, all that affects tourism, the ecosystem and biodiversity conservation, should they occur in the future, a framework to facilitate fast community resilience seems to be in place through the regional and national conservancies associations.

2.5 Conclusion and recommendations

The emergence of wildlife conservancies in Kenya has offered new ways in which tourism enhances supply of ecosystems of goods and services. The experiences with the Maasai Mara conservancies suggest that tourism has a major influence in biodiversity conservation and in the restoration of degraded ecosystem, thereby enhancing provision of ecosystem services. It is evident that through ecotourism practices, two potentially antagonist land-use practices, that is, wildlife conservation and livestock herding, now co-exist, thereby creating the right balance for sustainable provision of ecosystem services and community benefits. However, there are still persisting challenges related to the perception of unfair distribution of benefits and the way some conservation interventions relating to some ecosystem services such as new grazing management are handled. Furthermore, a common suppressed voice among many local residents suggests that zoning of grasslands and setting up of core conservation areas where cattle cannot be allowed to graze disadvantages the local people. The zoning mechanism could be re-looked into to avoid tensions between landowners and tourism investors so that zoning outcomes do not favour survival wildlife over the Maasai livestock, which too contributes to their wellbeing.

Another outstanding issue relates to how to extend benefits from nature to a greater majority of residents in the conservancies, particularly given that the working of the conservation model relates principally to the relationship between the landowner and the tourism investor. The landowner–tourist investor relationship potentially leaves out a significant proportion of people like women who traditionally are not landowners but who also need to be enlisted in enjoying the goods and services provided by nature. Therefore, through the conservancy-based tourism model to conservation and supply of ecosystem services, the Maasai Mara ecosystem has provided an innovative way to conservation in which tourism is central to its success that can be replicated in other places with similar land tenure situation.

In terms of research implications, this paper is an exploratory effort. Further studies need to be undertaken to understand issues highlighted here in more depth. The following issues are suggested for further studies: (i) issues of inclusion in benefit sharing, especially among non-land title holding members such as women. (ii) How core conservation zoning mechanisms and access to grazing area arrangements impact on the future stability conservancies. (iii) Fences and their effects on provision of ecosystem services and conservation in the conservancies. (iv) COVID-19 impacts, response and implications on the ability of the conservancies to cope with potential future uncertainties. (v) Comparative studies on these issues on the situation in conservancies in other regions in the country.

References

Adams, W.M., 2013. *Against extinction: The story of conservation*. Earthscan, UK and USA.

Allan, B.F., Tallis, H., Chaplin-Kramer, R., Huckett, S., Kowal, V.A., Musengezi, J., Okanga, S., Ostfeld, R.S., Schieltz, J., Warui, C.M. and Wood, S.A., 2017. Can integrating wildlife and livestock enhance ecosystem services in central Kenya?. *Frontiers in Ecology and the Environment*, *15*(6), pp. 328–335.

Anand, A. and Kim, D.H., 2021. Pandemic induced changes in economic activity around African protected areas captured through night-time light data. *Remote Sensing*, *13*(2), p. 314.

Bedelian, C. and Ogutu, J.O., 2017. Trade-offs for climate-resilient pastoral livelihoods in wildlife conservancies in the Mara ecosystem, Kenya. *Pastoralism*, *7*(1), pp. 1–22.

Bullock, J.M., Aronson, J., Newton, A.C., Pywell, R.F. and Rey-Benayas, J.M., 2011. Restoration of ecosystem services and biodiversity: Conflicts and opportunities. *Trends in Ecology & Evolution*, *26*(10), pp. 541–549.

Crowe, S., Cresswell, K., Robertson, A., Huby, G., Avery, A. and Sheikh, A., 2011. The case study approach. *BMC Medical Research Methodology*, *11*(1), pp. 1–9.

Dudley, N., 2008. *Guidelines for applying protected area management categories*. International Union for Conservation of Nature (IUCN), Gland, Switzerland.

Emboo River Camp Website, 2021. [online]. Available at https://www.emboo.camp/. Accessed on 13 August 2021.

Enonkishu Conservancy Website, 2021. [online]. Available at https://www.enonkishu.org/. Accessed on 13 August 2021.

Government of Kenya, 2013. Wildlife Conservation and Management Act 2013. [online]. Available at http://kenyalaw.org/kl/fileadmin/pdfdownloads/Acts/WildlifeConservationandManagementActCap376_2_.pdf. Accessed on 21 October 2021.

Government of Kenya, 2020. Impact of COVID-19 on tourism in Kenya: Measures taken and the recovery pathways. [online]. Available at http://www.tourism.go.ke/wp-content/uploads/2020/07/COVID-19-and-Travel-and-Tourism-Final-1.pdf. Accessed on 21 October 2021.

Hausmann, A., Slotow, R.O.B., Burns, J.K. and Di Minin, E., 2016. The ecosystem service of sense of place: Benefits for human well-being and biodiversity conservation. *Environmental Conservation*, *43*(2), pp. 117–127.

Ingram, J.C., Wilkie, D., Clements, T., McNab, R.B., Nelson, F., Baur, E.H., Sachedina, H.T., Peterson, D.D. and Foley, C.A.H., 2014. Evidence of payments for ecosystem services as a mechanism for supporting biodiversity conservation and rural livelihoods. *Ecosystem Services*, *7*, pp. 10–21.

Job, H., Becken, S. and Lane, B., 2017. Protected areas in a neoliberal world and the role of tourism in supporting conservation and sustainable development: An assessment of strategic planning, zoning, impact monitoring, and tourism management at natural World Heritage Sites. *Journal of Sustainable Tourism*, *25*(12), pp. 1697–1718.

Kiss, A., 2004. Is community-based ecotourism a good use of biodiversity conservation funds?. *Trends in Ecology & Evolution*, *19*(5), pp. 232–237.

Li, W., Buitenwerf, R., Munk, M., Amoke, I., Bøcher, P.K. and Svenning, J.C., 2020. Accelerating savanna degradation threatens the Maasai Mara socio-ecological system. *Global Environmental Change*, *60*, p. 102030.

Løvschal, M., Bøcher, P.K., Pilgaard, J., Amoke, I., Odingo, A., Thuo, A. and Svenning, J.C., 2017. Fencing bodes a rapid collapse of the unique Greater Mara ecosystem. *Scientific Reports*, *7*(1), pp. 1–7.

MEA (Millennium Ecosystem Assessment), 2005. *Ecosystems and human well-being* (Vol. 5).Island press, Washington, DC.

Mehta, H, 2021. Integrated Biodiversity, Pastoralism and Tourism Master Plan for NabosihoWilldife Conservancy, 2021. [online]. Available at https://www.youtube.com/watch?v=XTmnLpK3_5ITAPAS. Accessed on 31 July 2021.

MMWCA (Maasai Mara Wildlife Conservancies Association), 2021. [online]. Available at https://maraconservancies.org/. Accessed on 25 July 2021.

Msoffe, F.U., Ogutu, J.O., Said, M.Y., Kifugo, S.C., de Leeuw, J., Van Gardingen, P., Reid, R.S., Stabach, J.A. and Boone, R.B., 2019. Wildebeest migration in East Africa: Status, threats and conservation measures. *BioRxiv*, p. 546747.

Okello, M. M., Bonham, R. and Hill, T. (2014). The pattern and cost of carnivore predation on livestock in maasai homesteads of Amboseli ecosystem, Kenya:

Insights from a carnivore compensation programme. *International Journal of Biodiversity and Conservation*, *6*(7), pp. 502–521.

Okello, M.M. and Kiringe, J.W., 2004. Threats to biodiversity and their implications in protected and adjacent dispersal areas of Kenya. *Journal of Sustainable Tourism*, *12*(1), pp. 55–69.

Osano, P., de Leeuw, J. and Said, M., 2017. Case Study: Biodiversity- and wildlife-tourism-based Payment for Ecosystem Services (PES) in Kenya. In Namirembe S., Leimona B., van Noordwijk M. and Minang P., (eds.) *Co-investment in ecosystem services: Global lessons from payment and incentive schemes*. World Agroforestry Centre (ICRAF), Nairobi.

Osano, P.M., Said, M.Y., de Leeuw, J., Ndiwa, N., Kaelo, D., Schomers, S., Birner, R. and Ogutu, J.O., 2013. Why keep lions instead of livestock? Assessing wildlife tourism-based payment for ecosystem services involving herders in the Maasai Mara, Kenya. *Natural Resources Forum*, *37*(4), pp. 242–256.

Pullin, A.S., Bangpan, M., Dalrymple, S., Dickson, K., Haddaway, N.R., Healey, J.R., Hauari, H., Hockley, N., Jones, J.P., Knight, T. and Vigurs, C., 2013. Human well-being impacts of terrestrial protected areas. *Environmental Evidence*, *2*(1), pp. 1–41.

Sinclair, A.R.E., 1995. Serengeti past and present. In Sinclair, A.R.E. and Norton-Griffiths, M. (eds.) *Serengeti: Dynamics of an ecosystem* (pp. 3–30). University of Chicago Press, Chicago, USA.

Snyman, S. and Bricker, K.S., 2019. Living on the edge: Benefit-sharing from protected area tourism. *Journal of Sustainable Tourism*, *27*(6), pp. 705–719.

Spenceley, A., McCool, S., Newsome, D., Báez, A., Barborak, J.R., Blye, C.J., Bricker, K., Sigit Cahyadi, H., Corrigan, K., Halpenny, E. and Hvenegaard, G., 2021. Tourism in protected and conserved areas amid the COVID-19 pandemic. *Parks*, (27), pp. 103–118.

Stronza, A.L., Hunt, C.A. and Fitzgerald, L.A., 2019. Ecotourism for conservation?. *Annual Review of Environment and Resources*, *44*, pp. 229–253.

Tabor, K., Hewson, J., Tien, H., González-Roglich, M., Hole, D. and Williams, J.W., 2018. Tropical protected areas under increasing threats from climate change and deforestation. *Land*, *7*(3), p. 90.

Thompson, D.M., Serneels, S., Kaelo, D.O., Trench, P.C. (2009). Maasai Mara – Land Privatization and Wildlife Decline: Can Conservation Pay Its Way?. In Homewood, K., Kristjanson, P., Trench, P.C. (eds.) *Staying Maasai?*. Studies in Human Ecology and Adaptation, vol. 5. Springer, New York, NY. https://doi.org/10.1007/978-0-387-87492-0_3.

Veldhuis, M.P., Ritchie, M.E., Ogutu, J.O., Morrison, T.A., Beale, C.M., Estes, A.B., Mwakilema, W., Ojwang, G.O., Parr, C.L., Probert, J. and Wargute, P.W., 2019. Cross-boundary human impacts compromise the Serengeti-Mara ecosystem. *Science*, *363*(6434), pp. 1424–1428.

Waithaka, J., 2004. Maasai Mara – an ecosystem under siege: An African case study on the societal dimension of rangeland conservation. *African Journal of Range and Forage Science*, *21*(2), pp. 79–88.

Waithaka, J., 2020. The impact of COVID-19 pandemic on Africa's protected areas operations and programmes. *IUCN-WCPA Paper*. [online]. Available at https://www.iucn.org/sites/dev/files/content/documents/2020/report_on_the_impact_of_covid_19_doc_july_10.pdf. Accessed on 13 October 2020.

Walpole, M.J. and Goodwin, H.J., 2001. Local attitudes towards conservation and tourism around Komodo National Park, Indonesia. *Environmental Conservation*, *28*(2), pp. 160–166.

Walpole, M.J., Goodwin, H.J. and Ward, K.G., 2001. Pricing policy for tourism in protected areas: Lessons from Komodo National Park, Indonesia. *Conservation Biology*, *15*(1), pp. 218–227.

Weldemichel, T.G. and Lein, H., 2019. "Fencing is our last stronghold before we lose it all." A political ecology of fencing around the Maasai Mara National Reserve, Kenya. *Land Use Policy*, *87*, p. 104075.

Western, D., Russell, S. and Cuthill, I., 2009. The status of wildlife in protected areas compared to non-protected areas of Kenya. *PloS one*, *4*(7), p. e6140.

Western, D., Tyrrell, P., Brehony, P., Russell, S., Western, G. and Kamanga, J., 2020. Conservation from the inside-out: Winning space and a place for wildlife in working landscapes. *People and Nature*, *2*(2), pp. 279–291.

Western, D., Waithaka, J. and Kamanga, J., 2015. Finding space for wildlife beyond national parks and reducing conflict through community-based conservation: The Kenya experience. *Parks*, *21*(1), pp. 51–62.

Wunder, S., 2015. *Payments for environmental services: Some nuts and bolts*. CIFOR, Bogor, Indonesia.

3 Community-based conservation for sustainable ecosystem services and tourism

A case study of Murchison Falls conservation area, Uganda

Francis Mugizi, Jim Ayorekire, and Joseph Obua

3.1 Introduction

Community-based conservation (CBC) is a paradigm that has evolved since the 1980s and gained prominence in ecosystem services discourse over the years (Brooks et al., 2013). It is linked to management of resources and the attendant effects on sustainable livelihoods of local communities. Furthermore, CBC is an approach that incorporates the interests and views of local people (Brooks et al. (2013), addresses local ecological realities, and supports productivity of multi-functional landscapes and ecosystem services (Reyes-Garcia et al., 2013). In related contexts, CBC is an institution that simultaneously conserves biodiversity and enhances livelihoods of resource-dependent communities (Galvin et al., 2018). Biodiversity is important for ecotourism and provides a crucial link between conservation, ecosystem services and livelihoods.

There has been much debate on the merits of CBC, and Berkes (2007) discussed it from a global perspective, where he referred to it as a complex systems problem that needs to be addressed taking into account issues of scale, uncertainty and multiple institutional involvement. Perceived this way, community-based conservation offers a broad pluralistic approach to biodiversity protection and sustenance of ecosystem services (Galvin et al., 2018). From a regional perspective, CBC has evolved in Eastern and Southern Africa since the 1980s at field level implementation and policy reforms that led to the establishment of conservancies. For instance in Kenya, the conservancies were based on agreements between tourism operators and landholding communities, while in Namibia, communal conservancies were a centrepiece of national approach to community-based natural resource management that helped to sustain ecosystem services, promote tourism and enhance local community livelihoods (Nelson et al., 2021).

Despite many positive outcomes of community-based conservation highlighted above, some gaps remain to be addressed, for instance, limited implementation of community-based operational plans, negative perception of wildlife due to human–wildlife conflict, inadequate capacity to manage the impact of climate change on biodiversity and ecosystem services and continued habitat

DOI: 10.4324/b23145-5

fragmentation on community lands. A study by Hauser et al. (2021) revealed that despite the inclusion of local communities in CBC and balancing sustainable resource use and protection, there are some discontents with the contribution of CBC to rural development and biodiversity conservation. In addition, illicit activities in a number of protected areas continue to hinder conservation efforts. Moreover, few studies have empirically evaluated the impact of CBC on biodiversity conservation and rural development, thus making it a priority for future research.

Natural resource conservation and the attendant ecosystem services are fundamental to human well-being (Jax et al., 2013) and include regulation of natural systems, provision of food, wood, herbal medicines, recreation and education (Jones et al., 2016). Ecosystems, human and social capital have tri-directional relationship that provides sustainable benefits to society (Vinogradovs et al., 2020). Human capital protects ecosystems, while societies with impaired well-being experience decline in ecosystem services (Butler & Oluoch-Kosura, 2006). Social capital facilitates management of natural resources and is characterized by trust, community involvement and cohesion that sustain ecosystem services (Barnes-Mauthe et al., 2015).

Conservation of ecosystems and related environmental services should benefit communities that often bear the costs of wildlife damage (Costanza et al., 2017). To achieve consensus on this view, there is a need for increased understanding of the value of environmental services made possible by conservation (Macheka et al., 2021). While studies of tourism have progressed beyond understanding the relationship between tourism and conservation, the discourse on tourism-ecosystem services nexus lags behind (Pueyo-Ros, 2018) and environmental services have not been included in the national accounting system (Koko et al., 2020). Research on conservation, sustainable tourism and livelihoods requires a sound understanding of the underlying natural dynamics and socio-economic processes (Yin et al., 2021). To date, literature on the inextricable relationship between conservation, ecosystem services and tourism is limited (Pagdee & Kawasaki, 2021), and there is general paucity of research and baseline data to explain the nexus (Dewi et al., 2013). This gap in research, knowledge and literature exists for Uganda's conservation areas, including Murchison Falls Conservation Areas (MFCA), and provides the impetus for the study to examine the nexus between resource use, conservation, ecosystem services and tourism.

COVID-19 remains a global challenge viewed by tourism scholars as a watershed moment because of its global scale and widespread effects that continue to cause economic and social disruptions (Thurstan et al., 2021). From the conservation standpoint, it presents fundamental challenges, as well as opportunities, for current and future biodiversity conservation (Rogerson & Baum, 2020). In many countries, COVID-19 has caused serious perturbations that disrupted or entirely halted routine conservation monitoring and management activities. Increase in illegal hunting and illicit trade in wildlife products has been reported as well as global reduction in human movement and economic activity that has resulted in reduced wildlife disturbance as human populations retreated indoors (Thurstan et al., 2021). The global reduction in mobility has had significant negative consequences for the conservation sector, amid the postponement or cancellation of research, monitoring and decline in tourism revenue that supports conservation (Evans et al., 2020).

According to Eshun and Tonto (2014), community-based enterprises have emerged around conservation areas as a mutually reinforcing relationship between conservation, mitigating resource depletion and alleviating poverty. However, there is limited knowledge and evidence of the extent to which community-conservation relationships have been reinforced. At the same time, local communities continue to use resources in the buffer zones and protected areas, thus causing resource degradation (Cooper, 2018) and negatively affecting the biodiversity, ecosystem services and tourism (Pueyo-Ros, 2018). Given this background, it was imperative to carry out a study in MFCA guided by the following question: in what ways do community-based enterprises that are supported by Uganda Wildlife Authority benefit the adjacent local communities and reduce degradation of resources while contributing to conservation, ecosystem services and tourism in MFCA? The specific objectives were to (1) examine the approaches applied to achieve conservation and sustainable ecosystem services (2) assess ways in which community-based enterprises supported by tourism revenue sharing scheme contribute to conservation and ecosystem services (SES) and (3) analyse the inter-relationships between community-based enterprises (CBEs), conservation, ecosystem services and tourism.

This chapter is structured as follows: the literature review is presented in the next section, followed by a description of the methodology, presentation of results and the discussion. The chapter ends with conclusions, contribution to knowledge and future research direction.

3.2 Literature review

3.2.1 Conceptual considerations

The pillar concepts of the study considered in this chapter are CBCs, community-based enterprises, ecosystem services, conservation and livelihoods, as well as nature-based tourism.

Community-Based Enterprise is a business designed to improve the livelihoods of a community acting corporately in pursuit of a common good (Valchovska & Watts, 2013). CBE is managed to yield individual and group benefits (Peredo & Chrisman, 2006) and to contribute to poverty alleviation and natural resource conservation. It requires a good understanding of the socio-economic, environmental, social and cultural values that make CBEs successful (Peredo & Chrisman, 2006). Kiss (2004) postulated three types of CBE: (1) those wholly owned by a community, (2) those owned by families or groups in a community where all members pull their assets together to ensure that they benefit from the enterprise and (3) those partly owned by a community and government, NGO or private investors. Properly managed CBEs can ameliorate financial hardships of local communities by supplementing existing livelihood strategies and sources of income.

Ecosystem services are the interface between people and nature and include the benefits obtained from the contributions of the ecosystem's biophysical structure (Potschin & Haines-Young, 2016), processes and functions (Burkhard & Maes, 2017). Although they are indispensable, ecosystem services are not sustainably used, often left out of the national accounting system and rarely broached in tourism research and development (Koko et al., 2020). Valuing ecosystem services and incorporating them in decision-making processes are, therefore, crucial for ensuring their sustainable use (Geijzendorffer et al., 2015). Ecosystem services are classified into provisioning services such as food production, materials, energy and water supply, which are directly used by people; regulating services that include the way ecosystems control other environmental media or processes; cultural services that relate to the cultural or spiritual needs of people; and supporting services which include processes and functions that underpin the other three types of services, for instance, habitat protection, soil formation and nutrient recycling (Potschin & Haines-Young, 2016).

3.2.2 Conservation and livelihoods

Conservation of natural resources mainly takes place in protected areas. There is a close relationship between conservation and local community livelihoods because the primary goal of community-based conservation is to meet the community's needs (Cobbinah et al., 2015). The global environmental conservation community recognizes local communities as vital for the success of conservation (Wali et al., 2017). As such, it is important for communities to conserve and use natural resources sustainably for their subsistence needs and to support their livelihoods (Chao et al., 2018). In MFCA, community-based conservation is being achieved through a combination of approaches that include funding of CBEs to alleviate poverty, reduce pressure on the resources and maintain ecosystem services. Through the CBEs, Uganda Wildlife Authority (UWA) embraces a 'rights for responsibilities' strategy that grants local people access to harvest certain natural resources inside MFCA as well as gives responsibilities to monitor and regulate the harvests (Solomon et al., 2012). UWA hopes that this strategy will help to alleviate poverty, increase support for conservation and reduce illegal and uncontrolled exploitation of resources.

3.2.3 Nature-based tourism

One of the most vexing problems faced by conservation area managers is how to maintain the balance between resource use, conservation, ecosystem services and tourism (Anup, 2018). Many rural areas have been targeted for development of tourism-related CBEs to generate incomes and support conservation (Rastegar, 2019). To this end, Uganda Wildlife Authority has been implementing a revenue-sharing scheme in which 20% of the gate collections is channelled to support CBEs (Ahebwa et al., 2012). Although revenue sharing is an alluring concept for promoting conservation and local community development, there are economic and institutional shortcomings that need to be addressed (Tumusiime & Vedeld, 2012). For instance, the poverty levels, economic priorities and organizational capacity of the local people need to be factored in the planning and implementation framework of the revenue sharing scheme.

3.3 Theoretical framework

In view of the multiple concepts examined in this study, the social theory of risk and uncertainty and the conservation of resources theory are applied simultaneously. The social theory of risk and uncertainty applies to the concepts of conservation and ecosystem services (Peter, 2020). The theory views society as a self-organizing sub-system that is intertwined with environment (Japp & Kusche, 2009) and offers a broad perspective on the role of society in resource exploitation and degradation occasioned by unsustainable practices that compromise ecosystem services, livelihoods and tourism. This study is also underpinned by the *conservation of resources theory*, which applies to the concepts of community-based enterprises and tourism. The theory states that humans have the responsibility to conserve resources to sustain their livelihoods (Hobfoll et al., 2000). The resources include natural capital (Halbesleben et al., 2014), such as River Nile in MFCA, which provides fish for the local communities and biodiversity that supports nature-based tourism. The theory also offers fundamental insights into understanding the relationships between community-based enterprises, tourism and sustainable livelihoods, as well as the danger of unsustainable exploitation of resources that cause degradation and loss of ecosystem services (Holmgreen et al., 2017).

3.4 Methodology

3.4.1 Study area

Murchison Falls Conservation Area (1°42′N -02° 15′N and 31° 24°E-32°14′E) is located in northwestern Uganda and covers 3,893 km^2 (Uganda Wildlife Authority, 2014). It is Uganda's largest, oldest and most-visited protected area with 31% of all national park visits (MacKenzie et al., 2017). It comprises Murchison Falls National Park (MFNP), Bugungu Wildlife Reserve, Karuma Wildlife Reserve and Ajai Wildlife Reserve (AWR) (Figure 3.1). The vegetation is Sudanian and comprises forest, swamps, savannah grasslands and woodlands dominated by scattered *Combretum* and *Acacia* species (Uganda Wildlife Authority, 2014). The southern bank of the River Nile is ethnically diverse with over 56 ethnic groups. Human population density is over 111 persons/km^2 (Dell et al., 2020). The socio-economic activities include bushmeat hunting (Dell et al., 2020), agro-pastoralism and subsistence crop

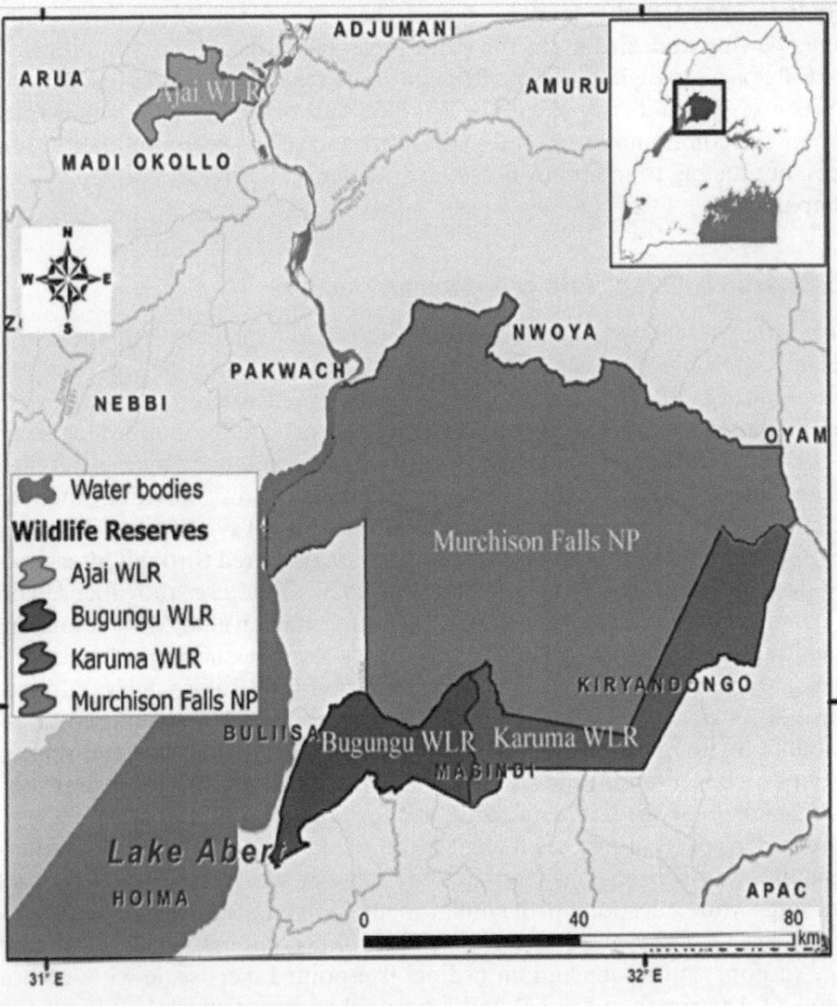

Figure 3.1 Location of the study area.

farming (Kizza et al., 2021). Below the western Rift valley escarpment, fishing in Lake Albert and the Albert Nile is the main occupation. Oil discovery and development in the Albertine graben have attracted job seekers and increased the population, especially around Lake Albert (Dowhaniuk et al., 2018).

The biodiversity consists of 145 tree species, 556 bird species, 114 species of mammals, 51 species of reptiles, 28 species of amphibians and 755 plant species (Plumptre et al., 2015). Mammals that attract visitors include lions, elephants, leopards, hippopotamus, buffalo, giraffe, Uganda kob and a range of primate species (Uganda Wildlife Authority, 2014). The delta area where River Nile flows into Lake Albert is an Important Bird Area, and the Victoria Nile stretch is an important breeding ground for the Nile crocodiles (Dendi & Luiselli, 2017). In 2018–2019, 104,000 tourists visited MFCA and tourism activities included the launch trip to the top of the Murchison Falls, game-viewing and birding in the delta area. The adjacent communities do not fully participate in tourism although some members are employed in the tourism service industry (Uganda Wildlife Authority, 2014). A small fraction of the communities operates transport and offers accommodation services, but the big tourism businesses are taken up by private individuals and companies.

3.5 Research design, data collection and analysis

A case study research design (Grossoehme, 2014) and qualitative mixed-methods approach encompassing theoretical and conceptual framework analysis (Muhumuza et al., 2013), questionnaire and key informant interviews were applied to collect data (Mohajan, 2018). The mixed-methods approach helps to overcome the limitations of applying a single method and gives a detailed insight into a problem that cannot be obtained by an individual method (Almeida, 2018). Furthermore, mixed methods permit systematic collection, organization and interpretation of information gathered through interviews and document reviews (Bracio & Szarucki, 2020). The case study focused on 15 community-based enterprises (CBEs) established around MFCA and are benefiting from Uganda Wildlife Authority's revenue-sharing scheme. The CBEs are spread across five districts bordering MFCA (Buliisa, Kiryandongo, Masindi, Nwoya and Pakwach). Twenty-five copies of a structured questionnaire with questions on participation in CBEs and tourism-related activities, conservation and sustainable ecosystem services were pretested with members of the CBEs and analysed for reliability (Taber, 2018); it generated a Cronbach alpha coefficient mean of 0.83 (SD = 0.016). The questionnaire was subsequently revised to improve clarity of wordings and meanings while questions with similar focus were merged. To achieve objectives 1 and 2, the revised questionnaire with a combination of open and fixed response questions and an ordinal five-point Likert scale with anchor points of 1 (strongly disagree) and 5 (strongly agree) (Wu & Leung, 2017), for measuring participation in CBES and conservation, was administered to 167 purposively selected respondents, by research assistants and one co-author who were familiar with the local languages. The lower end of the Likert scale (1) indicated weak participation in CBEs and support of conservation while the highest figure (5) indicated strong participation and a pro-conservation response.

To achieve objective 3, open-ended interviews were held with five MFCA staff members in order to gain a comprehensive understanding of the linkages between CBES, local community use of resources, conservation, ecosystem services and tourism. Literature was also reviewed to explain the interrelationships between community-based enterprises, conservation, ecosystem services and tourism in the context of the *theory of risk and uncertainty* and the *conservation of resources theory* based on Bradbury-Jones's (2014) approach. The questionnaire responses were edited, coded and entered in SPSS version 21 to create a data file and generate statistical summary. Data were subjected to chi-square tests to show the relationship between participation in CBEs, conservation and sustainable ecosystem services (SES). A preliminary insight on the data set was sought by Pearson's correlation analysis to determine the degree of interrelatedness between membership of CBE and participation in conservation and sustainability of ecosystem services. The analysis revealed a moderate correlation ($r = 0.50$) that confirmed the suitability of the data set for subsequent regression analysis.

3.6 Results and discussion

3.6.1 Demographic and socio-economic profile of the respondents

Out of 167 respondents that are involved in CBEs, 54.5% were males and 45.5% were females, 10.8% had no formal education, 59.3% had primary education, 24.6% had secondary school education while 5.4% had tertiary education. The majority (49.7%) were 31–50 years old, engaged in farming (97% multiple response), 66.7% in fishing and small-scale businesses (22%). Sixty-nine per cent earned less than Uganda shillings 200,000 (approx. USD 55) per month. Forty per cent indicated that they had 4–7 persons per household while 51% had more than 8 persons per household. The majority (91%) had lived near MFCA for more than ten years, 59.9% owned 1–3 hectares while 8.4% owned more than 12 hectares of land.

3.6.2 Conservation and sustainable ecosystem services in MFCA

Uganda Wildlife Authority employs various community-based conservation approaches including collaborative resource management, community outreach, corporate social responsibility, monitoring of wildlife outside the conservation areas and revenue sharing (Table 3.1). Interviews with key informants from Uganda Wildlife Authority (the institution managing MFCA) revealed that the MFCA General Management Plan (GMP) is not explicit on sustainability of ecosystem services. However, the role of MFCA in providing sustainable ecosystem services is viewed in terms of the ecosystem management approach, which is stated in the management plan prepared according to Uganda Wildlife Act 2019 and Uganda Wildlife Policy 2014 guidelines. More than that, the interviews revealed that increased oil exploration activities in MFCA are likely to negatively impact the ecosystem services and baseline studies on the status of ecosystems and allied environmental services are being undertaken for planning long-term ecosystem monitoring.

The application of community-based approaches (Table 3.1), such as collaborative resource assessment and community outreach has helped to build social capital and networking among the communities which allow environmental, social and economic issues in MFCA to be addressed. On top of that, approaches such as corporate social responsibility and revenue sharing help to divert their attention from engaging in illicit resource use that

Table 3.1 Community-based conservation approaches that also sustain ecosystem services

Approach	*Description of activities*
Collaborative resource management	• Regulated firewood collection, fishing, beekeeping
Community outreach	• Sensitization of local communities on conservation and ecosystems services values
Corporate social responsibility	• Providing support to local schools and churches to enlist support for conservation
Monitoring wildlife outside the conservation area	• Tracking wildlife movements and returning them to the conservation area • Monitoring land-use activities on the fringes of MFCA
Revenue sharing	• Channelling 20% of gate collections to support community-based enterprises and development initiatives
Energy conservation and tree planting	• Promoting on-farm planting as source of firewood • Promoting energy-saving technologies

compromises conservation and ecosystem services. Tourism as a downstream enterprise benefits from the positive ripple effect of the approaches. The study further established that communities are mainly engaged in conservation through CBE activities (Table 3.2) that indirectly contribute to ecosystem services because of the benefits they get from conservation such as sustainable water supply that supports fishing, habitat for bees, source of raw materials for making crafts, firewood, mushrooms and others (Table 3.2). CBEs such as duck and goat rearing, metal fabrication and motorcycle transport services provided alternative sources of income that diverted the local communities from engaging in destructive resource extraction activities, hence promoting conservation.

Despite the different conservation approaches applied in MFCA, a number of challenges to sustainable ecosystem services are experienced. These include local communities flouting guidelines on resource extraction and engaging in illicit activities, increased human population that exerts pressure on the conservation area (search of farmlands and firewood), political promises to grant parts of the conservation area to local people and inadequate awareness of conservation and ecosystems services values.

3.7 Contribution of CBEs to conservation and sustainable ecosystem services (SES)

The local communities living around MFCA have low education backgrounds and are generally with low incomes. As noted earlier, out of the 167 respondents, the majority have primary education (n = 96), own 1–3 acres of land (n = 83) with a monthly income of less than UGX 200,000 (approx.

Table 3.2 Community-based enterprises established around MFCA

Type of community-based enterprise	*Nature of business*
1 Resource users	• The resource user group regulates resource extraction from MFCA in conjunction with Uganda Wildlife Authority, e.g., firewood, mushroom and herbal medicines.
2 River fishing	• River Nile capture fisheries group established to regulate fishing, make local people value conservation and establish fish ponds for livelihoods.
3 Beekeeping and honey production	• Apiary management and honey production for improved livelihoods. The group digs trenches to stop elephants from crossing across to community farms.
4 Ecotourism	• Women's group involved in providing accommodation facilities and sale of handicrafts to tourists.
5 Duck rearing	• Rearing of ducks for production of eggs and meat sold to restaurants and lodges to discourage engaging in illegal fishing in the River Nile and poaching park animals for bushmeat.
6 Handicraft making	• Production of handicrafts sold mainly to tourists as a source of employment and income.
7 Goat rearing	• Goats are reared for income and as a source of meat to reduce poaching wildlife for bushmeat.
8 Metal fabricators	• Production of metallic door and window frames, chairs, table stands and other fabrications for hotels and household use.
9 Motorcycle transporters group	• Transportation of passengers as an alternative source of income to reduce reliance on resources such as firewood and thatch grass collected from the park.

USD 55, $n = 115$) and yet the majority ($n = 85$) have more than 8 persons per household (Table 3.3). The number of people that participated in CBEs, conservation and sustainable ecosystem services activities varied with the demographic and socio-economic characteristics. For example, more males than females participated in CBEs (80 males, 63 females) and activities related to conservation and sustainable ecosystem services (79 males, 63 females). Respondents aged 31–50 years participated more than those in the other age groups in the CBES ($n = 74$) and conservation and sustainable ecosystem services ($n = 72$). Those who had stayed for more than 11 years near MFCA participated more in CBEs ($n = 130$) and conservation and sustainable ecosystem services than those who had stayed for less than ten years. The implication of the above demographic characteristics is that the local communities living around MFCA have limited prospects for employment and on-farm

Table 3.3 Demographic/socio-economic characteristics and participation in CBEs, conservation and SES activities ($n = 167$)

Demographic characteristics		*Participation in CBEs*		*Participation in conservation and SES activities*	
		Yes	*No*	*Yes*	*No*
Gender	Male	80	10	79	11
	Female	63	14	63	14
Marital status	Married	113	13	109	17
	Single	14	3	13	4
	Divorced	1	2	1	2
	Separated	4	0	4	0
	Widowed	7	1	7	1
Age group	18–30	34	5	32	7
	31–50	74	9	72	11
	51–64	28	4	28	4
	65+	7	6	10	3
Level of education	No formal Education	14	4	17	1
	Primary	82	14	80	16
	Secondary	37	4	34	7
	Tertiary/University	8	1	8	1
Household size	Less than 3	11	2	11	2
	4–7	55	11	55	11
	8 and above	74	11	73	12
Size of landholding	1–3 acres	84	16	85	15
	4–7 acres	30	7	29	8
	8–11 acres	9	0	9	0
	12 acres and above	14	0	12	2
Period of stay near MFCA	Less than 1 year	1	1	1	1
	1–5 years	6	1	5	2
	6–10 years	6	0	6	0
	11 years and above	130	22	130	22
Average monthly household income in UGX (USD1 = UGX 3,600)	Less than 200,000	98	17	97	18
	200,001–400,000	24	3	24	3
	400,001–600,000	5	1	5	1
	600,001–800,000	5	2	6	1
	Above 800,001	9	0	8	1

SES = Sustainable ecosystem services.

incomes as they have low education and small landholdings in spite of the large household sizes that would provide farm labour. Therefore, MFCA management needs to sustain support to the communities' alternative sources of income to reduce pressure on the resources, sustain ecosystem services and tourism.

Responses to statements on participation in conservation and sustainable ecosystem services as well as tourism differed across the five districts that border MFCA as four statements were significant at $p < 0.05$ while two were not. The significant statements were about participation in CBEs, conservation and sustainable ecosystem services that enhanced the respondents' understanding of the value of conservation and ecosystem services in ($\chi^2 = 23.0, p < 0.05$), improved their knowledge of conservation and community-based tourism development ($\chi^2 = 25.87, p < 0.05$), enabled them to know tourism policy and how it guides community-based tourism development ($\chi^2 = 39.36, p < 0.01$) and learnt about wildlife laws that foster conservation for community-based tourism development ($\chi^2 = 46.32, p < 0.01$) (Table 3.4). There were no significant relationships between participation and improved on-farm resource conservation and improved human-wildlife conflict management, possibly because the local communities do not link their participation in CBEs to conservation of on-farm resources. Instead, they link their participation in the CBEs to conservation of resources in MFCA, which is expected, because it is the topical message in the environmental awareness delivered by the staff during community-based conservation outreach. It is also likely that the local community regards human–wildlife conflict management as MFCA business and not theirs and do not include it as part of the CBE activities that they undertake.

The regression analysis revealed a weak relationship between membership of CBE and participation in conservation and sustainable ecosystem services activities ($R^2 = 0.245, p < 0.05$). This implies that other factors could have motivated the respondents to join the CBEs as avenues for participating in conservation and sustaining ecosystem services. For instance, the prospects of getting financial benefit in form of increased incomes from CBEs that are supported under UWA's revenue-sharing scheme ($n = 67$ strongly agreed) and the desire to diversify sources of income ($n = 99$ strongly agreed) (Table 3.5) could have been a major motivation for them to join the CBEs rather than the zeal to conserve resources and sustain ecosystem services in MFCA. Interviews with the local communities revealed that the local community members yearned to exploit the protected resources which would negatively impact ecosystem services and tourism. However, Wali et al. (2017) noted that sustainable use of resources in protected areas is a potent conservation tool if local people derive social and economic benefits from such use. Benefits accruing directly to individuals often encourage pro-conservation behaviour than community-level benefits. A similar view was echoed by the MFCA Wardens in charge of community-based conservation, who noted that the CBEs are supported under UWA's revenue-sharing scheme to eradicate illicit harvesting of resources, secure conservation and increase the ancillary benefits.

In addition, the wardens reported that the majority of local community members living around MFCA are impoverished multi-ethnic immigrants with low level of education and high socio-economic expectations, which make them increasingly dependent on the conservation area's resources for subsistence. Furthermore, inadequate funding for community-based conservation programmes in MFCA that would ensure effective community participation in conservation and sustainable ecosystem services was a constraint identified during preparation of the UWA Community Conservation Policy 2019 (UWA, 2019). To overcome this challenge, one of the policy's strategic

Table 3.4 Chi-square test of participation in conservation, sustenance of ecosystem services and tourism across the study districts bordering MFCA

Conservation, SES and tourism statements	*Item responses*	*Kiryandongo n(%)*	*Pakwach n(%)*	*Masindi n(%)*	*Nwoya n(%)*	*Buliisa n(%)*	χ^2	*df*	*p*
Improved natural resource conservation on my farm	*Strongly agree*	24(17.8)	6(4.4)	11(12.6)	0(0)	12(8.9)	20.439	12	0.059
	Agree	14(10.4)	4(3.0)	5(3.7)	0(0)	7(5.2)			
	Neutral	0(0)	2(1.5)	0(0)	0(0)	1(0.7)			
	Disagree	7(5.2)	11(8.1)	10(7.4)	0(0)	11(8.1)			
	Strongly disagree	1(0.7)	2(1.5)	1(0.7)	0(0)	0(0)			
Increased appreciation of conservation and ecosystem services in MFCA	*Strongly agree*	25(10.9)	8(6.1)	16(12.1)	0(0)	12(9.1)	18.850	12	0.092
	Agree	18(13.6)	9(6.8)	10(7.6)	0(0)	10(7.6)			
	Neutral	2(1.5)	2(1.5)	0(0)	0(0)	2(1.5)			
	Disagree	0(0)	5(3.8)	5(3.8)	0(0)	7(5.3)			
	Strongly disagree	0(0)	1(0.8)	0(0)	0(0)	0(0)			
Better understanding of the value of conservation and ecosystem services in MFCA	*Strongly agree*	26(19.5)	8(6.0)	18(13.3)	0(0)	15(11.3)	23.003	12	0.028
	Agree	17(12.8)	12(9.0)	9(6.8)	0(0)	8(6.0)			
	Neutral	1(0.8)	1(0.8)	1(0.8)	0(0)	2(1.5)			
	Disagree	0(0)	2(1.5)	5(3.8)	0(0)	6(4.5)			
	Strongly disagree	0(0)	2(1.5)	0(0)	0(0)	0(0)			

Improved knowledge of conservation and community-based tourism development	*Strongly agree*	20(15.0)	8(6.0)	16(12.6)	0(0)	8(6.0)	25.875	12	0.011
	Agree	16(12.0)	7(5.3)	12(9.0)	0(0)	9(6.8)			
	Neutral	2(1.5)	0(0)	0(0)	0(0)	1(0.8)			
	Disagree	7(5.3)	4(4.5)	5(3.8)	0(0)	13(9.8)			
	Strongly disagree	0(0)	3(2.3)	0(0)	0(0)	0(0)			
Able to know tourism policy and how they guide community-based tourism development	*Strongly agree*	25(16.6)	16(9.3)	15(9.9)	1(0.7)	17(11.3)	39.360	20	0.006
	Agree	12(7.9)	4(2.6)	7(4.6)	4(2.6)	7(4.6)			
	Neutral	0(0)	1(0.7)	2(1.3)	0(0)	0(0)			
	Disagree	8(5.3)	7(4.6)	8(5.3)	6(4.0)	7(4.6)			
	Strongly disagree	0(0)	0(0)	1(0.7)	3(2.0)	0(0)			
Learnt about wildlife laws that foster conservation for community-based tourism development	*Strongly agree*	29(19.2)	14(9.3)	13(8.6)	2(1.3)	18(11.9)	46.329	20	0.001
	Agree	13(8.6)	8(5.3)	13(8.6)	6(4.0)	10(6.6)			
	Neutral	1(0.7)	0(0)	3(2.0)	0(0)	0(0)			
	Disagree	3(2.0)	4(2.6)	4(2.6)	4(2.6)	3(2.0)			
	Strongly disagree	0(0)	0(0)	0(0)	2(1.3)	0(0)			

Table 3.5 Reasons for participation in CBE

Reason statements	*Strongly agree*	*Agree*	*Neutral*	*Disagree*	*Strongly disagree*
(1) Contribute to community development (*n* = 145)	71	40	1	18	15
(2) Support conservation and tourism (*n* = 150)	96	43	4	5	2
(3) Enable me get support from UWA (*n* = 147)	67	65	3	8	4
(4) Diversify my income (*n* = 149)	99	35	1	6	8
(5) Increase chances of benefiting from tourism-related opportunities (*n* = 150)	111	31	3	3	2
(6) Know tourism policies that guide community development (*n* = 151)	72	34	5	36	4
(7) Be sensitized about wildlife laws to foster conservation (*n* = 151)	76	50	5	18	2
(8) Gain knowledge of tourism products in MFCA (*n* = 149)	52	50	5	34	8
(9) Know about tourism investment opportunities in MFCA (*n* = 149)	51	43	7	39	10

objectives focuses on investments in community-based enterprises as an approach for promoting conservation and maintaining ecosystem services.

Local communities need to participate effectively in conservation and maintenance of ecosystem services. Although the meaning of '*community participation*' in conservation discourses has been contested (Thomas, 2013), local participation is crucial because of the social and environmental costs of conservation they often bear. Experience from Kenya indicates that CBEs are used as vehicles for transforming local communities and achieving conservation of resources (Juma & Khademi-Vidra, 2019).

In MFCA, more CBEs should be supported because they are genuine means of building strong and interdependent links between resource use, conservation, ecosystem services and tourism. One of the most widely accepted principles of achieving this is through meaningful local community participation. Tosun (2006) classified community participation into spontaneous, coercive and induced participation, while Bahreldin and Ariga (2011) classified it as traditional community-driven and the legislative government-initiated forms of community participation. In the case of MFCA, a blend of traditional community-driven and induced participation should be promoted for two main reasons: firstly, it allows the local community to have a say in conservation of natural resources through a bottom-up approach and representation in the resource management committee. Secondly, UWA's support for the CBEs through the revenue-sharing scheme induces the local community to participate directly in conservation and in the project management committee of the scheme.

Local people living around MFCA participate in conservation through community-based natural resource management (CBNRM) agreement, which allows them to harvest regulated quantities of resources. CBNRM reforms the conventional protectionist conservation philosophy and is perceived as a remedy to natural resource degradation underlain by the assumption that when local communities derive benefits, they cultivate the spirit of ownership and use them sustainably and hence support biodiversity conservation (Qin et al., 2020). In MFCA, local communities are allowed to harvest limited quantities of resources such as firewood from the savanna woodlands and fish from the River Nile. Fishing is a common livelihood activity that protected area managers often permit local communities to engage in under CBNRM arrangement. For instance, in Kibale National Park (western Uganda), the managers permit local communities to fish inside the Park under CBNRM (Solomon et al., 2012). CBNRM is a lever for promoting conservation, generating income for the local people and contributing to poverty alleviation. Furthermore, it encourages support for conservation by building a critical mass of conservation-minded people and inculcating pro-conservation behaviours that help to reduce pressure on natural resources (Grilli & Curtis, 2019). Although CBNRM is lauded as a worthwhile conservation tool, exploitation of resources needs to be regularly monitored to prevent illicit behaviour and sustain the resource base (Qin et al., 2020).

3.8 The inter-relationships between CBEs, conservation, ecosystem services and tourism

Local communities living adjacent to conservation areas have cultural norms that guide exploitation of resources (Richards, 2018). The CBEs support local community livelihoods through harvesting of limited quantities of resources from MFCA. Grounded on contextual analysis of reviewed literature and the social theory of risk and uncertainty, it is clear that local communities can sustainably exploit resources without jeopardizing conservation, ecosystem services and tourism. Conservation of resources theory discourages human activities that cause environmental stress and compromise ecosystem services and tourism in conservation areas such as MFCA. Nonetheless, sustainable use of resources, which is integral to a healthy ecosystem, can motivate local communities to participate in conservation that also sustains ecosystem services (Vasseur & Hart, 2002). The theoretical framework underpinning this study applies a multi-conceptual approach that provides a clear line of sight and weaves together an understanding of the interrelationships between resource use, livelihoods, conservation, sustainable ecosystem services and

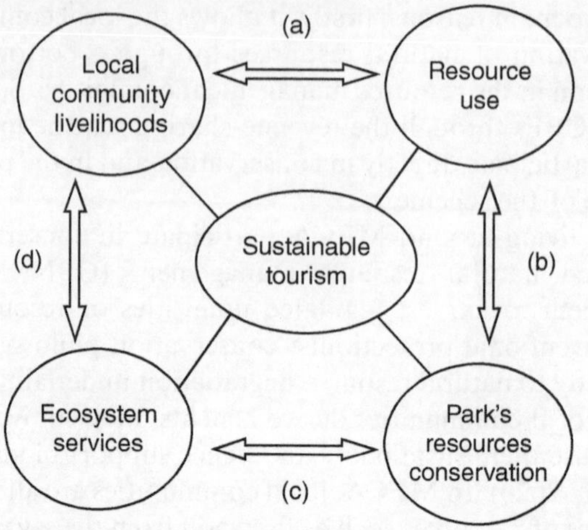

Figure 3.2 The interrelationships between local community, resource use, conservation, ecosystem services and sustainable tourism in a conservation area.

Source: Authors (2021).

tourism. Analysis of these concepts, within the theoretical framework and in the context of MFCA, generates the interrelationships illustrated in Figure 3.2 and indicated by arrows labelled (a) to (d).

The above diagram shows that the level of resource use by local communities can affect the availability of resources that sustain livelihoods (a) and unsustainable use of resources creates social and conservation risks (b), which make it imperative to understand the local community–natural resources relationship (Peter, 2020). Resource conservation determines the sustainability of ecosystem services and vice versa. Therefore, sustainable resource conservation should ensure the overall capacity of MFCA to provide ecosystem services (c) and that the socio-economic benefits from tourism should not diminish. Furthermore, ecosystem services and local community livelihoods (d) are interdependent, and involvement of local communities in activities that jeopardize ecosystem services threatens the provisioning and regulatory benefits derived from ecosystems. At the intersection of the four concepts is sustainable tourism, which is a major activity in a conservation area such as MCFA that depends on and is affected by the environmental and socio-economic dynamics and processes. From the above perspectives, the role of UWA is to regulate resource use, work in partnership with local communities through the CBEs, and coordinate development and management of tourism in MFCA through the application of a blend of community-based conservation approaches that balance local community needs and livelihoods with conservation and sustainable ecosystem services.

Figure 3.2 is also reminiscent of Stone and Stone's (2019) systems thinking approach, in which the nexus between the four concepts represents a 'sustainable tourism system'. They applied the systems thinking approach to illuminate the understanding of the relationships between protected area, tourism and community livelihoods in Botswana's Chobe national park, which is analogous to MFCA. They define a 'system' as a set of connected and interrelated [concepts] and 'systems thinking' as the ability to understand the whole rather than part of it. In their view, the essential properties of a system taken as a whole derive from the interactions of its parts which, in this study, are the local community livelihoods, resource use, conservation and ecosystem services.

3.9 Conclusions and contribution to knowledge

Uganda Wildlife Authority, through .the revenue-sharing scheme, has supported the local communities that live around MFCA to engage in CBEs that apply a combination of community-based conservation approaches to sustain their livelihoods and ecosystem services. Review of documents has revealed that the General Management Plan of MFCA is not explicit on management of the conservation area for provision of ecosystem services. Instead, ecosystem services are viewed as an integral part of the management of MFCA's entire ecosystem, which is primarily aimed at supporting wildlife-based tourism.

It is clear from this study that CBEs provide avenues for local communities' participation in conservation activities that sustain ecosystem services and enhance knowledge of wildlife laws and tourism policies that guide community-based tourism development. Similarly, it has become clear that there is a link between the demographic and socio-economic characteristics of the local communities living around MFCA and their participation in CBEs, conservation and sustainable ecosystem services. For example, the number of males and females who participate is more than those who do not. At the same time, the number of males is higher than that of females who participate in conservation-related activities. The number of respondents in the 31–50 years age group participate more than those in the other age groups. However, for sustainability of conservation and ecosystem services, the management of MFCA should target the youth in the age group 18–30 years to increase their participation as the future lies in their hands.

The formation of CBEs has enhanced understanding of the value of conservation and ecosystem services among the local communities living around MFCA in the five study districts. In view of this, the management of MFCA should support formation of more CBEs in other adjoining areas to spread and strengthen community-based conservation initiatives that support ecosystem services and tourism. In MFCA, optimal interrelationships between local community livelihoods, sustainable resource use, conservation and provision of ecosystem services can support sustainable tourism. However, Uganda Wildlife Authority needs to articulate these concepts in the management plans of conservation areas and embed them in policies for implementation as pillars of sustainable tourism.

This chapter provides a formidable theoretical framework that weaves together and deepens the understanding of the nexus between resource use, conservation, ecosystem services and tourism in MFCA that was not available before. Furthermore, the chapter is a new milestone that adds originality to research on the interrelationships between these concepts that crucially inform the practice of community-based conservation. Therefore, future research in this area should focus on generating empirical evidence based on valuation of ecosystem services and elucidating how it is inextricably linked to tourism in conservation areas. Such information is needed to refocus the management plans that integrate the concept of sustainable ecosystem services with conservation and nature-based tourism.

Acknowledgement

This study was supported by Makerere University under the Research and Innovations Fund (RIF) provided by the Government of Uganda. We gratefully acknowledge the participation of local communities in the study areas and Uganda Wildlife Authority.

References

Ahebwa, W.M., van der Duim, R., & Sandbrook, C., 2012. Tourism revenue sharing policy at Bwindi Impenetrable National Park, Uganda: A policy arrangements approach. *Journal of Sustainable Tourism*, *20*(3), pp. 377–394.

Almeida, F., 2018. Strategies to perform a mixed methods study. *European Journal of Education Studies*, *5*(1), pp. 137–151

Anup, K. C. (2018). Tourism and its role in environmental conservation. *Journal of Tourism and Hospitality Education*, *8*, pp. 30–47.

Bahreldin, I. & Ariga, T., 2011. Evaluation of two types of community participation in development projects: A case study of the Sudanese Neighbourhood of Al-Shigla. *Journal of Architecture and Planning. Architecture Institute of Japan (AIJ)*, *76*(670), pp. 2369–2379.

Barnes-Mauthe, M., Oleson, K.L., Brander, L.M., Zafindrasilivonona, B., Oliver, T.A., & van Beukering, P., 2015. Social capital as an ecosystem service: Evidence from a locally managed marine area. *Ecosystem Services*, *16*, pp. 283–293.

Berkes, F., 2007. Community-based conservation in a globalized world. *Proceedings of the National Academy of Sciences*, *104*(39), pp. 15188–15193.

Bracio, K. & Szarucki, M., 2020. Mixed methods utilisation in innovation management research: A systematic literature review and meta-summary. *Journal of Risk and Financial Management*, *13*(11), p.252.

Bradbury-Jones, C., Taylor, J., & Herber, O., 2014. How theory is used and articulated in qualitative research: Development of a new typology. *Social Science & Medicine*, *120*, pp. 135–141.

Brooks, J., Waylen, K.A., & Mulder, M.B., 2013. Assessing community-based conservation projects: A systematic review and multilevel analysis of attitudinal, behavioral, ecological, and economic outcomes. *Environmental Evidence*, *2*(1), pp. 1–34.

Burkhard, B. & Maes, J., 2017. Mapping ecosystem services. *Advanced Books*, *1*, p. e12837.

Butler, C.D. & Oluoch-Kosura, W., 2006. Linking future ecosystem services and future human well-being. *Ecology and Society*, *11*(1), pp. 1–17.

Chao, F., Yunli, B., Linxiu, Z., Shuai, W., & Xue, Y., 2018. Coupling conservation and livelihoods for sustainable management of protected areas in East Africa. *Journal of Resources and Ecology*, *9*(3), pp. 266–272.

Cobbinah, P.B., Black, R., & Thwaites, R., 2015. Biodiversity conservation and livelihoods in rural Ghana: Impacts and coping strategies. *Environmental Development*, *15*, pp. 79–93.

Cooper, R. (2018). *Current and projected impacts of renewable natural resources degradation on economic development in Uganda. K4D Emerging Issues Report*. Institute of Development Studies, Brighton, UK.

Costanza, R., De Groot, R., Braat, L., Kubiszewski, I., Fioramonti, L., Sutton, P., Farber, S., & Grasso, M., 2017. Twenty years of ecosystem services: How far have we come and how far do we still need to go? *Ecosystem Services*, *28*, pp. 1–16.

Dell, B.M., Souza, M.J., & Willcox, A.S., 2020. Attitudes, practices, and zoonoses awareness of community members involved in the bushmeat trade near Murchison Falls National Park, northern Uganda. *PloS One*, *15*(9), p. e0239599.

Dendi, D. & Luiselli, L., 2017. Population surveys of Nile crocodiles (Crocodylus niloticus) in the Murchison Falls National Park, Victoria Nile, Uganda. *European Journal of Ecology*, *3*(2), pp. 67–76.

Dewi, S., Van Noordwijk, M., Ekadinata, A., & Pfund, J.L., 2013. Protected areas within multifunctional landscapes: Squeezing out intermediate land use intensities in the tropics?. *Land Use Policy*, *30*(1), pp. 38–56.

Dowhaniuk, N., Hartter, J., Ryan, S.J., Palace, M.W., & Congalton, R.G., 2018. The impact of industrial oil development on a protected area landscape: Demographic and social change at Murchison Falls Conservation Area, Uganda. *Population and Environment*, *39*(3), pp. 197–218.

Eshun, G. & Tonto, J.N.P., 2014. Community-based ecotourism: Its socio-economic impacts at Boabeng-Fiema Monkey Sanctuary, Ghana. *Bulletin of Geography. Socio-economic Series*, (26), pp. 67–81.

Evans, K.L., Ewen, J.G., Guillera-Arroita, G., Johnson, J.A., Penteriani, V., Ryan, S.J., Sollmann, R., & Gordon, I.J., 2020. Conservation in the maelstrom of Covid-19 – A call to action to solve the challenge, exploit opportunities and prepare for the next pandemic. *Animal Conservation*. *10*(1), 1–4

Galvin, K.A., Beeton, T.A., & Luizza, M.W., 2018. African community-based conservation. *Ecology and Society*, *23*(3), pp. *23*(3), pp. 1–32.

Geijzendorffer, I.R., Martín-López, B., & Roche, P.K., 2015. Improving the identification of mismatches in ecosystem services assessments. *Ecological Indicators*, *52*, pp. 320–331.

Grilli, G. & Curtis, J., 2019. Encouraging pro-environmental behaviors: A review of methods and approaches. ESRI Working Paper No. 645 December 2019.

Grossoehme, D.H., 2014. Overview of qualitative research. *Journal of Health Care Chaplaincy*, *20*(3), pp. 109–122.

Halbesleben, J.R., Neveu, J.P., Paustian-Underdahl, S.C., & Westman, M., 2014. Getting to the "COR" understanding the role of resources in conservation of resources theory. *Journal of Management*, *40*(5), pp. 1334–1364.

Hauser, V.H., Engen, S., Muñoz, L., & Fauchald, P., 2021. Assessing a nationwide policy reform toward community-based conservation of biological diversity and ecosystem services in the Alpine North. *Ecosystem Services*, *49*, p. 101289.

Hobfoll, S.E. & Shirom, A. (2000) Conservation of resources theory: Applications to stress and management in the workplace. In Golembiewski, R.T., (Ed.), *Handbook of Organization Behavior*, Marcel Dekker, New York, 57–80.

Holmgreen, L., Tirone, V., Gerhart, J., & Hobfoll, S. E. (2017). Conservation of resources theory. *The Handbook of Stress and Health: A Guide to Research and Practice*, *2*(7), pp. 443–457.

Japp, K.P. & Kusche, I. 2009. Systems theory and risk. In O.Z. Jenns (Ed.), *Social Theory of Risk and Uncertainty: An Introduction*, pp. 76–96. John Wiley & Sons, New York.

Jax, K., Barton, D. N., Chan, K. M., De Groot, R., Doyle, U., Eser, U., …, & Wichmann, S., 2013. Ecosystem services and ethics. *Ecological Economics*, *93*, pp. 260–268.

Jones, L., Norton, L., Austin, Z., Browne, A.L., Donovan, D., Emmett, B.A., Grabowski, Z.J., Howard, D.C., Jones, J.P., Kenter, J.O., & Manley, W., 2016. Stocks and flows of natural and human-derived capital in ecosystem services. *Land Use Policy*, *52*, pp. 151–162.

Juma, L.O. & Khademi-Vidra, A., 2019. Community-based tourism and sustainable development of rural regions in Kenya; perceptions of the citizenry. *Sustainability*, *11*(17), p.4733.

Kiss, A., 2004. Is community-based ecotourism a good use of biodiversity conservation funds?. *Trends in Ecology & Evolution*, *19*(5), pp. 232–237.

Kizza, D., Ocaido, M., Mugisha, A., Azuba, R., Nalule, S., Onyuth, H., Musinguzi, S.P., Nalubwama, S., & Waiswa, C., 2021. Knowledge, attitudes and practices on bovine trypanosomosis control in pastoral and agro pastoral communities surrounding Murchison Falls National Park, Uganda. *Tropical Animal Health and Production*, *53*(2), pp. 1–11.

Koko, I.A., Misana, S.B., Kessler, A., & Fleskens, L., 2020. Valuing ecosystem services: Stakeholders' perceptions and monetary values of ecosystem services in the Kilombero wetland of Tanzania. *Ecosystems and People*, *16*(1), pp. 411–426.

Macheka, M. T., Maharaj, P., & Nzima, D. (2021). Choosing between environmental conservation and survival: Exploring the link between livelihoods and the natural environment in rural Zimbabwe. *South African Geographical Journal*, *103*(3), pp. 358–373.

MacKenzie, C.A., Salerno, J., Hartter, J., Chapman, C.A., Reyna, R., Tumusiime, D.M., & Drake, M., 2017. Changing perceptions of protected area benefits and problems around Kibale National Park, Uganda. *Journal of Environmental Management*, *200*, pp. 217–228.

Mohajan, H.K., 2018. Qualitative research methodology in social sciences and related subjects. *Journal of Economic Development, Environment and People*, *7*(1), pp. 23–48.

Muhumuza, M., Sanders, M., & Balkwill, K. (2013) A theoretical framework for investigating ecological problems associated with biodiversity conservation in national parks: A case of the Rwenzori Mountains National Park, Uganda. *Open Journal of Ecology*, 3(2), pp. 196–204. doi: 10.4236/oje.2013.32023

Nelson, F., Muyamwa-Mupeta, P., Muyengwa, S., Sulle, E., & Kaelo, D., 2021. Progress or regression? Institutional evolutions of community-based conservation in eastern and southern Africa. *Conservation Science and Practice*, *3*(1), p. e302.

Pagdee, A. & Kawasaki, J., 2021. The importance of community perceptions and capacity building in payment for ecosystems services: A case study at Phu Kao, Thailand. *Ecosystem Services*, *47*, p. 101224.

Peredo, A.M. & Chrisman, J.J., 2006. Toward a theory of community-based enterprise. *Academy of Management Review*, *31*(2), pp. 309–328.

Peter, S., 2020. Integrating key insights of sociological risk theory into the ecosystem services framework. *Sustainability*, *12*(16), p.6437.

Plumptre, A. J., Ayebare, S., Mugabe, H., Kirunda, B., Sekisambu, R., Mulondo, P., & Mudumba, T., 2015. *Biodiversity Surveys of Murchison Falls Protected Area*, p. 26. Wildlife Conservation Society, Kampala, Uganda.

Potschin, M., Haines-Young, R., Fish, R., & Turner, R. K. (Eds.). (2016). Routledge Handbook of Ecosystem Services. Routledge, London and New York.

Pueyo-Ros, J., 2018. The role of tourism in the ecosystem services framework. *Land*, *7*(3), p. 111.

Qin, H., Bass, M., Ulrich-Schad, J. D., Matarrita-Cascante, D., Sanders, C., & Bekee, B. (2020). Community, natural resources, and sustainability: Overview of an interdisciplinary and international literature. *Sustainability*, *12*(3), pp. 1–14.

Rastegar, R., 2019. Tourism development and conservation, do local resident attitudes matter? *International Journal of Tourism Sciences*, *19*(3), pp. 181–191.

Reyes-Garcia, V., Ruiz-Mallen, I., Porter-Bolland, L., Garcia-Frapolli, E., Ellis, E.A., Mendez, M.E., Pritchard, D.J., & Sanchez-Gonzalez, M.C., 2013. Local understandings of conservation in southeastern Mexico and their implications for community-based conservation as an alternative paradigm. *Conservation Biology*, *27*(4), pp. 856–865.

Richards, G., 2018. Cultural tourism: A review of recent research and trends. *Journal of Hospitality and Tourism Management*, *36*, pp. 12–21.

Rogerson, C.M. & Baum, T., 2020. COVID-19 and African tourism research agendas. *Development Southern Africa*, *37*(5), pp. 727–741.

Solomon, J., Jacobson, S.K., & Liu, I., 2012. Fishing for a solution: Can collaborative resource management reduce poverty and support conservation? *Environmental Conservation*, *39*(1), pp. 51–61.

Stone, M.T. & Stone, L.S., 2019. Understanding the relationship between protected areas, tourism and community livelihoods – A systems thinking approach. In M. Mkono (Ed.), *Positive Tourism in Africa*, pp. 36–52. Routledge, London.

Taber, K. S. (2018). The use of Cronbach's alpha when developing and reporting research instruments in science education. *Research in Science Education*, *48*(6), pp. 1273–1296.

Thani, P.R., Kc, R., Sharma, B.K., Kandel, P., & Nepal, K., 2019. Integrating biodiversity conservation and ecosystem services into operational plan of community forest in Nepal: Status and gaps. *Banko Janakari*, *29*(1), pp. 3–11.

Thomas, Y., 2013. Ecotourism development in Ghana: A case of selected communities in the Brong-Ahafo Region. *Journal of Hospitality Management and Tourism*, *4*(3), pp. 69–79.

Thurstan, R. H., Hockings, K. J., Hedlund, J. S., Bersacola, E., Collins, C., Early, R., ..., & Bunbury, N. (2021). Envisioning a resilient future for biodiversity conservation in the wake of the COVID-19 pandemic. *People and Nature*, *3*(5), pp. 990–1013.

Tosun, C., 2006. Expected nature of community participation in tourism development. *Tourism Management*, *27*(3), pp. 493–504.

Tumusiime, D.M. & Vedeld, P., 2012. False promise or false premise? Using tourism revenue sharing to promote conservation and poverty reduction in Uganda. *Conservation and Society*, *10*(1), pp. 15–28.

Uganda Wildlife Authority, 2014. *Murchison Falls Protected Area General Management Plan 2013-2023*. Ministry of Tourism, Wildlife and Antiquities, Kampala, Uganda.

Uganda Wildlife Authority, 2019. *Uganda Wildlife Authority*, Community Conservation Policy, Kampala, Uganda.

Valchovska, S. & Watts, G., 2013. *Community-based rural enterprise in the UK – Model development and success factors*. Report of the research project "Scaling the Rural Enterprise" – Bridging the Urban Rural Divide (BURD) funded by EPSRC, UK.

Vasseur, L. & Hart, W., 2002. A basic theoretical framework for community-based conservation management in China and Vietnam. *The International Journal of Sustainable Development & World Ecology*, *9*(1), pp. 41–47.

Vinogradovs, I., Villoslada, M., Nikodemus, O., Ruskule, A., Veidemane, K., Gulbinas, J., Morkvenas, Ž., Kasparinskis, R., Sepp, K., Järv, H., & Klimask, J., 2020. Integrating ecosystem services into decision support for management of agroecosystems: Viva Grass tool. *One Ecosystem*, *5*, p. e53504.

Wali, A., Alvira, D., Tallman, P., Ravikumar, A., & Macedo, M. (2017). A new approach to conservation: Using community empowerment for sustainable well-being. *Ecology and Society*, *22*(4), pp. 1–13.

Wu, H. & Leung, S.O., 2017. Can Likert scales be treated as interval scales? A simulation study. *Journal of Social Service Research*, *43*(4), pp. 527–532

Yin, C., Zhao, W., Cherubini, F., & Pereira, P., 2021. Integrate ecosystem services into socio-economic development to enhance achievement of sustainable development goals in the post-pandemic era. *Geography and Sustainability*, *2*(1), pp. 68–73.

4 The Weak Link in Managing for Ecological Integrity and Responsible Tourism

The Case of Visitor Conduct in Lake Nakuru National Park, Kenya

Fredrick Nyongesa Kassilly

4.1 Introduction

The International Union for Conservation of Nature and Natural Resources (IUCN) states that national parks are designated for three primary management goals with equal emphasis on each, namely preservation of species and genetic diversity, maintenance of environmental services, and tourism and recreation (Suh & Harrison, 2005). Ever since the establishment of the first national park in the world (Yellowstone National Park) in 1872, managers of national parks and conservation biologists have for long questioned whether the primary goals of national park management are practically and sustainably compatible. Empirical evidence reveals that recreational goals often conflict with preservation goals. The fact that managing a park for ecological integrity as a biodiversity reservoir often runs counter to managing it for exploitation as a recreation facility has generated interest among researchers seeking an understanding of the effects of recreational use of national parks on animals and their habitats. The result has been the birth and development of a new basic and applied field of study known as recreation ecology.

Recreation ecology is defined as the scientific study of visitor impacts and their effective management, which mainly focuses on impacts of outdoor recreation on natural or semi-natural environments including national parks (Liddle, 1997). Over the past few decades, recreation ecology has grown into an increasingly coherent body of knowledge, with new approaches of study including more accurate prediction techniques (Leung, 2012). Notably though, it still suffers from a lack of adequate attention commensurate with societal concerns about recreation impacts on wildlife and its habitat as there are still only a few scientists in the world devoted primarily to this field. Currently, research in the field of recreation ecology is much more developed in North America, Europe, and Australia than in other areas of the world, but it nevertheless remains a key emerging field of global importance with a primary research focus being the anthropogenic use–impact relationships with respect to wild animals and their habitat (Stankey, McCool & Stokes, 1984). Alongside recreational ecology is an emerging concept in tourism referred to as responsible tourism, which

DOI: 10.4324/b23145-6

emphasizes responsible behavior and actions among all tourism players, including tourists/park visitors (Mihalic et al., 2021).

Lake Nakuru National Park is classified as a premier park in Kenya and is among the country's most popular national parks with high visitation rates by both local and foreign tourists (Kassilly, 2000). Although empirical evidence reveals that some visitor behaviors in the park translate to animal harassment and also pose danger to the ecological integrity of the wildlife habitat, no known study has addressed the subject, and information on visitor behaviors in the park remains scanty and undocumented. The need for research in recreation ecology in Kenyan national parks cannot be overemphasized. Currently, development of an effective national park visitor management strategy is hampered by several factors including inadequate information on the specific visitor problem behaviors and their impacts on wild animals and their habitat, drivers of the problem behaviors, and the limitations of the current on-site messages in managing visitor conduct within the park. Adequate knowledge of visitor conduct and its influence on the welfare of wildlife and its habitat remain a weak link in efforts by Kenya Wildlife Service (KWS) to manage the park for ecological integrity and responsible tourism. The purpose of this study was to fill this knowledge gap.

4.2 Statement of the Research Problem

All over the world, wildlife protected areas including national parks are the major global mechanism for conservation of biodiversity and remain popular destinations for recreationists seeking to experience nature and appreciate stunning scenic landscapes. Other social benefits of recreational wildlife viewing include increased physical and mental health, enhanced social bonding, and a feeling of a great sense of place. Although visiting national parks can also enhance conservation directly via fees and other income for the parks, conduct of park visitors is known to have both direct and indirect negative impacts on wildlife and their habitats, which presents management challenges for national parks in pursuit of their dual objective of providing recreational opportunities for people and preserving natural environments. The enduring management challenge, more so for heavily visited national parks, has over time necessitated development of effective visitor management programs, which have globally become an integral part of park management.

Marion (2012) points out that although managers of national parks recognize that some degree of resource impairment is an inevitable consequence of visitor presence in parks, thereby implying a minimum degree of acceptable disturbance to parks resulting from recreational use, empirical evidence reveals that more often than not, the Limits of Acceptable Change (LAC) in individual parks are exceeded, thereby necessitating the development of visitor conduct management protocols in national parks worldwide.

Wildlife is Kenya's foremost natural resource and remains the country's prime tourist attraction. The country's 23 national parks are among the world's most visited wildlife protected areas. Lake Nakuru National Park

Table 4.1 Code of conduct for visitors to Kenya's National Parks

S/No.	*Description*
1	Respect the privacy of wildlife, this is their habitat
2	Beware of the animals, they are wild and can be unpredictable
3	Don't crowd around the animals or make sudden noises or movements.
4	Don't feed the animals, it upsets their diet and leads to human dependence
5	Stay in your vehicle at all times, except at designated picnic or walking areas
6	Keep below the maximum speed limit (40 kph/25 mph)
7	Keep quiet, noise disturbs the wildlife and may antagonize your fellow visitors
8	Never drive off-road, this severely damages the habitat
9	When viewing wildlife, stay not closer than 20 meters
10	Leave no litter and never discard burning objects or leave fires unattended
11	Stay over or leave before dusk
12	Night game driving is not allowed

Source: Kenya Wildlife Service records.

stands out as a popular local and foreign tourist destination in the country with its fair share of high visitor numbers and high incidents of problem visitor conduct (Kassilly, 2000) despite a clear code of conduct comprising "Dos" and "Don'ts" for visitors to the country's national parks (Table 4.1) in addition to on-site messages meant to guide visitor conduct within national parks.

Despite existence of empirical evidence of problem behaviors among visitors to Lake Nakuru National Park and despite the realistic need for informed strategic interventions to protects wild animals and their habitat from park visitors, research in recreation ecology within the park has suffered great neglect by researchers and no known study has sought to systematically document the problem behaviors among the park's visitors, identify drivers of such behaviors or examine the effectiveness of on-site messages in regulating visitor conduct in the park. The need for a study of visitor conduct as the weak link in managing Lake Nakuru National Park for ecological integrity as a biodiversity reservoir and responsible tourism as a recreation facility cannot be overemphasized.

4.3 Significance of the Study

The study provides insight information on the relationship between recreational use of Lake Nakuru National Park and its impacts on wildlife and the physical park environment. Such information is crucial for devising management strategies to ensure a sustainable balance between managing the park for purposes of human recreation and for preservation of biodiversity. Furthermore, study findings have relevance to other wildlife protected areas in Kenya and elsewhere.

The study is both a basic and an applied research in a field of study that largely remains virgin in Kenya. By applying a purely social science approach, the study is a methodological contribution to applied wildlife management in Kenya and serves as a foundational study upon which similar work can be conducted in future. Study findings constitute an important reference source to managers and resource planners of wildlife protected areas, students and researchers in the fields of conservation biology, wildlife management, tourism management and recreation ecology.

4.4 Objectives of the Study

4.4.1 General Objective

The general objective of the study was to identify problem visitor behaviors in Lake Nakuru National Park and determine the effectiveness of on-site messages in managing visitor conductor within the park.

4.4.2 Specific Objectives

(i) To identify the spectrum of visitor behaviors that cause problems to wildlife in Lake Nakuru National Park.
(ii) To identify the spectrum of visitor behaviors that cause problems to the wildlife habitat in Lake Nakuru National Park.
(iii) To identify drivers of problem behaviors among visitors in Lake Nakuru National Park.
(iv) To determine the effectiveness of on-site messages in managing visitor conduct within Lake Nakuru National Park.

4.4.3 Research Questions

(i) What visitor behaviors are problems to wildlife in Lake Nakuru National Park?
(ii) What visitor behaviors are problems to the wildlife habitat in Lake Nakuru National Park?
(iii) What are the drivers of problem behaviors among visitors in Lake Nakuru National Park?
(iv) How effective are on-site messages in managing visitor behavior in Lake Nakuru National Park?

4.5 Literature Review

4.5.1 Visitor Management in National Parks

Visitor management is defined as an administrative action oriented toward maintaining the quality of park resources and visitor experiences (Watkins, 2013). Worldwide, recreation and tourism activities in national parks continue to show trends of increasing visitation (Cordell, 2008). Associated with this rise in visitation are anthropogenic disturbances that result in negative impacts on wildlife and its habitat, which in turn raise concerns as to whether recreation and tourism activities in protected areas can be managed sustainably (Suh & Harrison, 2005). Available literature on recreation ecology suggests that there is a variety of management alternatives that might be applied to guide visitor use and minimize resulting impacts in national parks (Manning et al., 1999).

Visitor management practices may be classified along a spectrum according to the directness with which they act on visitor behavior (Chavez, 1996). Whereas direct management practices act directly on visitor behavior, leaving little or no freedom of choice, indirect management practices attempt to influence the decision factors upon which visitors base their behavior (McCool & Christensen, 1996). As an example, a direct management practice for reducing campfire-related impacts would be a regulation to prohibit campfires while an indirect management practice would be an education program informing visitors of the undesirable ecological and aesthetic impacts of campfires and encouraging them to carry and use alternative sources of heat for cooking (McCool & Christensen, 1996). Management of visitor behaviors through information and education plays an important role in natural resource management within wildlife protected areas (Buckley & Littlefair, 2007).

According to Mason (2008), the main ways of managing visitors in national parks include controlling the number of visitors, adapting the resource to handle the number of visitors, or changing the visitor's behavior. Mason (2005) classifies visitor management techniques as either "hard" or "soft," depending on the program a protected area chooses to implement, and explains that hard management techniques involve creating rules and regulations to physically manage the resource, while soft techniques involve promoting education and learning in the hope of modifying the visitor's behavior. Watkins (2013) notes that use of "soft" techniques like education programs that teach visitors the impact of their activities, on-site boards/messages that have rules of conduct, and providing lots of information to visitors can change their behavior, thereby managing the effects they have on the park environment.

Generally, indirect visitor management practices are favored when and where they are believed to be effective (McCool & Christensen, 1996). Emphasis on indirect management practices, however, has not been uniformly endorsed because some indirect management practices may be insufficiently effective in cases where some visitors, for example, may ignore management efforts to influence the decision factors that guide their behavior, thereby hampering attainment of management objectives (Shindler & Shelby, 1993). Information/education programs remain the most popular indirect visitor management practice (Abbe & Manning, 2007) despite having varying degrees of application to a variety of recreation management problems (Hendee & Dawson, 2003). Duncan and Martin (2002) report that information may have limited effectiveness in managing two types of problem behaviors, namely those that are deliberately illegal (e.g., capture of live animals within parks) or unavoidable (e.g., disposal of human waste) although it can adequately address the other three types of problem behaviors, namely careless actions (e.g., littering), unskilled actions (e.g., selecting an inappropriate campsite), and uninformed actions (e.g., using dead snags for firewood).

4.6 Research Methodology

4.6.1 Study Area

The study was conducted in Lake Nakuru National Park, which is one of the premier national parks in Kenya (Figure 4.1). It lies between latitudes 0° 17′ and

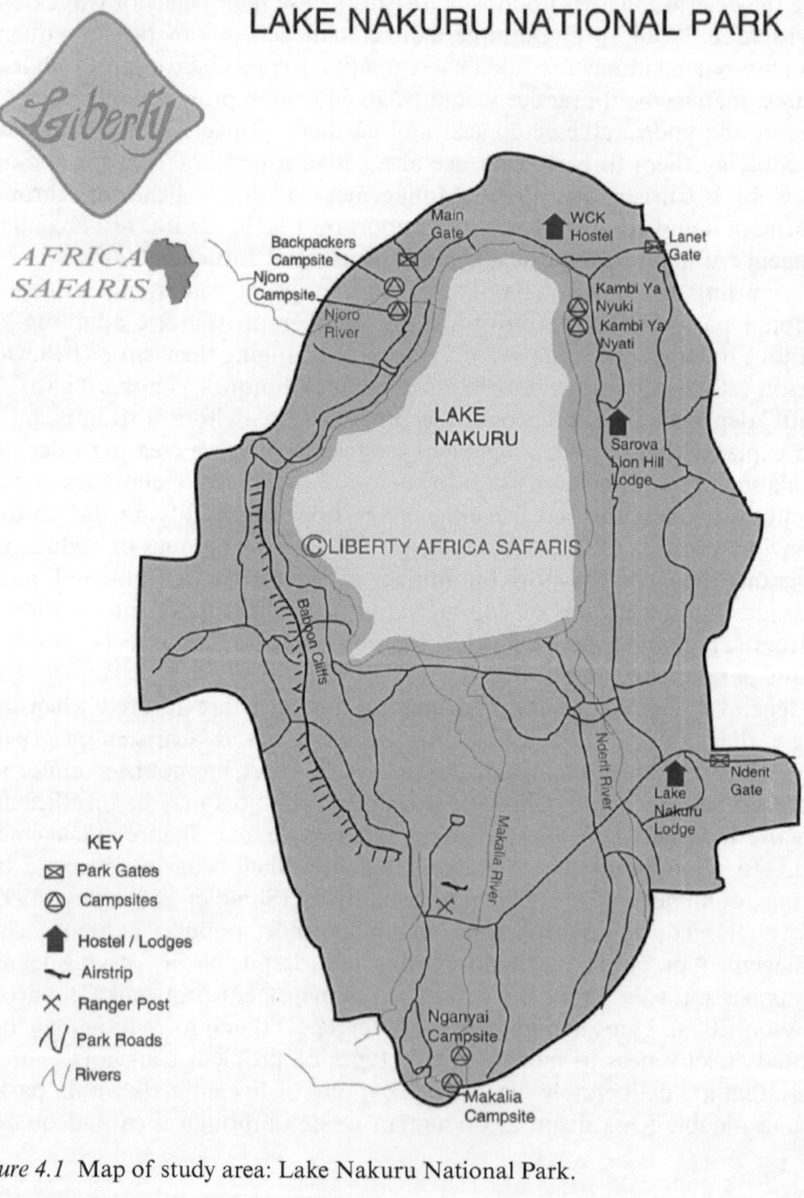

Figure 4.1 Map of study area: Lake Nakuru National Park.

0° 30′ South and between longitudes 36° 2′ and 36° 9′ East and between the 760 mm and 1015 mm isohyets. It lies at the floor of the Great Rift Valley, measures 187.9 square kilometers, and was the first national park in Africa to be gazette primarily for birds namely the greater and lesser flamingoes in 1968. At the center of the national park is the pea-shaped Lake Nakuru. Rivers Njoro, Makalia, Lamudiac, and Nderit all empty into the lake. Human settlement (rural and urban), ranching and agriculture outline the bounds of the park.

4.6.2 Research Design

This study employed a descriptive research design. Descriptive research determines and reports the way things are; it describes data and characteristics about the population and phenomena being studied. As Cooper and Schindler (2003) asserted, a study concerned with finding out who, what, when, where, and how of a phenomenon is a descriptive study. The study was accomplished in two parts. Part one involved an observational study on the influence of on-site park messages on visitor behaviors in Lake Nakuru National Park. Part two of the study involved collection of primary data on problem visitor behaviors in the park using a self-administered, closed-ended, drop-and-collect questionnaire developed in accordance with Dilman's Total Design Method (Fillion, 1978).

The target population (Kothari, 2008) for part one of the study comprised the visitor population in Lake Nakuru National Park during the study period while the target population for part two comprised the park's 200 non-commissioned security officers charged with the responsibility of patrolling and monitoring visitor conduct within thc park.

4.6.3 Data Collection Procedures

Part one of the study involved the researcher individually making observations on visitor compliance or non-compliance with four on-site park messages guiding their behavior as follows:

(i) Plea Message Without Interpretation
"Do Not Come Out Of Your Vehicles At This Point"
(ii) Plea Message With Interpretation
"Do Not Come Out Of Your Vehicles At This Point.
This Is To Avoid Encounters With Dangerous Animals"
(iii) Sanction Message Without Interpretation
"Visitors Are Penalized KES 1000 For Irresponsible Waste Disposal"
(iv) Sanction Message With Interpretation
"Visitors Are Penalized KES 1000 For Irresponsible Waste Disposal.
This Is To Hire Extra Labor To Clean Up The Place"

The first two messages were alternately put at a popular animal watering point where visitors usually come out of their vehicles to get a better and closer view of the wildlife. The last two messages were alternately put at a popular picnic site where visitors usually take their lunches or snacks and where they hold group parties. Observations were made at each site on five randomly selected days over a four-week period in August (2015). On each of these days, the researcher unobtrusively observed a systematic random sample of visitors (every nth visitor) and recorded whether or not they complied with the on-site park regulation (Park et al., 2008). The researcher's dressing and behavior was generally consistent with park visitors to conceal his identity (Burrus-Bammel & Bammel, 1984). Observations were made during peak use hours and fair weather when use levels were high enough to ensure that the observer was incognito. Visitors were followed at a discrete distance of up to 50 meters and the observer mingled with them at each observation site. Observations on compliance or non-compliance were based on 100 individually recorded visitor behaviors at each site. Where possible, additional information noted about observed visitors included gender (male/female), age (child/adult), and solo visitor/member of a group.

In part two of the study, a self-administered, closed-ended, drop-and-collect questionnaire was individually served on each of the 100 randomly selected non-commissioned security personnel (officers) with a clear register of the selected respondents maintained for purposes of sending reminders to them about the need to return the questionnaires on time. A polite introductory letter accompanied the questionnaire and not only explained the objectives of the study but also assured respondents of confidentiality of the information received from them. Through purposive sampling, only those officers who have served in the park for a minimum of five years participated in the study. From a sampling frame of 200, a total of 100 officers were randomly selected and served with the research questionnaire.

Responses to the questionnaire items on problem behaviors were based on a three-point Likert scale format ranging from "Major Problem" (Occurs Regularly), "Minor Problem" (Occurs Only Sometimes), and "Not A Problem" (Does Not Occur) for ten described visitor behaviors with respect to the wild animals and their habitat, respectively. A "Not Sure" option was provided for respondents who were undecided about a given questionnaire item. Respondents were further requested to indicate what they believed was the major driver of problem behavior among park visitors from a set of five (5) known common drivers of such behaviors.

4.6.4 Data Processing, Analysis, and Presentation

All returned questionnaires were first checked for completeness and accuracy. The researcher then coded the respective responses as per the questionnaire items using the Statistical Package for Social Sciences (SPSS) to derive the desired output on the adopted Likert scale. The data collected was analyzed using descriptive statistics (frequencies, means, percentages) as well as inferential statistics. Where appropriate, Analysis of Variance (ANOVA) was performed to compare statistical means. Responses to questionnaire items on problem visitor behaviors were analyzed as follows: "Major Problem" was assigned a score of 1, "Minor Problem" was assigned a score of 2, "Not A Problem" was assigned a score of 3, and "Not Sure" was assigned a score of 4. A questionnaire item with a missing response was assigned a score of zero and excluded from analysis (Wilson, 1976; Bath, 1993). The mean score on each item was used to categorize the item into "major problem," "minor problem," "not a problem," and "not sure/indeterminate" (Wilson, 1976; Bath, 1993). Individual item means were used to calculate the overall cluster mean. Internal consistency among parallel responses to questionnaire items within each cluster of problem visitor behaviors (wildlife versus the habitat) was determined using Cronbach's alpha values (Bath, 1993). A value of 0.70 or higher is usually considered acceptable (Nunally, 1970).

4.7 Results

4.7.1 Questionnaire Return Rate

The study recorded 100% questionnaire return rate. Of the 100 respondents, 80 (80%) were males and 20 (20%) were females. A majority (70%) were within the 20–39 age bracket while 30% were within the 40–59 age bracket.

4.7.2 Problem Visitor Behavior with Respect to Wildlife in Lake Nakuru National Park

All (10) described behaviors were evaluated to be problem cases (Table 4.2). Three of them (visitors touching/attempting to touch animals, throwing items at animals to draw their attention, and visitor vehicles colliding with animals within the park) were evaluated to be "minor problem" cases (means core of 2.08) (Table 4.3). Seven (70%) of the described behaviors (offering food to animals, making noise to draw attention of animals, crowding around certain animals, persistently following certain animals, crowding at drinking sites, disturbing resting animals, disturbing feeding animals) were evaluated to be "major" problem cases (mean score of 1.62) (Table 4.4). The overall score for the described visitor behaviors was 1.75, which translates to visitor behavior being a "major problem" to wildlife in Lake Nakuru National Park. Cronbach's alpha value for all the cluster items was 0.81.

4.7.3 Problem Visitor Behavior with Respect to the Wildlife Habitat in Lake Nakuru National Park

All (10) described visitor behaviors were evaluated to be problem cases (Table 4.5). "Major" and "minor" problem cases were co-dominant. Five of the described visitor behaviors (destroying vegetation within the park, lighting unnecessary fires at campsites, dropping litter in rivers within the park, dropping litter in the lake within the park, relieving themselves at undesignated

Table 4.2 Prevalence of visitor behaviors evaluated to be problems with respect to wildlife in Lake Nakuru National Park

Visitor behavior	*Major problem (%)*	*Minor problem (%)*	*Not a problem (%)*	*Not sure (%)*	*Mean score*	*Prevalence ranking*
Crowding around certain animals	64	30	6	0	1.42	1
Persistently following certain animals	56	36	8	0	1.52	2
Making noise to draw attention of animals	54	30	12	4	1.58	3
Disturbing feeding animals	56	32	8	4	1.60	4
Offering food to animals	48	40	10	2	1.66	5
Disturbing resting animals	64	26	10	0	1.72	6
Crowding at animal drinking sites	42	36	20	2	1.82	7
Visitor vehicles colliding with animals	38	30	22	10	2.04	8
Throwing items at animals to draw their attention	28	46	26	0	2.06	9
Touching/ attempting to touch animals	22	66	10	2	2.14	10
Overall evaluation					1.75	Major problem
Cronbach's alpha value for items					0.81	Acceptable

places within the park) were evaluated to be "Minor problems" (mean score 2.32) (Table 4.6). Another five (dropping litter at wrong places within campsites, driving off roads, dropping litter on roads within the park, dropping litter at wrong places at picnic sites within the park, disembarking from

Table 4.3 Visitor behaviors identified as minor problems with respect to wildlife in Lake Nakuru National Park

Visitor behavior	*Major problem (%)*	*Minor problem (%)*	*Not a problem (%)*	*Not sure (%)*	*Mean score*	*Remark*
Touching/attempting to touch animals	22	66	10	2	2.14	Minor problem
Throwing items at animals to draw their attention	28	46	26	0	2.06	Minor problem
Visitor vehicles colliding with animals	38	30	22	10	2.04	Minor problem
Overall evaluation					2.08	Minor problem
Cronbach's alpha value for items					0.94	Acceptable

Table 4.4 Visitor behaviors identified as major problems with respect to wildlife in Lake Nakuru National Park

Visitor behavior	*Major problem (%)*	*Minor problem (%)*	*Not a problem (%)*	*Not sure (%)*	*Mean score*	*Remark*
Offering food to animals	48	40	10	2	1.66	Major problem
Making noise to draw attention of animals	54	30	12	4	1.58	Major problem
Crowding around certain animals	64	30	6	0	1.42	Major problem
Persistently following certain animals	56	36	8	0	1.52	Major problem
Crowding at animal drinking sites	42	36	20	2	1.82	Major problem
Disturbing resting animals	64	26	10	0	1.72	Major problem
Disturbing feeding animals	56	32	8	4	1.60	Major problem
Overall evaluation					1.62	Major problem
Cronbach's alpha value for items					0.90	Acceptable

Table 4.5 Prevalence of problem behaviors with respect to the wildlife habitat in Lake Nakuru National Park

Visitor behavior	*Major problem (%)*	*Minor problem (%)*	*Not a problem (%)*	*Not sure (%)*	*Mean score*	*Prevalence ranking*
Dropping litter at wrong places at camp sites	54	36	10	0	1.26	1
Dropping litter at picnic sites within the park	42	46	12	0	1.58	2
Driving off the roads	36	62	2	0	1.66	3
Disembarking from vehicles at undesig-nated places	44	46	10	0	1.66	4
Dropping litter on roads within the park	38	50	10	2	1.76	5
Dropping litter in rivers within the park	10	68	18	4	2.04	6
Dropping litter in the lake within the park	16	60	18	6	2.14	7
Relieving themselves at undesignated places	10	40	50	0	2.40	8
Destroying vegetation within the park	10	40	32	18	2.48	9
Lighting unnecessary fires at camp sites	6	36	56	2	2.54	10
Overall evaluation					2.00	Minor problem
Cronbach's alpha value for items					0.88	Acceptable

vehicles at undesignated places within the park) were evaluated to be "major problems" (mean score 1.58) (Table 4.7). The overall score for the described visitor behaviors was 2.00, which translates to visitor behavior being a "minor problem" to the wildlife habitat in Lake Nakuru National Park. Cronbach's alpha value for all the cluster items was 0.88.

Table 4.6 Visitor behaviors identified as minor problems with respect to the wildlife habitat in Lake Nakuru National Park

Visitor behavior	*Major problem (%)*	*Minor problem (%)*	*Not a problem (%)*	*Not sure (%)*	*Mean score*	*Remark*
Destroying vegetation within the park	10	40	32	18	2.48	Minor problem
Lighting unnecessary fires at camp sites	6	36	56	2	2.54	Minor problem
Dropping litter in rivers within the park	10	68	18	4	2.04	Minor problem
Dropping litter in the lake within the park	16	60	18	6	2.14	Minor problem
Relieving themselves at undesignated places	10	40	50	0	2.40	Minor problem
Overall evaluation					2.32	Minor problem
Cronbach's alpha value for items					0.89	Acceptable

Table 4.7 Visitor behaviors identified as major problems with respect to the wildlife habitat in Lake Nakuru National Park

Visitor behavior	*Major problem (%)*	*Minor problem (%)*	*Not a problem (%)*	*Not sure (%)*	*Mean score*	*Remark*
Dropping litter at wrong places at camp sites	54	36	10	0	1.26	Major problem
Driving off the roads	36	62	2	0	1.66	Major problem
Dropping litter on roads within the park	38	50	10	2	1.76	Major problem
Dropping litter at picnic sites within the park	42	46	12	0	1.58	Major problem
Disembarking from vehicles at undesignated places	44	46	10	0	1.66	Major problem
Overall evaluation					1.58	Major problem
Cronbach's alpha value for items					0.85	Acceptable

4.7.4 Drivers of Problem Behavior among Visitors in Lake Nakuru National Park

Table 4.8 presents findings on the drivers of problem behaviors among visitors in Lake Nakuru National Park. Respondents ranked "lack of knowledge about park regulations" the number one driver of problem behaviors among park visitors (41%) followed by "mutual imitation of what fellow visitors were doing" (21%), "lack of understanding park regulations" (13%), a "don't care attitude" (10%), "visitor lack of control over body functions" (8%), and visitors' temerity to make the visit more rewarding" (7%).

4.7.5 Effectiveness of On-Site Park Messages in Managing Visitor Conduct in Lake Nakuru National Park

A total of 400 park visitors were observed, of whom 230 were males and 170 were females. Solo visitors were more compliant than group visitors. Out of 60 clearly identified solo visitors, 54 (90%) complied with the park regulations, compared to 74 (30%) of the 240 clearly identified group visitors. The least compliant were school parties. Of the 140 identified visitors in school uniforms, only 35 (25%) complied with the on-site park messages regarding visitor conduct.

The effectiveness of on-site park messages in managing visitor conduct in the park was between 75% and 41%, with a mean of 60.75% (Table 4.9). Generally, sanction messages were significantly ($P < 0.05$) more effective in influencing visitor conduct in the park than plea messages with mean compliance rates of 70.00% and 51.50%, respectively. For both message types, inclusion of an interpretive component (explanation) resulted in a significant ($P < 0.005$) increase in compliance rates. The sanction message with the explanation recorded 10% more compliance while the plea message with the explanation recorded 21% more compliance.

Table 4.8 Prevalence of drivers of problem behaviors among visitors in Lake Nakuru National Park

Driver	*Percent endorsement*	*Rank*
Visitors' lack of knowledge about park regulations	41	1
Visitors' imitation of what fellow visitors are doing	21	2
Visitors' lack of understanding of park regulations	13	3
Visitors' "Don't care attitude"	10	4
Visitors' lack of control over body functions	8	5
Visitors' temerity to make the visit more rewarding	7	6

Table 4.9 Compliance ratings of visitor behaviors for on-site park messages

Message content	*Message type*	*Compliance rating (%)*	*Rank*
"Visitors Are Penalized KES 1000 For Irresponsible Waste Disposal. This Is To Hire Extra Labor To Clean Up The Place"	Sanction message with interpretation	75	1
"Visitors Are Penalized KES 1000 For Irresponsible Waste Disposal"	Sanction message without interpretation	65	2
"Do Not Come Out Of Your Vehicles At This Point. This Is To Avoid Encounters With Dangerous Animals"	Plea message with interpretation	62	3
"Do Not Come Out Of Your Vehicles At This Point"	Plea message without interpretation	41	4

4.8 Discussions

4.8.1 Questionnaire Return Rate

The high questionnaire return rate is associated with the fact that respondents were officers of a disciplined unit (KWS), and they took the task of answering the questionnaire seriously in line with their work ethic. It is plausible to explain the high return rate with the fact that the questionnaire items were about issues the respondents were familiar with from their routine duties of patrolling the park and enforcing the "visitor code of conduct." Furthermore, the two reminders sent to the respondents before the questionnaires were collected, and the anonymity associated with the questionnaire encouraged respondents to provide the required information.

4.8.2 Scale Reliability

The Cronbach's alpha values for item clusters investigated in the study are high and all fall within acceptable ranges. As measures of internal consistency, the values demonstrate that the individual questionnaire items within the clusters are highly closely related as a group. Hence, inter-item correlation within the two clusters was appreciably high, which translates to a high-scale reliability in the study.

4.8.3 Problem Visitor Behaviors to Wildlife Welfare in Lake Nakuru National Park and the Need for Responsible Tourism

Study findings reinforce the view advanced by Fitter (1986) and Knight and Cole (1991) that preservation of nature and provision of recreational opportunities within the same land area will always inevitably conflict and demonstrate further the tricky balance between the need to maintain the ecological integrity of wildlife protected areas amidst the need to preserve them as facilities for recreation, as highlighted by Canfield et al. (1999) and Ewert (1999). Further, the findings not only lend credence to the argument by Marion (2012) that protected area recreationists often pose a real and significant threat to the very wildlife resource they so much cherish. From a management perspective, the study presents a case of a realistic need for efforts to protect the park's wildlife from recreationists through programs aimed at minimizing their disturbances on wildlife. This study not only defines reported problem visitor behaviors to the wildlife as largely intentional and therefore avoidable but also classifies the resulting disturbances as harassment based on the criteria by Knight and Cole (1995).

This study recognizes the basis for prevention of harassment of birds by recreationists in Lake Nakuru National Park on the basis of evidence from available literature, which cites visitor harassment as the reason for not only nest desertion (White & Thurrow, 1985), reduction in ground and shrub nesting among birds at campsites (Blakesley & Reese, 1988), and lower reproduction success (Bunneil et al., 1981) but also responsible for elevated heart rates, increased energy expenditure in disturbance flights coupled with decreased energy acquisition, and reduced vigor, which may ultimately result in increased sickness, disease, and potentially death (Hutchins & Geist, 1987). The study maintains that harassment of herbivores and carnivores in Lake Nakuru National Park should be checked as informed by Hutchins and Geist (1987) that they suffer decreased productivity, displacement, and change in behavior in response to being disturbed during feeding, resting, being persistently followed or being crowded by recreationists as was reported in this study. Furthermore, animal harassment as was reported in this study results in displacement of animals which forces them to seek refuge in unfamiliar, less secure, and less ecologically favorable sections of their habitat, which poses a threat to their survival (Morgantini & Hudson, 1979; Taylor & Knight, 2003) including forcing some prey species to accidentally move in the way of predators and natural enemies in their bid to escape the harassment (Muthee, 1992). Gill and Sutherland (2000) equate avoidance to use of certain areas by animals due to human harassment with loss of habitat.

Animal harassment through noise from visitors' shouts as reported in this study and from vehicle engines should be avoided because it distresses wildlife, resulting in changes in their activity patterns to escape from the noise sources (Morgantini & Hudson, 1979; Finnessey, 2012). Furthermore, persistently high noise levels can reduce a habitat's suitability for wildlife because it is associated with elevated stress hormone levels in animals, reproductive failure, and increased energy use coupled with interruptions in their natural activity patterns and habitat use, which may result in increased exposure to predators and decreased parental care (Bowles, 1995).

From a human dimension context, this study advocates effective visitor management to avoid a situation where visitor harassment of animals as was reported in this study forces them to escape the harassment by moving out of protected areas into surrounding human-occupied dispersal areas as reported by Canfield et al. (1999), where they cause damage to property, destroy crops, compete with livestock for water and pasture, prey on livestock, threaten human lives and generally become a nuisance in the human-wildlife interface areas surrounding national parks (Knight & Cole, 1991), which ultimately escalates human-wildlife conflicts within such areas.

This study presents a strong case for responsible tourism in Lake Nakuru National Park to avert cases where visitors interfere with animal welfare by interfering with their natural feeding patterns through visitors offering food to them, which makes them prone to poisoning (Knight and Cole (1991) or by persistently interfering with their normal feeding activities by closely following them as was reported in the study. A case example is presented by Muthee (1992) who reports that day hunters in Maasai Mara Game Reserve resorted to night hunting due to constant harassment by visitors during day time. Related studies by Yarmology, Bayer and Geist (1988) found that animal harassment caused them to resort to grazing and browsing at night and seek cover more frequently during the day at the expense of feeding, which negatively impacts on their welfare and which the management of Lake Nakuru National Park should avoid.

4.8.4 Problem Visitor Behaviors to the Wildlife Habitat in Lake Nakuru National Park and the Need for Responsible Tourism

Findings on problem visitor behaviors within the context of the ecological integrity of the wildlife habitat as reported in this study closely mirror those by Finnessey (2012) that, generally, wildlife habitat disturbance occasioned by problem visitor behaviors within protected areas is associated with degradation of ecosystems (aquatic and terrestrial), vegetation destruction, air and water pollution, noise pollution, trampling, irresponsible solid waste disposal, and littering. Further, study findings corroborate undocumented observations that on the whole, park visitors in Kenya generally partake in irresponsible solid waste disposal and littering, which not only pollutes the wildlife habitat but also causes general degradation of the physical environment and water bodies.

The study posits that there is urgent need to promote responsible tourism in Lake Nakuru National Park in its efforts to ensure ecological integrity of the terrestrial and aquatic wildlife habitat in the park. Special attention should be paid to recreationists who not only damage vegetation and make unnecessary noises within the park but also leave behind lots of garbage including plastic paper bags, beer cans, water bottles, food wrappings, and left-over food items, which are eaten by animals. Cases of plastic paper bags being found in dung of herbivores as reported for Maasai Mara Game Reserve (Kassilly, 2000) should be avoided.

Generally, destruction of vegetation, which was one of the problem visitor behaviors in this study, is common among visitors to wildlife protected areas in Kenya and like elsewhere outside Kenya results from vegetation being crushed, sheared off, bruised, uprooted, de-barked; cutting off of flowers, leaves, and branches (Turton, 2005); root damage; cutting trees for firewood; and harvesting of wild flowers (Bridle & Kirkpatrick, 2003). Within wildlife protected areas in Kenya, visitors normally target vegetation associated with medicinal values for humans and livestock, including the aloe plant, the neem tree, and *Salvadora persica*, commonly referred to as "mswaki" because of its oral cleaning properties, and *Mondia whytei*, commonly referred to as "mukombero" (Kassilly, 2000). Visitor conduct management in Lake Nakuru National Park should ensure such destruction of vegetation is averted.

Often, off-road driving as a problem visitor behavior as reported in this study is undertaken to satisfy an incessant penchant among visitors to see it all in a day (Opala, 1996) and in pursuit of the famous "big five" (Eltringham, 1984; Western, 1992) or to offer a close view of animals' nesting, feeding, drinking, or resting sites causes trampling whose undesirable effects on vegetation include reduced vigor and regeneration, loss of ground cover, breakage and bruising of stems, and changes in species richness (Cooper, De Lacy & Jago, 2007). On soils, off-road driving results in loss of organic matter, soil compaction, and scarring, which reduces soil air and water permeability, increases run-off, and accelerates soil erosion during rains (Finnessey, 2012). Ensuring ecological integrity of Lake Nakuru National Park requires that off-road driving be avoided.

4.8.5 Drivers of Problem Behavior among Visitors in Lake Nakuru National Park

This study recognizes that part of managing problem behavior among visitors to lake Nakuru National Park in its endeavor to develop a culture of responsible tourism among the park visitors involves identifying the root causes of their problem behavior. The fact that lack of knowledge and understanding of park regulations accounts for a majority of problem behaviors reveals the urgent need for effective visitor information, education, and awareness to manage visitor conduct in Lake Nakuru National Park. Such information needs to be accessible to the visitors prior to the park visit because while in the park, most visitors are too keen on watching the wildlife and may not pay much attention to anything else. The regulations should be circulated in diverse ways, including being put on the KWS website (for both local and foreign visitors), through pamphlets in KWS stations countrywide (for local visitors) and tour company offices (for foreign visitors) to facilitate pre-visit access by potential visitors. Additionally, a reminder of the same regulations should be availed to the visitors at the point of entry. Whereas the study stresses that management of Lake Nakuru National Park should pay attention to all drivers reported in the study, special attention should be paid to visitors driven by the "don't care attitude" to flout park regulations especially youths from schools and colleges and those who disregard regulations because they want to make the most of their money while within the park for attitude change. Management of Lake Nakuru National Park needs to consider constructing washrooms across the park to prevent cases of visitors being forced to relieve themselves in the bush. The fact that visitor propensity to flout park regulations is higher among groups than solo visitors is a matter the park management should pay attention to. Available literature (Duncan & Martin, 2002) classifies visitor problem into five basic types along a spectrum that applies to the Kenyan situation as well, where at the ends of the spectrum are those behaviors that are deliberately illegal (e.g., capture of live animals within parks) or unavoidable (e.g., disposal of human waste at undesignated places) for which information may have limited effectiveness. However, the other three types of problem behaviors, namely careless actions (e.g., littering), unskilled actions (e.g., selecting an inappropriate campsite), and uninformed actions (e.g., using dead snags for firewood) may be considerably more amenable to information/education programs (Garrett & Martin, 2002). Where relevant and feasible, management of Lake Nakuru National Park needs to consider unique aspects of a given visitor problem behavior in the development of its visitor management strategy.

4.8.6 Effectiveness of On-Site Messages in Managing Visitor Behavior in Lake Nakuru National Park

From a management perspective, the study demonstrates the potential of on-site park messages in contributing to, if appropriately worded and sited, the park's overall visitor conduct strategy. This is more so given that wildlife management agencies face decreasing budgets for on-site personnel to provide face-to-face communication to recreationists, as reported by Winter, Cialdin and Bator (1998). The finding that message wording has a bearing on its effectiveness corroborates the view advanced by Winter (2008) that message content has a big influence on message effectiveness in controlling visitor conduct in wildlife protected areas and agrees with Marion and Reid (2009) that messages with a moral appeal and those that provide a cause and effect relationship of visitor actions register higher conformity ratings than those without them. In a way, the study reveals that to some extent, park visitors employ a rational approach in interpreting instructional messages within the facility and avoid behaviors that are associated with penalties or sanctions. This explains higher compliance rates for sanction messages (which tell visitor the penalty for engaging in prohibited activities) than plea messages (which tell visitors what to do). The role of appropriateness in message design including wording in influencing visitor conduct as revealed by this study corroborates findings by Park et al. (2008) that due attention needs to be paid to message attributes, including content for effective visitor conduct management in national parks. The management of Lake Nakuru National Park is encouraged to develop appropriately worded on-site messages for visitor conduct management in its endeavor to promote responsible tourism.

On the whole, study findings demonstrate the importance of public information, education, and awareness creation in managing visitor conduct in recreation areas through on-site messages. A plausible explanation for higher compliance rates for messages with interpretive components than those without them borrows from the view advanced by Christensen and Dustin (1989) that the interpretive component communicates the significance or meaning of the prohibited action in a way that instills not only an understanding and appreciation but also a sense of responsibility and self-regulation in visitors to make them do the right thing even when they are not being seen. This study advocates for use of the education and information approach in managing visitor conduct in Lake Nakuru National Park. In so doing, it is cognizant of the views advanced by Duncan and Martin (2002) that the approach may have limited effectiveness in managing problem behaviors that are deliberately illegal (e.g., capture of live animals within parks) or unavoidable (e.g., disposal of human waste at undesignated places) but more effective for problem behaviors that are classified as careless actions (e.g., littering), unskilled actions (e.g., selecting an inappropriate campsite), and uninformed actions (e.g., using dead snags for firewood) as reported by Garrett and Martin (2002). Management of Lake Nakuru National Park needs to consider unique aspects of a given visitor problem behavior in the development of its visitor management strategy.

4.9 Conclusions, Recommendations, and Suggestions for Further Work

4.9.1 Conclusions

Based on the findings, the study concludes that:

1 Problem behaviors to the wildlife and its habitat characterize visitor behavior in Lake Nakuru National Park.
2 Visitors touching/attempting to touch animals, throwing items at animals to draw their attention, and visitor vehicles colliding with animals within the park were evaluated to be "minor problem" situations with respect to wildlife in the park.
3 Offering food to animals, making noise to draw attention of animals, crowding around certain animals, persistently following certain animals, crowding at drinking sites, disturbing resting animals, disturbing feeding animals) were evaluated to be "major problem" situations with respect to the wildlife in the park.
4 Destroying vegetation within the park, lighting unnecessary fires at campsites, dropping litter in rivers within the park, dropping litter in the lake within the park, visitors relieving themselves at undesignated places within the park were evaluated to be "minor problem" situations with respect to the wildlife habitat in the park.
5 Dropping litter at wrong places within campsites, driving off roads, dropping litter on roads within the park, dropping litter at wrong places at picnic sites within the park, disembarking from vehicles at undesignated places within the park were evaluated to be "major problem" situations with respect to the wildlife habitat in the park.

6 Visitor behavior in Lake Nakuru National Park is a bigger problem to the welfare of animals than it is to the welfare of their habitat.
7 Youths in schools and colleges are more defiant to visitor conduct regulations than adult visitors to Lake Nakuru National Park.
8 Problem behaviors among visitors in Lake Nakuru National Park are largely caused by visitor ignorance of the code of conduct, defiance of regulations on the part of visitors, and by mutual imitation among visitors in the park.
9 On-site sanction messages have greater influence on visitor conduct than plea messages in Lake Nakuru National Park.
10 Inclusion of an interpretive segment (explanation) in on-site park messages makes them more effective in controlling visitor conduct in Lake Nakuru National Park.

4.9.2 Recommendations

1 There is need for the management of Lake Nakuru National Park to develop a visitor management program to protect both wildlife and the physical environment in the park. Such a program should incorporate an elaborate public education and information strategy to enhance pre-visit awareness among the publics on the need to minimize negative visitor impacts on the park's wildlife and physical environment.
2 On-site park messages have potential for managing visitor conduct in Lake Nakuru National Park and should be part of the overall visitor management strategy.

4.9.3 Suggestions for Further Work

1 Given that the study was focused on Lake Nakuru National Park and given that Kenya has several national parks with unique features including visitation rates, this study has served as a foundation upon which further research should be done. It is suggested that similar and more elaborate studies be conducted in other national parks in the country to provide a national view of problem visitor behaviors.

References

Abbe, J.D. and R.E. Manning. (2007). Wilderness day use: Patterns, impacts, and management. *International Journal of Wilderness*, *12*(3), 21–25.

Bath, A.J. (1993). *Fire Management in Yellowstone*. Ph.D. Thesis, University of Calgary.

Blakesley, J.A. and K.P. Reese. (1988). Avian use of campground and non-campground sites in riparian zones. *The Journal of Wildlife Management*, *52*, 399–402.

Bowles, A.E. (1995). Responses of wildlife to noise. Pages 109–156. In R.L. Knight and K.J. Gutzwiller (Eds.), *Wildlife and Recreationists: Coexistence through Management and Research*. Island Press, Washington DC.

Bridle, K.L. and J.B. Kirkpatrick. (2003). Impact of nutrient additions and digging for human waste disposal in natural environments, Tasmania, Australia. *Journal of Environmental Management*, *69*, 299–306.

Buckley, R., & Littlefair, C. (2007). Minimal-impact education can reduce actual impacts of park visitors. *Journal of Sustainable Tourism*, *15*(3), 324–325.

Bunneil, F.L., D. Dunbar, L. Koza, and G. Ryder. (1981). Effects of disturbance on the productivity and numbers of white pelicans in British Columbia: Observation models. *Colonial Water birds*, *4*, 2–11.

Burrus-Bammel, L.L. and G. Bammel. (1984). Application of unobtrusive methods. In J.D. Peine (Ed.), *Proceedings of a Workshop on Unobtrusive Techniques to Study Social Behavior in Parks*. National Park Service Southeast Regional Office, Natural Science and Research Division.

Canfield, J.E., L.J. Lyon, J.M. Hills, and M.J. Thompson. (1999). Ungulates. Pages 6.1–6.25. In G. Joslin and H. Youmans, coordinators. *Effects of Recreation on Rocky Mountain Wildlife: A Review for Montana*. 307 pp.

Chavez, D.J. (1996). *Mountain Biking: Issues and Actions for USDA Forest Service Managers*. Research Paper PSW-RP-226. Albany, California: Pacific Southwest Research Station, USDA Forest Service.

Christensen, H. and D. Dustin. (1989). Reaching recreationists at different levels of moral development. *Journal of Park and Recreation Administration*, *7*, 72–80.

Cooper, A. and P.S. Schindler. (2003). *Business Research Methods* (8th ed.). McGraw-Hill, Boston.

Cooper, C., T. De Lacy and L. Jago (Eds.). (2007). *Impacts of Recreation and Tourism on Plants in Protected Areas in Australia*. Catherine Pickery & Wendy Hill, Australia

Cordell, H.K. (2008). The latest on trends in nature-based outdoor recreation and tourism. *Forest History Today*, Spring, 4–10.

Duncan, G.S. and S.R. Martin. (2002). Comparing the effectiveness of interpretive and sanction messages for influencing wilderness visitors' intended behavior. *International Journal of Wilderness*, *8*, 20–25.

Eltringham, S. (1984). *Wildlife Resources and Economic Development*. John Wiley and sons, New York.

Ewert, A.W. (1999). Outdoor recreation and natural resource management. An uneasy alliance. *Parks and Recreation*, *39*(2), 1–9.

Fillion, F.L. (1978). Increasing the effectiveness of mail surveys. *Wildlife Society Bulletin*, *6*, 135–141.

Finnessey, L. (2012). *The Negative Effects of Tourism on National Parks in the United States of America*. Unpublished B.Sc Thesis, Johnson & Wales University.

Fitter, R. (1986). *Wildlife for Man. How and Why We Should Conserve Species*. Collins, London.

Garrett, S.D. and S. Martin. (2002). Comparing the effectiveness of interpretive and sanction messages for influencing wilderness visitors' intended behavior. *International Journal of Wilderness*, *2*, 20–24.

Gill, J.A. and W.J. Sutherland. (2000). Predicting the consequences of human disturbance from behavior. Pages 51–64. In L.M. Goosling and W.J. Sutherland (Eds.), *Behaviour and Conservation*. Cambridge University Press, Cambridge.

Hendee, J.C. and C.P. Dawson. (2003). *Wilderness Management: Stewardship and Protection of Resources and Values* (3rd ed.). Fulcrum Publishing, Golden.

Hutchins, M. and V. Geist. (1987). Behavioral considerations in the management of mountain dwelling ungulates. *Mountain Research and Development*, *7* 135–144.

Kassilly, F.N. (2000). *Human Dimensions in Wildlife Resource Management in Kenya: A Study of People-Wildlife Relations Around Two Conservation Areas*. Dr.rer.nat Dissertation, University of Natural Resources and Applied Life Sciences, Vienna.

Knight, R.L. and D.N. Cole. (1991). Effects of recreational activity on wildlife in woodlands. *Transactions of the North American Wildlife and Natural Resources Conference (USA)*, *56*, 228–247.

Knight, R.L. and D.N. Cole. (1995). Factors that influence wildlife responses to recreationists. Pages 71–79. In R.L. Knight and H.K. Cordall (Eds.), *Wildlife and Recreationists: Coexistence through Management and Research*. Island Press. Washington DC.

Kothari, C.R. (2008). *Research Methodology. Methods and Techniques* (2nd ed.). New Age International, New Delhi.

Leung, Y.F. (2012). Recreation ecology research in East Asia's protected areas: Redefining impacts? *Journal for Nature Conservation*, 20(6), 349–356.

Liddle, M.J. (1997). *Recreation Ecology: The Ecological Impact of Outdoor Recreation and Ecotourism*. Chapman and Hall, London.

Manning, R., W. Valliere, B. Wang, S. Lawson, and J. Treadwell. (1999). *Research to Support Visitor Management at Statue of Liberty/Ellis Island National Monuments*. University of Vermont School of Natural Resources, Burlington, Vermont.

Marion, J.L. (2012). *Recreation Ecology Research Findings: Implications for Wilderness and Park Managers*. USGS Patuxent Wildlife Research Center, USA.

Marion, J. and S. Reid. (2009). Minimizing visitor impacts to protected areas: The efficacy of low impact education programs. *Journal of Sustainable Tourism*, *15*(1), 5–27.

Mason, P. (2005). Visitor management in protected areas of the periphery: Polar perspectives. *Tourism and Hospitality Planning & Development*, *2*, 171–190.

Mason, M. (2008). Complexity theory and the philosophy of education. *Educational philosophy and theory*, *40*(1), 4–18.

McCool, S.F. and N.A. Christensen (1996). Alleviating congestion in parks and recreation areas through direct management of visitor behavior. In D.W. Lime (Ed.), *Congestion and Crowding in the National Park System: Guidelines for Management and Research*. (MAES Misc. Pub. 86-1996). Department of Forest, St. Paul, MN

Mihalic, T, S. Mohammadi, A. Abbasi, and L.D. David. (2021). Mapping a sustainable and responsible tourism paradigm: A bibliometric and citation network analysis. *Sustainability*, *13*, 853.

Morgantini, L.E. and R.J. Hudson. (1979). Human disturbance and habitat selection in elk. Pages 132–139. In M.S. Boyce and L.D. Hayden-Wing (Eds.), *North American elk: Ecology, Behavior and Management. Proceedings of a Symposium on Elk Ecology and Management*. University of Wyoming, Laramie, Wyoming.

Muthee, L. (1992). Ecological impacts of tourism use on habitats and pressure point species. Pages 18–38. In C.G. Gakahu (Ed.), *Tourist Attitudes and Use Impacts in Maasai Mara Game Reserve*. Wildlife Conservation International. Nairobi.

Nunally, J.C. (1970). *Introduction to Psychological Measurement*. McGraw-Hill Publications, New York.

Opala, K. (1996). *The Magic of the Mara*. Swara magazine, July-August.

Park, L.O., R.E. Manning, J.L. Marion, S.R. Lawson, and C. Jacobi. (2008). Managing visitor impacts in parks: A multi-method study of the effectiveness of alternative management practices. *Journal of Park and Recreation Administration*, *26*, 97–121.

Shindler, B. and B. Shelby. (1993). Regulating wilderness use: An investigation of user group support. *Journal of Forestry*, *91*(22), 41–44.

Stankey, G.H., S.F. McCool, and G.L. Stokes. (1984). Limits of acceptable change: A new framework for managing the Bob Marshall Wilderness complex. *Western Wildllands*, *10*(3), 33–37.

Suh, J. and S. Harrison. (2005). *Management Objectives and Economic Value of National Parks: Preservation, Conservation and Development*. Discussion Paper No. 337, May 2005, School of Economics, The University of Queensland.

Taylor, A.R. and R.L. Knight. (2003). Wildlife responses to recreation and associated visitor perceptions. *Ecological Applications*, *13*, 951–963.

Turton, S.M. (2005). Managing environmental impacts of recreation and tourism in rainforests at the Wet Tropics of Queensland World Heritage Area. *Geographical Research*, *43*, 140–151.

Watkins, S.G. (2013). *Trend Report: Visitor Management Policies in protected Areas and National Parks*. School of Hospitality and Tourism Management, University of Guelph.

Western, D. (1992). Ecotourism: The Kenya challenge. Pages 49–62. In C.G. Gakahu and B.E. Goodie (Eds.), *Ecotourism and Sustainable Development in Kenya*. Wildlife Conservation International, Nairobi.

White, C.M. and T.L. Thurrow. (1985). Reproduction of ferruginous hawks exposed to controlled disturbance. *The Condor*, *87*(1), 14–22.

Wilson, G.D. (1976). Attitudes. In: G.D. Wilson and H.J. Essence (Eds.), *A Text Book of Human Psychology*. MTP Press Ltd, Lancaster, England.

Winter, P.L. (2008). Park signs and visitor behavior: A research summary. *Park Science*, *2*, 34–35.

Winter, P.L., R.B. Cialdin, and R.J. Bator. (1998). An analysis of normative messages in signs at recreation settings. *Journal of Interpretive Research*, *3*, 39–47.

Yarmology, C., M. Bayer, and V. Geist. (1988). Behaviour responses and reproduction in mule deer, *Odocoileus heminious*, following experimental harassment with an all-terrain vehicle. *Canadian Field Naturalist*, *102*, 425–429.

Part III

Management of Cultural Ecosystems Services

5 Conceptual Understanding of Cultural Ecosystem Services in Tourism

Aman Kumar

5.1 Introduction

5.1.1 Ecosystem Services

Ecosystem services investigations have been the part of study, and its findings are utilized in policy planning and decision making related to tourism. Integration of different parts of ecosystem can be included for conservation, and resident's well-being is also the part of ecosystem services assessment and study (Margules and Pressey, 2000). Services which are catered by the human being from ecosystem like water purification, fuelwood, tourism, recreation and aesthetic aspect contribute to human being benefits (Millennium Ecosystem Assessment, 2003). It provides conservation action and protection of human well-being, which contributes to the social relevance of conservation assessment (Balvanera et al., 2001; Knight et al., 2006). Perhaps the ecosystem services have been defined many times (Millennium Ecosystem Assessment, 2003) but has not acquired a standard definition and has many different names like ecosystem function, ecological function, ecological system or environmental services (Egoh et al., 2007). Many authors claim that ecosystem services are provided by the natural system, while others claim that it is the human-modified natural system (Costanza et al. 1997; Millennium Ecosystem Assessment, 2003).

Ecosystem services are broadly categorized into four parts: (i) provisioning services (ii) regulating services (iii) cultural services and (iv) supporting services. Provisioning services include food, fuel and wood from the ecosystem; regulating services include services of gas regulation, water purification, biological processes, floods, climate and soil erosions; cultural services form the part of spiritual, aesthetic, educational oriented and recreational; supporting services includes biodiversity in the ecosystem, nutrients, habitats and production (Millennium Ecosystem Assessment, 2005).

Various parts of ecosystem services related to the soil have been studied, including that the soil is one of the relevant natural resources in the nature which provides a series of goods and services and needs to be included in the soil ecosystem services evaluation and policy making decisions (Adhikari and Hartemink, 2016). It purely depends upon the properties of the soil;

DOI: 10.4324/b23145-8

the degradation in the soil properties might lead to soil erosions, landslides, serious environmental problems and instability in the biodiversity (Godfray et al., 2010).

The conceptual understanding of ecosystem and ecosystem services is crucial to understand their benefits on human being. Conceptual comprehensiveness plays an essential part in decision making for various aspects related to the ecosystem. The classification and definition of ecosystem services bring the essential function of decision making in relation to ES and ecosystem (Fisher et al., 2009). The Millennium Assessment (MA) in context to ES, involving 1300 scientists and featuring outcome that out of 24 ES 15 ES are in the level to decline, will have a negative impact in the future on welfare of human being. It further emphasized on this part that the scientific community should involve to find out ES evolving concepts and their validity, used by various stakeholders (Millennium Ecosystem Assessment, 2005).

Biodiversity attributes in relation to the ecosystem services include processes which are important to comprehend biodiversity in ecosystem services. The relationship between biodiversity and ecosystem services is to know about the loss of biodiversity due to increase in provisioning of ecosystem services. Now it is increasingly adopted to frame management strategies for protection of the biodiverse areas. Earlier researchers worked on the relationship of biodiversity and ecosystem services related to the nature of interaction and impact of environment, different species and ecosystem functions (Harrison et al., 2014). Population dynamics and its impact on the ecosystem functioning by proposing the concept of Service Providing Unit (SPU) explain the ecological units which provide ecosystem services (Luck et al., 2003). The study brings the ecosystem service providers (ESP) and mixed the concept of ESP with SPU to assess various levels of ecosystem like population, community or functional level. The linking of ecosystem provisioning and biodiversity brings the benefit of protection and conservation of area and ecological restoration (Kremen, 2005).

Economic global development has been putting pressure on resource's depletion, and it's time for the decision-making bodies, that is, government and business communities to emphasize on this part (Millennium Ecosystem Assessment, 2005). MA has differentiated conceptually the sub-group assessment (SGA) to assess the role of ecosystem for human well-being (Church et al., 2017). This is conducted by way of provision and connected values of ecosystem which are produced from it (Chan, 2012a). In 2012, the Intergovernmental Science-Policy Platform on Biodiversity and Ecosystem Services (IPBES) was established, which emphasized the assessment at global as well as sub-global level and in four continental regions (Diaz et al., 2015). There is a lack of study on the systematic assessment of tourism highlighted in SGA due to the fact that tourism is a complex type of human recreation (Hall, 2005), and it has impacts on economy, environment and society (Duffy, 2015; Hunt et al., 2015). In Brundtland, the commission includes the same ecosystem services as well as the sustainable development aims (Duffy, 2015; Saarinen and Rogerson, 2014). The ecosystem concept found its role in interdisciplinary resource management and is often present in the discussion of natural as well as social sciences in relation to environmental protection sustainability (Costanza et al., 1997, 2014). The researcher's community focuses upon the development of standardized methods of identification, model and value of ecosystem services for the policy decision maker at various spatial levels (Ash et al., 2010). Human beings are dependent on the services received from the ecosystem, which are used merely as resources for consumption, but the differentiation between services and benefits of ecosystem services brings the change in the utilization of ecosystem services, along with protection and conservation from their exploitation (Millennium Ecosystem Assessment, 2005). Ecosystem services not only provide human well-being through their services and benefits but also help in social and cultural re-generation (Bhagwat, 2009). The tourism concept can be developed in three ways in connection to ecosystem services: (i) the community understands the value of the nature assessment; (ii) tourism depends upon the ecosystem resources for its growth and development and (iii) the services in tourism actively come from the ecosystem and form the particular tourism product (Busby and Rendle, 2000). Many literature on recreation and tourism concentrate on the part of cultural ecosystem services (CES) (Bryce et al., 2016; Fish et al., 2016; Hernandez-Morcillo et al., 2013). The ecosystem services at urban area of Stockholm, Bolund and Hunhammar (1999) emphasized on the urban ecosystem and its impact and benefits of provisions to the residents of the settlement. Valuation of the ecosystem services has been neglected in relation to the cost-benefit analysis (CBA), which can bring that if the monetary or non-monetary values are included in the ecosystem effectively result in enhanced benefits of infrastructure development and conservation-related projects. In this study, the urban ecosystem comprises seven parts including lawns, cultivated lands, wetlands, lakes, trees, forests in the urban part and water streams. Benefits of the ecosystems rated to urban ecosystem include the air filtration, reduction in the level of noise pollution, regulation of micro-nutrients, drainage of rainwater, treatment related to sewage, and cultural and recreational value. The contribution of ecosystem services for urban structure designing and resource preservation requires the awareness of ecosystem services by city planners and decision makers.

Daily (1997) asserted that the ecosystem is the relationship and process among the natural ecosystem and different species to build and raise the human being life. Zhang et al. (2007) conducted a study on ecosystem services (ES) and ecosystem dis-services (EDS) related to agriculture, which led to a discussion on dis-services of agriculture ecosystem by decreasing the productivity level of agriculture production and enhancing the cost of production among unwanted species, for example, herbivory, struggle for water and nutrition. The fertility of the soil enhanced by micro- or macroinvertebrates (Thomas, 1999) claims that pests in agriculture fields like herbivores and seed eaters along with a variety of pathogens decrease production and simultaneously enhance the cost of production in agriculture. More use of pesticides to control these pests increases the resistance among some pests against pesticides and generates negative health impact on human beings along with the other species.

5.2 Tourism and Ecosystem Services

Tourism is conceptualized in three manners in context to ES. First, the tourism sector expresses it as people emphasized their value towards nature-based services. People provide value individually based on economic assessment or a combination of social and cultural values realized through tourism (Church et al., 2017; Duffy, 2015). Tourism is economically benefitted from the participation of ecosystem services towards the well-being of people (Millennium Ecosystem Assessment, 2005). Second, tourism is dependent on a variety of ecosystem services and is also found to bring a change in ES. It can have a negative impact on the terrestrial ecosystem by introducing mass tourism as the ecosystem services not only helps cultural services but also contributes to provisioning and regulating services. Third, tourism produces its services from the ecosystem and brings well-being to people. Tourism not only gets the benefits from ecosystem services but also shows its impacts on the ecosystem through particular tourism product that is offered, for example, farm tourism (Busby and Rendle, 2000; Church et al., 2017). Ecotourism is highly conceptualized in the study of sub-global assessment of ES, in which 8 out of 14 ecosystem services have been studied. Few studies are based on animal species hunting and others are reliant on agriculture tourism or farm tourism and economies. The assessment did not include the part of ecosystem used as tourism product which can proceed to negative as well as positive results for ecosystem and on people benefits as well (Church et al., 2017; Hunt et al., 2015).

Benefits of tourism on health using ES for cognitive capacity restoration, stress relief and revitalization are derived only when people participate in nature-based tourism activities. The nature, tourism and well-being connection interpret how culture ecosystem services are produced for people and what it means to them (Willis, 2015).

Provision of biophysical environment for tourism along with ecosystem services for land, water and air is important, which provides roads, accommodations, gas regulations and climate for tourism. These biophysical environment and ecosystem services together generate numerous direct and indirect effects on the tourism sector besides raw material provisioning. Degradation of these resources might lead to loss for tourism sustainability in the long run (Simmons, 2013; Whinam and Chilcott, 1999). Ecosystems are culturally, socially important in tourism for people who can have rest and relaxation, recreational activities as well as psychological benefits. The aesthetic part of ecosystem in the form of landscapes, rivers and other natural products enhances the opportunities for nature-dependent activities for the people (Simmons, 2013).

Nature-generated tourism activities are the combinations of various features of landscapes including mountains, rivers, forests and wildlife. The ecosystem service-related perspective of users of landscapes (tourists, local community and businessmen) depends upon their attitude toward it. The interaction between landscapes and users brings the possibilities to manage and to take decisions related to it, which promote the destination facilities and services along with destination image (Bachi et al., 2020).

5.3 Cultural Ecosystem Services in Tourism

Millennium ecosystem assessment has defined cultural ecosystem services (CES) as non-material benefits providing requirements for psychological benefits to people. CES of tourism provides its benefits for human being in the form of psychological well-being, which plays a significant role to motivate and satisfy the need of the human being to be a part of tourism decision making and management. The concept and framework of CES in tourism bring more sense to understanding tourism, CES and well-being connection and emphasized on overall management of natural resources in tourism. Tourism management includes various literature point of views related to CES challenges and psychological benefits for human beings and part of destination values, which act as motivation for that destination for the tourists. Tourism conceptualization is mixture of not only the part of activities but also includes the psychological aspect related to experience and mental state of mind leading to clear understanding of tourism management practices of natural destinations (Bieling and Plieninger, 2013; Chan, 2012b; Mannell and Iso-Ahola, 1987; Willis, 2015).

In tourism, the benefits of CES are harder to explain due to its nature of tangibility and intangibility. Benefits of intangible ecosystem services in tourism for tourists are difficult to assess (Andersson et al., 2014; Leyshon, 2014; Smith and Ram, 2017). Landscapes provide the natural (Lee and Han, 2002), cultural (Rossler, 2006), rural (Arriaza et al., 2004) and wildlife aspects in tourism, and change in these services generates change in demand of tourism. Therefore, a clear understanding of CES is required not only in conceptual form but also for the tourism sector in general. The study of landscape, tourism and ecosystem services reveals that four major interaction factors are included, namely spirituality, emotional, existence value and intellectual domain, and empirically analysed generate the Common International Classification of Ecosystem Services (CICES) developed by the European Environment Agency (EEA) to assess the ecosystem services valuation. For visitors, the experience of landscapes is more emotional than other CES (Smith and Ram, 2017).

Tourism activities of cultural ecosystem services produced by parks and areas which are under protection indicate the impacts on human health and well-being. There is increased impact on ecosystem by tourism activities, which lessen the positive impact on ecology. Consideration of these impacts for the planning and decision making regarding CES for natural resources is essential to be beneficial to human well-being and health (Taff et al., 2019). Simply participating in nature-based tourism activities relates to happiness by self-perception and positive state of mind (Tarrant, 1996).

Studies are related to CES like rural tourism along with the community-based tourism (CBT) with CES (Gao and Wu, 2017). Heritage tourism related to CES includes motivational part of tourists (Mccain and Ray, 2003) and use of CES as tool of research for the assessment of tourist satisfaction (Smith and Ram, 2017). Many other studies related to CES focus upon the subjective part like human well-being, recreation and formation of tourism product identity (Pyke et al., 2016). The mapping aspect of CES is also studied; like in Chile the outcomes are used for decision making for tourist destinations (Nahuelhual et al., 2016). Other study includes Italy for CES to make its outcome as a base to policy of land use (Zoderer et al., 2016), photo base study approach to identify the tourists' perception for the choice of landscapes of South Brazil and in Vineyard landscapes (Bastarz and Biondi, 2011).

CES provides tangible as well as intangible benefits of services to visitors in tourism, which are based on only its marketable nature for tourism but lack in formation of standardization in its definition to measure its impacts to be utilized for decision making and planning. The formation of various governmental organizations on biodiversity and ecosystem services has enhanced the opportunity to conceptualize the concept of CES. Various methods to account for the cultural ecosystem services based on different aims have been found in literature. Most of the literature indicating the CES indicators are not clearly defining them. This problem of less clarity of CES indicators can be improved and conceptualized in more precise way by including stakeholders to bring more visibility (Hernandez-Morcillo et al., 2013).

The CES and Heritage at coastal region of Mexico through social media highlight that local residents are more inclined than foreign tourists towards aesthetic values of the destination and bird watching as found out through the number of photographs uploaded in social media. The analysis of these social media put its implication for the management and planners for nature conservation and decision making at destinations (Ghermandi et al., 2020).

The CES in urban green area brings benefits of improvement in the quality of life of locals. The use of CES in relation to land use and management in monetary and non-monetary value indicate the changeover at different green infra and managerial stage, for example, learning about the environment brings low monetary but high level of non-monetary value; unknown place brings the low intensity of management than CES providing tourism at place. Hence, it is necessary that land strategy be included in planning ecosystem services with land use and management intensity (Langemeyer et al., 2015).

Mountains provide CES to local residents as well as to tourists. The change at global level has generated challenges to the decision makers and policy makers for appropriate methods to evaluate the CES. The model of perception related to the aesthetic aspect of landscape pattern by using the photographs survey is conducted, and outcome depicts that respondent provided more emphases for aesthetic values of Alpine landscapes for its pattern of settlement, infrastructure and farming land use. Aesthetic value is more for the highest altitude ecosystem services and lowest for near to street and valley part showing little view. This map that comes out of study can be used in future by stakeholders for decision making for ecosystem services and landscape planning (Schirpke et al., 2016).

Albania brings its cultural heritage and tourism along with the food, dance, communist era to be utilized for marketing. The tourism of Albania has the potential to utilize its resources of CES to achieve its economic developmental objectivity. The records of visitors highlight that visitors are willing to invest in nature and cultural ecosystem services (CES) (Seidl, 2014).

Assessment of CES provided by the biodiversity in context to demand and supply framework in African mammals' reveals through investigation that supply of large species of mammals increases the hope of tourists and does not affect the expectation to watch specific species. Predator species are in more demand than others, and the demand and supply in four protected areas (taken under study) are similar. The traits of specific tropical part increase the expectation of the tourists to see particular species. The demand and supply assessment across the wildlife ecosystem brings the framework to conceptualize the CES of wildlife tourism and utilizes it for decision making in wildlife tourism and its conservation part (Arebieu et al., 2017).

The use of CSE in urban settlement provides specific ecosystem services to the residents and community like another ecosystem. The specific part of ecosystem services provision like food and soil erosion control is less important in context to urban region, and it provides services including values of recreation and educational part. The assessment of urban planning by using specific indicators of the cultural ecosystem services shows that there is a lack in conceptual clarity on the part that effects the decision making and planning for urban area. It brings the need to conceptualize the specific indicators of CES for urban planning (Rosa et al., 2015).

CES provides material and non-material benefits to the community but lack in providing potential benefits like infrastructure development along with failing to control and reduce the environmental impacts created by human beings. Stakeholders only discuss the facilities and local impact. If the combination includes stakeholders, then part of conceptual understanding of ecosystem services with local understanding will improve and bring the responsibility on local people to manage and preserve the natural environment (Bullock et al., 2018).

CES benefits are not clearly conceptualized due to intangibility and non-market ability. An analysis of management aspect of coastal and watershed cultural ecosystem services was conducted by environment managers of Hawaii. Only 10% of the managers are able to show specific policy related to CES and through interview reveals that maximum part of CES management focus for the benefits to local community ignoring the spatial scale. Identification of CES managerial characteristics leads to indicator development to notice the changes in ecosystem service and human well-being (Pleasant et al., 2014).

5.4 Methodology

This study is based on secondary data. Sixty-four research papers of reputed journals have been studied. Research papers related to ecosystem services, cultural ecosystem services and tourism were used to understand the concept of ecosystem services (ES), cultural ecosystem services (CES) in the context of tourism.

5.5 Case Studies Discussion

The case study of Bachi et al. (2020) of Mote Verde, which is the district of Camanducia in the state of Southeast of Brazil. The super humid climate of the region varies in summer from 17 degree to 25 degree Celsius and in winter varies from −5 degree to 10 degree Celsius. The high altitude provides Ombrophilous forest and Araucaria, which is also known as pine tree. This landscape attracts many tourists by its mixture of vegetations, trees and mountains, and received "A" classification from the Ministry of Tourism. Photography questionnaires were used to assess the characteristics of landscapes in relation to cultural ecosystem services along with the content analysis. The outcome is visualized in three parts related to management of landscapes, opportunities and challenges in it and finally the planning at spatial level for short or long term.

1 Results related to management aspects come out as CES do not find equality of landscapes, rather it has gathered around small urban region in Monteverde district, which indicates that local and tourists sign equal value to destination.
2 Results of cultural ecosystem services (CES) that are produced by different landscapes of Monteverde district provide various benefits to people. This knowledge can be used for spatial planning of the landscapes, mapping and modelling of the CES of the landscapes in response to principles of sustainable development could be achieved the Sustainable Development Goals (SDGs).
3 Residents of the Monteverde can participate in decision making, designing and planning of values of attachments related to physical part of the area. Zoning will promote aesthetic and cultural values part of the landscapes of the area based on sustainability principles and help in marketing and promotion strategies to motivate the visitors to select this area for their visit.

Managing the services related to the cultural ecosystem study by Pleasant et al. (2014) discusses the various ecosystem services and their benefits to tourists as well as local people. Hawaii region has a rich history related to culture and varied forms of CES, which is dependent on the managers who are managing these CES for various stakeholders. The hypotheses related to ecosystem services by following the guidelines of MEA are being tested, including (i) selection of particular ecosystem services are prioritized to be considered for the management by various types and organizations of decision makers (ii) establishing the relatedness between ecosystem services and human well-being, and (iii) How these services are managed by the decision makers at the perspective of individual stage on the bases of environmental decision making.

The results show that in Hawaii, emphases have been provided to the management of CEC to secure local community to save it from various threats to culture and climate change. It is found that the CES profits are significant for locals as well as for tourists. Various policies are focusing on the management of these CES, but on the socioeconomic front there is lack in guiding source in policy. Results of the study indicate that there is a lack of accounting for these CES which renders its hard impact on management of CES. Different ecosystem services in relation to its beneficial role for the people under the proposal of MEA have been found with certain important contrast for the Hawaii region. In Hawaii, the ecosystem services which are not clearly defined by the market, for example, culture, the stakeholders pointed out the first priority for these services management and it will help to involve the local community to bring the changeover in management planning into an effective policy-related decision.

The relationship between the CES with socio-cultural values and well-being was studied by Bullock et al. (2018) by taking case study of Ireland's coastal area of Dublin. It indicates that the study of socio-cultural values can be obtained at two levels: at output and process levels, including participants' view on it and for decision making. Second, the constructivism approach led to more discussions and debates on the value system related to socio-culture aspects with the ecosystem services and generate more concrete insight on this CES aspects rather than looking at it only in monetary term. This approach will generate the qualitative valuation while the analytical deliberative approach produces the quantitative values in terms of its process. In the study, the deliberative value formation (DVF) approach is used to lead to the outcome that cultural as well as aesthetic values in relation to the community are significant. This approach produces that nature plays its role in the well-being of the people and brings to light the fact that the benefits are getting held slow down by the people's perception of pollution. The longer series of deliberative discussions on ecosystem services benefits and challenges could bring a better conceptual understanding of ecosystem services (ES) and their related problems and solution for the community.

Smith and Ram's (2017) study discusses developing a new research tool to assess the benefits of landscapes for tourists and local community by CES established by the MEA. It includes a sample of 876 respondents, clear that it not only led to assess the benefits but also conceptualize the CES-related aspects. Fifteen variables were taken into factor analysis. The four benefits come out from CES of landscapes in the form of interconnection between the tourists and landscapes like spiritual, existed values, intellectual and emotional contacts. As for the theoretical and managerial decision making, this study brings the assessment tool, which is helpful for the assessment of intangible benefits produced by landscapes to tourists and local residents, and provide the aid to support for tourism industry professional for decision making related to CES benefits connected to the landscape.

5.6 Conclusion

Cultural ecosystem services (CES) are playing a significant role in tourism and in providing various material and non-material benefits to visitors as local community. Various elements of CES have been investigated in connection to CES like coastal region and urban region along with the well-being aspect related to community and socio-cultural values. Millennium Ecosystem Assessment (MA) has categorized the ecosystem services into four sections, which is also relevant for tourism nature-based product development and decision making. The inclusion of CES in the tourism services and facilities development requires an utmost understanding of the part of ecosystem services (ES) generally and cultural ecosystem services (CES) specifically. As the ES are an important component of tourism, they bring the combination of tourism products to tourists in one way and also enhance the degradation of the ecosystem services (e.g., producing mass tourism or unplanned tourism in the region) if not controlled at planning and decision level. There is a need to conceptualize the understanding of the assessment process of ecosystem services by the tourism product developers, decision makers and tourism service providers. The different challenges related to CES must be conceptualized to utilize them for the tourism benefits generation. The psychological elements of visitors (e.g., behaviour, attitude, perception) to be made part of better conceptual understanding of CES in connection to tourism benefits. Community-based tourism (CBT) should also be conceptualized in relation to CES of the area for the tangible and intangible benefits generated from tourism activities for community as well as for tourist or visitors. The indicators of CES need to be researched under different aspects related to tourism and framed in a clear manner so that the tourism planners and decision makers can put proper conceptual understanding regarding each and every indicator to be the part of CES-related tourism development. More findings regarding the conceptuality of ES and CES associated with tourism phenomena related to various natural geographical locations are still awaited.

5.7 Recommendations

The following recommendations are made on the basis of the study:

1 The concept of ecosystem services in relation to tourism is less explored. So, there is a need to explore these various latent indicators in relation to tourism.
2 Cultural ecosystem services are not explored fully in all aspects related to tourism. Need to explore them for more conceptual understanding of the CES.

3 ES and CES are not conceptualized in relation to tourism up to its potential. They need to be studied for their various other aspects. So, that it can be better conceptualized and have theoretical and managerial implications.

References

Adhikari, K. and Hartemink, E., 2016. Linking soil to ecosystem services – A global review. *Geoderma*, Volume 262, pp. 101–111.

Andersson, E., et al., 2014. Cultural ecosystem services as a gateway for improving urban sustainability. *Ecosystem Services*, Volume 12, pp. 165–168.

Arebieu, U., et al., 2017. Mismatch between supply and demand in wildlife tourism: Insights for assessing cultural ecosystem services. *Ecological Indicator*, Volume 78, pp. 282–291.

Arriaza, M., et al., 2004. Assessing the visual quality of rural landscapes. *Land and Urban Planning*, Volume 69, No. 1, pp. 115–125.

Ash, N., et al., 2010. *Ecosystem and human well-being: A manual for assessment practitioners*. Washington, DC: Island Press.

Bachi, L., et al., 2020. Cultural Ecosystem Services (CES) in landscapes with a tourist vocation: Mapping and modeling the physical landscape components that bring benefits to people in mountain tourist destination in southeastern Brazil. *Tourism Management*, Volume 77, p. 104017.

Balvanera, P., et al., 2001. Conserving biodiversity and ecosystem services. *Science*, Volume 291, p. 2047.

Bastarz, C. and Biondi, D., 2011. Applicacao do Metodo Q para a Valoracao da paisagem de Morretes, Parana, Brasil, como Subsidio ao Planejamento do Turismo. *Turismo Em Analise*, Volume 22, No. 3, pp. 651–680.

Bhagwat, S., 2009. Ecosystem services and sacred natural sites: Reconciling material and non-material values in nature conservation. *Environmental Values*, Volume 18, pp. 417–427.

Bieling, C. and Plieninger, T., 2013. Recording manifestations of cultural ecosystem services in the landscapes. *Landscape Research*, Volume 30, pp. 649–667.

Bolund, P. H. and Hunhammar, S., 1999. Ecosystem services in urban areas. *Ecological Economics*, Volume 29, pp. 293–301.

Bryce, R., et al., 2016. Subjective well-being indicators for large scale assessment of cultural ecosystem services. *Ecosystem Services*, Volume 21B, pp. 258–269.

Bullock, C., et al., 2018. An exploration of the relationship between cultural ecosystem services, socio-cultural values and well-being. *Ecosystem Services*, Volume 31, pp. 142–152.

Busby, G. and Rendle, S., 2000. The transition from tourism on farm to farm tourism. *Tourism Management*, Volume 21, No. 6, pp. 635–642.

Chan, K., et al., 2012a. Where are culture and social in ecosystem services? A framework for constructive engagement. *BioScience*, Volume 62, pp. 744–756.

Chan, K., et al., 2012b. Rethinking ecosystem services to better address and negate cultural values. *Ecological Economics*, Volume 74, pp. 8–18.

Church, A., et al., 2017. Tourism in sub-global assessment of ecosystem services. *Journal of Sustainable Tourism*, Volume 25, No. 11, pp. 1529–1546.

Costanza, R., et al., 1997. The value of world's ecosystem services and natural capital. *Nature*, Volume 387, pp. 253–260.

Costanza, R., et al., 2014. Changes in the global value of ecosystem services. *Global Environmental Change*, Volume 26, pp. 152–158.

Daily, G., 1997. *Nature's services*. Washington, DC: Island Press.

Diaz, S., et al., 2015. A Rosetta Stone for nature's benefits to people. *PLOS Biol*, Volume 13, No. 1, p. E1002040.

Duffy, R., 2015. Nature-based tourism and neoliberalism: concealing contradictions. *Tourism Geographies*, Volume 17, No. 4, pp. 529–543.

Egoh, B, et al., 2007. Integrating ecosystem services in conservation assessments: A review. *Ecological Economics*, Volume 63, No. 2007, pp. 714–721.

Fish, R., et al., 2016. Cultural ecosystem services: A novel framework for research and critical engagement. *Ecosystem Services*, Volume 21B, pp. 208–217.

Fisher, B., et al., 2009. Defining and classifying ecosystem services for decision making. *Ecological Economics*, Volume 68, pp. 643–653.

Gao, J. and Wu, B., 2017. Revitalizing traditional villages through rural tourism: A case study of Yuanjia village, Shaanxi province, China. *Tourism Management*, Volume 63, pp. 223–233.

Ghermandi, A., et al., 2020. Social media-based analysis of cultural ecosystem services and heritage tourism in a coastal region of Mexico. *Tourism Management*, Volume 77, p. 104002.

Godfray, H., et al., 2010. Food security: The challenge of feeding 9 billion people. *Science*, Volume 327, No. 5967, pp. 812–818.

Hall, C., 2005. *Tourism, rethinking the social sciences of mobility*. Harlow: Prentice-Hall.

Harrison, P. A., et al. (2014). Linkages between biodiversity attributes and ecosystem services: A systematic review. *Ecosystem Services*, Volume 9, No. 1, 191–203.

Hernandez-Morcillo, M., et al., 2013. An empirical review of cultural ecosystem service indicators. *Ecological Indicators*, Volume 29, pp. 434–444.

Hunt, C., et al., 2015. Can ecotourism deliver real economic, social, and environmental benefits? A study of the Osa Penninsula, Costa Rica. *Journal of Sustainable Tourism*, Volume 23, No. 3, pp. 339–357.

Knight, A., et al., 2006. An operational model for implementing conservation action. *Conservation Biology*, Volume 20, No. 2, pp. 408–419.

Kremen, C., 2005. Managing ecosystem services: What do we need to know about their ecology?. *Ecology Letters*, Volume 8, pp. 468–479.

Langemeyer, J., et al., 2015. Contrasting values of cultural ecosystem services in urban area: The case of park Montjuic Barcelona. *Ecosystem Services*, Volume 12, pp. 178–186.

Lee, K. and Han, S.-H., 2002. Estimating the use and preservation values of national parks' tourism resources using a contingent valuation method. *Tourism Management*, Volume 23, No. 5, pp. 531–540.

Leyshon, C., 2014. Cultural ecosystem services and the challenge for cultural geography. *Geography Compass*, Volume 8, No. 10, pp. 710–725.

Luck, G., et al., 2003. Population diversity and ecosystem services. *Trend in Ecology and Evolution*, Volume 18, pp. 331–336.

Mannell, R. C. and Iso-Ahola, S. E., 1987. Psychological nature of leisure and tourism experience. *Annals of Tourism Research*, Volume 14, pp. 314–331.

Margules, C. and Pressey, R., 2000. Systemetic conservation planning. *Nature*, Volume 405, pp. 243–253.

McCain, G. and Ray, N., 2003. Legacy tourism: The search for personal meaning in heritage travel. *Tourism Management*, Volume 24, pp. 713–717.

Millennium Ecosystem Assessment, 2005. *Millennium ecosystem assessment: Ecosystem and human well-being 5*. Washington, DC: Land Press.

Millennium Ecosystem Assessment, 2003. *Ecosystem and human wellbeing: A framework for assessment*. Washington, DC: World Resources Institute.

Nahuelhual, L., et al., 2016. Mapping social values of ecosystem services: What is behind the map?. *Ecology and Society*, Volume 21, No. 3, pp. 1–24.

Pleasant, M., et al., 2014. Managing cultural ecosystem services. *Ecosystem Services*, Volume 8, pp. 141–147.

Pyke, S., et al., 2016. Exploring well-being as a tourism product resource. *Tourism Management*, Volume 55, pp. 94–105.

Rosa, D. L., et al., 2015. Indicators of cultural ecosystem services for urban planning: A review. *Ecological Indicators*, Volume 61, pp. 74–89.

Rossler, M., 2006. World heritage cultural landscapes: Ä UNESCO flagship programme 1992-2006. *Landscape Research*, Volume 31, No. 4, pp. 333–353.

Saarinen, J. and Rogerson, C., 2014. Tourism and the millennium development goals: Perspectives beyond 2015. *Tourism geographies*, Volume 16, No. 1, pp. 23–30.

Schirpke, U., et al., 2016. Cultural ecosystem services of mountain regions: Modelling the aesthetic value. *Ecological Indicators*, Volume 69, pp. 78–90.

Seidl, A., 2014. Culture ecosystem services and economic development: World Heritage and early efforts at tourism in Albania. *Ecosystem Services*, Volume 10, pp. 164–171.

Simmons, G. D., 2013. Tourism and ecosystem services in New Zealand. In: Dymond J. R. (ed.), *Ecosystem services in New Zealand – conditions and trends*. New Zealand: Manaaki Whenua Press, Lincoln, pp. 343–348.

Smith, M. and Ram, Y., 2017. Tourism, landscapes and cultural ecosystem services: A new research tool. *Tourism Recreation Research*, Volume 42, No. 1, pp. 113–119.

Taff, B. D., et al., 2019. The role of tourism impacts on cultural ecosystem services. *Environments*, Volume 6, No. 4, p. 43.

Tarrant, M., 1996. Attending to past outdoor recreation experiences: Symptom reporting and changes in affect. *Journal of Leisure Research*, Volume 28, No. 1, pp. 1–17.

Thomas, M. B. (1999). Ecological approaches and the development of "truly integrated" pest management. *Proceedings of the National Academy of Sciences*, Volume 96(11), 5944–5951.

Whinam, J. and Chilcott, N., 1999. Impacts of tramping on alpine environments in central Tasmania. *Journal of Environmental Management*, Volume 57, pp. 205–220.

Willis, C., 2015. The contribution of cultural ecosystem services to understanding the tourism-nature-wellbeing nexus. *Journal of Outdoor Recreations and Tourism*, Volume 10, pp. 38–43.

Zhang, W., et al., 2007. Ecosystem services and dis-services to agriculture. *Ecological Economics*, Volume 64, pp. 253–260.

Zoderer, B., et al., 2016. Identifying and mapping the tourists perception of cultural ecosystem services: A case study from an Alpine region. *Land Use Policy*, Volume 56, pp. 251–261.

6 Impact of Cultural Ecosystem Services on Livelihoods in the North Vidzeme Biosphere Reserve, Latvia

Anda Mežgaile, Agita Livina, Mahender Reddy Gavinolla, and Iveta Druva-Druvaskalne

6.1 Introduction

The United Nations Educational, Scientific and Cultural Organization (UNESCO) Man and Biosphere (MAB) Program is aimed at promoting, conserving, and ensuring the sustainable use of natural resources (Ishwaran et al., 2008). Biosphere Reserves (BR) are "areas of terrestrial ecosystems that aim to promote strategies that integrate the conservation of biodiversity with its sustainable use" (UNESCO, 1970). Often, the biosphere concept is considered as a local solution for the planetary problems (Heinze et al., 2020), in which tourism is considered as an important dimension of BR management (UNESCO, 2002; Moreno-Llorca et al., 2020). Ecosystem services (ES) are one of the major aspects of biosphere management that have emerged in the recent past (Bridgewater & Babin, 2017). ESs are "the benefits humans derive from ecosystems" (MEA, 2005), and these include regulatory, provisioning, cultural, and support services. This approach influences human well-being, culture, and global economy (Bolzonella et al., 2019; Ma et al., 2020). For example, recreational fishing, agriculture, and cultivation-related activities supported by tourism for the communities in and around the biosphere are the major sources of livelihood (Bires & Raj, 2019). Further, the global agenda of sustainable development goals 2030 insists on the conservation and promotion of natural resources and biodiversity (Rosa, 2017), in which MAB is one of the areas that contributes (Pool-Stanvliet & Coetzer, 2020). It is important to note that the conservation is complementary to improved socio-economic conditions associated with tourism and other ES of the destination communities (Fletcher, 2009; Bires & Raj, 2019), as it leads to the recognition, belongingness, ownership, and stewardship among the community members (Chitakira et al., 2012). In this regard, the concept of biosphere reserve contributes significantly to nature conservation and sustainable management of ES while enhancing the well-being and livelihood of the communities (UNESCO, 2017).

However, the interdependence of ES and society is complex (Pahl-Wostl, 2007; Negev et al., 2019) and complicates the goal of sustainable development due to the deterioration of ES (Guerry et al., 2015) from the climate change, pollution (de la Vega-Leinert et al., 2012), and numerous anthropogenic

DOI: 10.4324/b23145-9

activities (Hugé et al., 2020). A change in the ESs in turn affects livelihoods, employment, migration, and conflict (Rodríguez-Robayo et al., 2020). Most importantly, it influences the livelihood of the people living in and around the biosphere (Schaaf, 2009; Bires & Raj, 2020). For instance, a study conducted as part of the UNESCO's, MAB program, in order to understand the impact of climate change on the loss of revenue from recreational and tourism-based ES at Lac Saint-Pierre Biosphere Reserve in Québec, Canada, has several implications on the livelihood of the people in the biosphere area (He et al., 2019).

Several studies examined the role of the biosphere reserve in the restoration of ESs. As an example, one of the studies conducted on role of the MAB program in the restoration of ESs in the African countries revealed that the program contributes to the protection of cultural ecosystem services and enhanced livelihood (Azadi et al., 2021). A case analysis elucidated the benefits to the stakeholders and trade-offs of ESs of the biosphere reserve (Heinze et al., 2020). A significant number of studies examined numerous areas of ESs in the context of BR, such as mapping benefits and use of ESs (Delgado-Aguilar et al., 2017), poverty alleviation (Duan & Wen, 2017), and community perception on wildlife conservation, livelihood, and poaching control (Epanda et al., 2019), sustainable livelihood framework (Huluka & Wondimagegnhu, 2019), change in the cultural landscape of BR (Rescia et al., 2010), connecting conservation to the tourism through community participation (Florian & Hubert, 2019), linking medicinal plant conservation with livelihood (Maikhuri et al., 2017), livelihood from timber products and coffee forest biosphere reserve (Beyene et al., 2020), contribution to the livelihood in the Himalayan biosphere reserve (Yadav et al., 2019), the welfare of the local community (Pour et al., 2017). The above-mentioned studies were mainly carried out in general on the ESs in the BR. Nevertheless, a few studies examined the ESs of BR from the tourism point of view, which includes livelihood diversification from agriculture to tourism (Bires & Raj, 2020), local economic development and livelihood (Mondino & Beery, 2019), stakeholders' perspectives on ecotourism as a tool to the conservation awareness and sustainable local development (Mondino & Beery, 2019), stakeholder discourse on the practice of sustainable tourism development principles in BR (Lyon et al., 2017), visitor experience on nature-based tourism in BR (Moreno-Llorca et al., 2020), visitor perception on ESs in BR (Viirret et al., 2019), and tourism-based livelihood (Praptiwi et al., 2021).

Despite these studies contributing to the ESs in the BR, research in the context of cultural ecosystem services (CES), particularly, the impact of CES on livelihood is scant. Auxiliary to this result of the above-mentioned studies varies from one BR to another, due to the change in the socio-economic, political, demographic, and cultural context of the place. Nevertheless, several studies have shown that the ESs in BR contributes to well-being and livelihood. Studies in the context of livelihood and well-being analysis will be an initial step to reduce the pressure on BR, thus contributing to the effective management of BR. In this background, a special aspect of the study is to know the role of cultural heritage and CES in the development of a low population density area and in areas with a shrinking population. In this regard, the study is aimed at understanding the contribution of CES to the livelihood in the North Vidzeme Biosphere Reserve (NVBR) in Latvia. The NVBR is selected as a study site since it is one of the popular tourist attractions and the only biosphere reserve located in the northern part of Latvia. This study is conducted by using the mixed-method approach that combines the case study, semi-structured interviews with the community members, and analysis of published reports to understand the interconnected phenomenon of cultural ecosystem services and livelihood. The outcome of this exploratory study will provide insights and an improved understanding of the implications, issues, and challenges of cultural ecosystem services to the livelihood and well-being of the biosphere reserves. Further, this piece of research provides a contribution to the existing body of literature and insights to the biosphere reserve management authority for the sustainable management of cultural ecosystem services on livelihoods.

6.2 Literature Review

The direct links between ecosystem services and households' well-being have been long discussed, as ES contribute to household livelihoods and rural regions particularly (Robinson et al., 2019). It has been found, that people in natural resource-dependent communities with diverse livelihoods and ways of life consider CES as crucial to well-being (Ruskule et al., 2018; Elwell et al., 2020). Willis in his study in the Jurassic coast in the United Kingdom concluded that CES and psychological well-being plays an important role in tourist motivation and satisfaction (Willis, 2015). Chakraborty et al. (2020) emphasize the difference of CES for tourists and livelihoods. In addition, society and local community members can rely on income streams that come from CES (Wisely et al., 2018), such as recreation and ecotourism, athletics, education, traditional knowledge, and cultural values.

Parks and protected areas are recognized for the important ES or benefits to provide society, including tourism and recreation activities. ES, which is commonly defined as the contributions of ecosystem structure and function to human well-being, has emerged as a key concept for managing the biosphere reserves – known as living labs (Kermagoret & Dupras, 2018). The biosphere reserves include such diverse landscapes as river landscapes, forests, and wetlands, which allows humans to practice diverse cultural activities that shape their well-being, for example, cultural festivals, religious sanctuaries, and rituals, making their landscape bodies highly appreciated landscape elements. CES are non-material benefits people obtain from ecosystems through spiritual enrichment, cognitive development, reflection, recreation, and aesthetic experiences. This includes cultural diversity, spiritual-religious values, knowledge systems, educational values, inspiration, aesthetic values, social relations, sense of place, cultural heritage values, recreation, and ecotourism (MEA, 2005). However, many of these services can only be provided if people visit protected areas, nature parks, and biosphere reserves through tourism opportunities or live there (Livina & Reddy, 2017; Taff et al., 2019). In different parts of the world, humans and ecosystems have co-evolved, which has led to the development and refinement of local and traditional knowledge and management strategies through constant adaptation and learning (Folke et al., 2005).

Recent literature on ES is increasingly emphasizing the links between biodiversity and ES, well-being, and nature (Cortinovis & Geneletti, 2018; Bradbury et al., 2021; Misiune et al., 2021; Pan et al., 2021). One of the important parts of the latest research in ES, especially in terms of CES, is to focus on values and benefits that are obtained from people interacting with their livelihoods and vice versa. A stronger focus is to incorporate the importance of place, nature attachment, and attachment to the local environment. Furthermore, the literature on ecosystem services has underlined that socio-cultural and environmental benefits are undervalued as relative in economic decisions and they are often intangible, implicit, unstated, difficult to express, and poorly represented in public policy processes (Kaltenborn et al., 2020). However, it is pointed out that the studies show multiple relationships between ES and well-being, but only a few studies actually try to quantify these relationships. Despite these challenges, researchers from broad disciplines have begun to address this need by defining frameworks and methodologies to quantify benefits people receive from ecosystems (Cruz-Garcia et al., 2017; Taff et al., 2019; Bradbury et al., 2021).

Although it has been accepted that well-being is intricately linked to nature, there is increasing agreement that there is a range of positive relationships between biodiversity attributes and ecosystem services of social and cultural importance (Kaltenborn et al., 2020; Bradbury et al., 2021). To explain how local people experience their life situation, both subjective (individually experienced satisfactions, linked to particular contexts and situations) and objective (without references to individuals, for example, environmental elements like forests, sea, and wetlands) factors need to be taken into account (Kaltenborn et al., 2020). It has been discussed, that a single ecosystem service, recreation-based on natural environments is perceived differently by tourists and local resource users. Such differences show up due to the dynamics of recreation-seeking activities of the tourists and the livelihoods-based activities of local users (Chakraborty et al., 2020). Bradbury et al. (2021) indicate that habitat conversion, conserving, and restoring sites typically benefit human prosperity as well.

Following are the several empirical studies and case discussions that provide an overview on the best practices of CES in protected areas, which are connected to BR.

6.3 Best Practices of Cultural Ecosystem Services in Protected Areas

Bieling (2014) through 14 short stories by residents of Swabian Alb biosphere reserve (Germany) reveals rich evidence of connections to identity, heritage values, inspiration, aesthetic values, and recreation, and underlines that non-material benefits are created by locals. It has been found that farmers in Kenyan food plains highly valued the aesthetic role of wetlands, pointing to the diversity of fauna and flora; fishers and farmers in Chile, highly valued birds are a proxy for terrestrial biodiversity; fishers and farmers in Columbia perceived aesthetic values of place as very important to well-being (Elwell et al., 2020).

In the Maguri-Motapung Wetlands of Assam, India, there were 29 key ES identified through household survey, five of the CES-tourism, educational and research, recreational visit, bird watching, and spiritual/inspirational value. In focus group discussions with local people, the top-ranked services were fishing, tourism (because it provides employment as guides or hotel employees), and habitats of biodiversity. They identified seven livelihood strategies of which five were linked to ES, but only one to cultural ecosystem services – tourism (tourist guide, bird watching guide). Findings showed that local people are highly dependent on ES and a decline in ES could negatively impact livelihoods (Bhatta et al., 2016).

In the United States, in 2017, 331 million park visitors spent an estimated $18.2 billion in local regions while visiting National Park Service lands across the country that supported a total of 306 thousand jobs, $11.9 billion in labor income, $20.3 billion in value-added, and $35.8 billion in economic output in the national economy (Thomas et al., 2018). Due to the COVID-19 pandemic, in 2020 amount of park visitors decreased by 28%, compared with 2017. In 2020, 237 million park visitors spent an estimated $14.5 billion in local gateway regions while visiting National Park Service lands across the country that supported a total of 234 thousand jobs, $9.7 billion in labor income, $16.7 billion in value-added, and $28.6 billion in economic output in the national economy (Thomas & Koontz, 2021). Thus, recreational ecosystem services of 15 German national parks were estimated; the consumer surplus of recreation varies from EUR 385.3–621.8 million (considering multiple destination trip bias) and EUR 1.690–2.751 billion (without multiple-destination trip bias), respectively (Mayer & Woltering, 2018).

In Bangladesh, coastal communities have been dependent on coastal ES since historical times through cultural interactions, carving out livelihood benefits from the surroundings. Using qualitative framework research method, it was found that from tourists' point of view, recreational and CES are "nature out there," but from livelihood-based point of view of the locals – there is stronger spiritual connection and greater diversity of place attachment, also by livelihood acquisition. Interaction with surrounding environments of locals shows the connection between recreation and livelihoods practices that are coupled with local ecological knowledge. Stakeholders are quite worried about the health and the beauty of particular landscape-coastal ecosystems because these are vital economic assets for their income and livelihoods (Chakraborty et al., 2020). The authors recommend using the approach of engaging local communities in sustainable tourism planning and resource protection, which can empower them with better livelihood options (Vaeliverronen et al., 2017; Wisely et al., 2018; Chakraborty et al., 2020).

In the Brazilian Amazon, they took research to assess the links between recreational ES and the benefits for the well-being of traditional livelihoods. Results showed that associations of Brazilian Amazon between ES and extractive activities of Brazil nut and rubber are very weak. Qualitative analysis of case studies allowed us to understand that where there are multifunctional livelihoods, recreational ecosystem services are indeed helping to enhance social values of non-timber forest product extractives that otherwise would be suppressed by prevailing "cattle ranching" lifestyles (Ribeiro et al., 2018).

Kermagoret and Dupras (2018) mentioned in their study in the Manicouagan-Uapishka World Biosphere Reserve, northern Quebec, Canada, there is a need to select ES that is significant and relevant for the particular study area for which ecological and economic data are available. They selected seven key ES and two of them correspond to CES: recreation and tourism, cultural heritage, and cultural diversity. They explain cultural awareness in depth that it includes values that humans place on the keeping of historically significant landscapes and forms of land use. Cultural heritage and cultural diversity are linked to indigenous traditional practices (Kermagoret & Dupras, 2018).

In the Natura 2000 Network in Greece, 22 potential proxy indicators of non-material benefits that people may get from nature in Natura sites were obtained. Results of the research indicated hot spot Natura sites for CES values and supply. They also developed risk analysis for proposed wind farm development within these protected areas, 26% were identified with a serious and high risk of degradation of their aesthetic value (Vlami et al., 2021).

A case study from Nepal proves that willingness to pay for CES in forest areas differs according to management modality, economic status, and proximity to forest area (Acharya et al., 2021), which impacts the livelihoods of local people. Overall, respondents in different sub-groups (rich, poor, nearby, distant) mean willingness to pay for bequest value ranged from US$ 3.5 to US$ 8.0/HH/year for all scenarios; for aesthetic values (including inspirational aspects and pleasure) on average they were willing to pay US$ 2.2 to US$4.6/HH/year (Acharya et al., 2021).

The latest research on nature-based tourism from Germany shows that 81% of respondents participated in nature-based tourism, on average participated in 63 leisure activities, 29 day-trips, and seven short overnight trips, with economic significance 151 EUR billion, including travel costs, on-site expenses, and opportunity costs. The most frequently mentioned motivation to participate in nature-based recreation was, firstly, to experience nature and, secondly, to cultural offers in the sight (Hermes et al., 2021).

In particular, there is a lack of quantitative research of the livelihood benefits that may be received from CES, and research is needed to better understand the relationship between protected areas, especially, biosphere reserves and human health and well-being, including impacts on livelihoods (Plieninger et al., 2013; Taff et al., 2019).

6.4 Methodology

The methodology consists of four steps (see Figure 6.1). Authors integrated insight from different non-economic studies, including questionnaires, interviews, and focus group discussions. The study includes semi-structured interviews (June 2021) of six experts: entrepreneurs, tourism specialists in the local municipality, spatial planner/landscape architects of Vidzeme planning region, representative of Nature Conservation Agency, and head of local cultural center. Semi-structured interviews consisted of six questions: significant natural and cultural values, most popular natural and cultural values by locals, characteristics of local residents, the impact of the NVBR to local people and their income, most significant leisure experience from the NVBR, main suggestions and proposals from locals and from business people how they started the local business, what are the responsiveness of locals.

The survey tool for residents of the NVBR was developed in early 2006 by researchers I. Druva-Druvaskalne and Agita Līviņa and surveys were carried out in 2007. The repeated survey of residents on environmental, economic, and social issues was conducted in April–May 2018 by involving the tourism students from Vidzeme University of Applied Sciences. The survey consisted of three parts. This study has used analysis of the statements on environmental, economic, social, and institutional aspects, understanding of term sustainability and information of most popular tourist places and objects as well as about most prospective tourism objects and places, and profile information of respondents. The survey was conducted face-to-face with residents; the total respondents were 308 from all local municipalities of the NVBR.

The authors conducted an online Google survey on well-being and active aging for residents in Latvia. The survey included seven questions, profiles of respondents, and it was opened from April 2021 to June 2021. The introductory part of the survey form included information on data processing, storing, access, and also confirmation to take part in the survey. The authors collected 93 responses, from the 65.6% people in urban areas and 34.4% in rural areas. The proportion is close to the distribution of residents in urban and rural areas in Latvia.

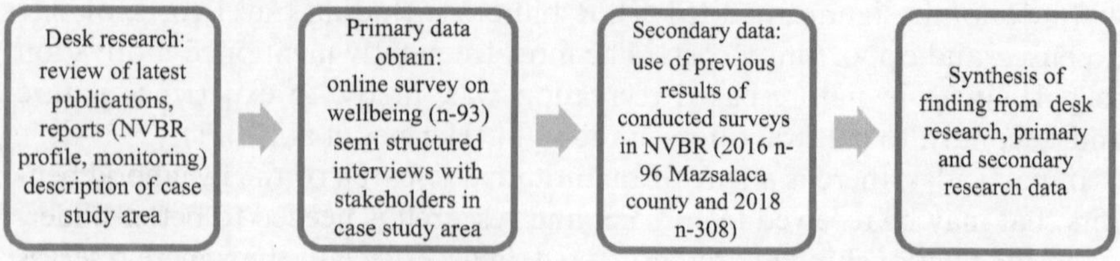

Figure 6.1 Research design framework.

6.5 Study Area

The NVBR is a classic example of a resource-rich rural region facing difficulties to find the best strategy for management. The NVBR embrace Salaca River basin, which includes the Salaca River and Lake Burtnieks along with their tributaries, and the marine area to a depth of 10 m covering 457,600 hectares of land and 16,750 hectares of sea, approximately about 6% of all area of Latvia and is home to 2.6% of the population of Latvia (50,333 inhabitants on January 1, 2021 (PMLP, 2021)). The NVBR was established in 1997 and already on December 15, 1997, was recognized as a protected territory of international importance in the framework of the UNESCO program. It includes not only objects of nature but also man as one of the elements – internationally significant nature and landscape values are preserved, ensuring sustainable social and economic development. The conservation objective of the territory is to protect natural values and landscapes of international significance, promoting sustainable economic and social development. The area includes 24 nature reserves, 1 nature park, and 2 protected Natura 2000 marine areas (Dabas aizsardzības pārvalde, 2021).

The NVBR represents two natural regions of Latvia: the Coastal Lowland and the Middle Latvia Lowland. The highest elevation in the Reserve is 127 m above sea level. The present landscape, formed at the end of the last glacial period approximately 10,000 years ago, is characterized by forests, moraines, drumlins, kettle lakes, rivers, wetlands, raised bogs, semi-natural grasslands, coastal meadows, and agricultural lands. Forests cover more than 45% of the BR, while wetlands and inland waters cover approximately 10%. There are 63 lakes larger than 3 ha in size within the Reserve. Of 63 biotopes found in Latvia that are important at the European Union level, 37 are represented in the NVBR, additionally of 61 EU Bird Directive species nesting in Latvia, 48 do so in the NVBR. The NVBR abuts Latvia's border with Estonia, sharing important wetland areas with neighboring Estonian districts. From 2005 to 2008 in the project "Biodiversity Protection in North Vidzeme Biosphere Reserve," Microgrant program was developed to support economic development activities for local communities in forestry, agriculture, and tourism.

6.6 Research Findings

Mostly, when there are discussions about economic growth, gross domestic product values are used, which measures economic production of a country but can only capture the well-being of society through material wealth. Thus, increased use of the Human Development Index, which assesses the long-term progress of the country in three basic dimensions of human development: a long and healthy life, access to knowledge, and a decent standard of living. Latvia's Human development index value for 2019 was 0.866 (37 out of 189 countries, which put the country in the very high human development category (UNDP, 2020)).

In survey results of NVBR in 2007 and 2018 (aim: to identify the understanding of the population living in the territory of the ZBR about sustainable development (living and working)) already show significant parts of CES – a sense of place, recreational, cultural diversity, and knowledge systems. More than 72% agree that they are proud to live in NVBR; more than 91% think that it is necessary to protect traditions through cultural experiences; 67% agree that protected area increases the value of prosperity; 63% think that access to education is good or very good; and 69% agree that tourism activities help to decrease unemployment and helps to balance the economic situation. To the question, "Which are the most prospective tourism development territories in the NVBR?," territories around water bodies (lakes, rivers, coast) and around the more populated areas (cities, parishes) were provided as answers (Druva-Druvaskalne & Slara, 2007; Druva-Druvaskalne, 2008; Druva-Druvaskalne & Livina, 2019). This is consistent with recent research on urban agglomeration planning to provide smart and high-quality access to cultural ecosystem services for residents from urban areas (Baro et al., 2016; Cortinovis & Geneletti, 2018; Almenar et al., 2021; Pan et al., 2021). NVBR area is very attractive for tourists, has rich natural and cultural values and recreation and ecotourism activities, but it is not managed as a united tourism destination (Arklina et al., 2020).

Although it is difficult to describe the local resident of the NVBR because the area is large and each community has its own individual characteristics; experts characterized the NVBR residents as kind, polite, hospitable, independent, strong, and proud of themselves for living in biosphere reserve. Main sources of income, agriculture, forestry, small business, tourism, and hospitality, paid employment in governmental institutions or other organizations. They actively take part in small business grant projects to receive funding for their (mostly creative or service-oriented) business idea realization to create workplaces for local people in the NVBR cities and rural areas. They organize and participate in cultural events, enjoys nature trails, cycling and hiking routes, water tourism, and local, small accommodations and choose to buy local products in their leisure time.

The most essential branches of the economy that are developed in the territory of the NVBR are agriculture, timber industry, food processing, fish processing, and construction works. There are 310 ecological farms, mainly involved in crop farming and dairy farming. In Table 6.1, statistical data are compiled from State Regional Development Agency about total revenue per capita budgeted by the municipality, municipal budget real estate tax revenue per capita, population density, and territorial development level index. From 2009 to 2021, July NVBR consisted fully and partly from ten counties. As you can see, despite the decrease in population density, in the last eight years there has been an increase in total revenue per capita budgeted by the municipality and municipal budget real estate tax revenue per capita (VRAA, 2021).

The past years have demonstrated growth in the tourism industry – both in terms of tourist accommodation, the diversity of services provided, and through the improvement and development of the tourism infrastructure.

Table 6.1 Review of economic development in the NVBR

County	*Total revenue per capita budgeted by the municipality, EUR*		*Municipal budget real estate tax revenue per capita, EUR*		*Population density, people/km²*		*Territorial development level index**	
	2011	*2019*	*2011*	*2019*	*2011*	*2019*	*2013*	*2019*
Alojas	802	1,285	27	55	9.5	7.8	−0.766	−0.658
Burtnieku	616	1,161	40	86	12	10.8	0.057	−0.049
Kocenu	920	1,535	35	67	14	12.4	0.083	0.093
Limbazu	775	1,448	40	88	16.5	14.9	−0.152	0.155
Mazsalaca	709	1,242	27	63	9.3	7.6	−0.574	−0.295
Nauksenu	764	1,487	52	106	8.1	6.5	−0.214	−0.123
Rujienas	922	1,458	23	48	17.4	14.6	−0.594	−0.438
Salacgrivas	701	1,276	38	85	14.6	12.2	0.088	0.085
Strencu	950	1,333	24	53	11.1	8.81	−0.790	−0.879
Valkas	1,524	1,448	28	59	11.5	10.8	−0.949	−0.504

* Territorial development level index determines the level of socio-economic development of territories, which consists of eight indicators.
Source: VRAA, 2021.

This has resulted in an increased number of tourists and people spending their time in the NVBR, visiting museums, and attending cultural and other community events. The number of accommodation services increased from 75 in 2007 to 194 in 2021 (with 1,059 rooms and 2,854 beds). Assessment of Sustainable Development Profile indicators of the NVBR (Arklina, 2018) showed that indicators of CES have increased compared to 2008. For example, in biological diversity (size of indicator species population of salmon (Salmo salar), wolf (Canis lupus), lynx (Lynx lynx)), tourist accommodations, but decrease in rural bird index, population, social events organized by NVBR administration, which provides recreation, sense of place aesthetic, cultural, social relations, and educational cultural ecosystem services.

In spring 2018, the survey results of the NVBR residents show that 70% of respondents provided an explanation of what they understood by sustainable development. The most active explanators were respondents aged 60–69 (71%), 45–59 years (68%) with secondary education (61%) or higher education (53%) who live more than 20 years (74%) or lifetime (82%) in a given area. 56% of respondents understood sustainable development as slow, long-term development as well as thinking today for tomorrow. Some citations from the explanations:

> "Live today thinking about tomorrow." "Development with a purpose." "Sustainable development is a development that is good both now and in the future."

A total of 23% of respondents emphasized the significance of nature value conservation: *"In nature and human life, everything must be balanced, it must not be wasted, it must be helped to clean up."* Three percent mentioned saving of resources: *"Correct allocation of money to prospective areas", "Redeployment of resources to long-term objectives."* Eight percent has a narrow, specific understanding of sustainable development: *"Development of the municipality and preservation of the site." "Interaction between the municipality and residents."*

Table 6.2 shows compared results of social dimension statements in 2007 and 2018. In all statements, there is a decline in answering "I don't know." It shows an increase in general awareness of residents in the area. The most dramatic increase in disparities is on the issue of priorities for economic prosperity and housing or nature conservation. 62.7% (48.1% in 2007) of respondents agree that the need for good jobs and housing is more important than nature conservation. The number of respondents who disagreed with this statement has fallen by nearly 10% over a decade. The majority of residents are proud to live in the NVBR, and they support the preservation of local cultural traditions such as dance and choirs. 77.9% support the statement that the NVBR provides long-term protection of cultural and natural resources.

In spring 2020, when COVID-19 pandemic lockdown started for the first time, tourism specialists in NVBR recognized that tourists more than usually were interested in nature-based tourism and cultural activities; also (and very important), demand of opportunity to experience and to buy local products increased. New partnerships and cooperation were developed, as well as new products and services. Secondly, it was recognized that travel distance of tourists increased, which meant that they were consuming more local services and products (staying overnight or more than one night). The number of tourists in rural accommodations increased. Main requested activities were hiking and cycling routes (Green railways, Baltic Coastal hiking route, nature trails in bogs and near lakes), which underlined the will for tourists to escape from the daily routine and experience nature and culture through all CES (recreation and ecotourism, aesthetics, educational value, cultural heritage, and sense of place, inspiration, social relations, and spiritual value).

Latest collection of Tourism Statistics in Latvia 2021 (CSB, 2021) says that there were 45,145 visitors, of which 10,554 were non-residents. The number of nights spent in total – 14,482. According to statistics, the average expenditure per day for residents for one day travel was 26.9 EUR, but for non-residents – 50.3 EUR. It means that visitors spent approximately 1,461 363 EUR (by residents 930,497 EUR and by non-residents 530,866 EUR) in local services (shops, accommodations, services, etc.). It is necessary to take into account accommodation services that are offered in online collaborative economy platforms Airbnb, Booking, Expedia, and TripAdvisor. In Latvia in 2019, guests spent 1.5 million nights booked through one of the four digital platforms, 2.7% of them in the Vidzeme region (Eurostat, 2021).

According to expert interviews, main nature and cultural values in NVBR are the River Salaca, Lake Burtnieks, coastline, bogs, Vidzeme Rocky seashore, the landscape itself, grasslands, cites, craftsmen, bakers, blacksmiths, potters,

Table 6.2 Results of social dimension statements

Statement	*Agree, % 2007*	*Agree, % 2018*	*Disagree, % 2007*	*Disagree, % 2018*	*Don't know, % 2007*	*Don't know, % 2018*
I'm proud to live in a specially protected nature area.	69.8	72.1	10.3	10.4	19.9	17.5
I think more important to me as a local resident is the need for good jobs and housing, not for nature protection.	48.1	62.7	37.2	27.9	14.7	9.4
I participate in events, seminars, and festivities organized by the NVBR.	29.1	26.9	39.8	62.3	31.1	10.7
I'm happy to answer the tourist's/stranger's question, I help out.	85.7	87.7	8.5	9.1	5.8	3.2
Strangers, tourists are disturbing me when they come here, asking questions.	11.3	12.7	81.0	83.1	7.7	4.2
I don't plan to change my place of life by moving to town, a bigger city.	60.0	61.4	26.2	28.9	13.8	9.7
We need to preserve our traditions by promoting self-action choirs and collective dance activities.	87.7	91.9	5.2	3.9	7.1	4.2
The existence of the NVBR provides a long-term protection of natural and cultural resources.	74.8	77.9	5.4	5.8	19.8	16.2

Source: Survey results, 2018.

folk dance, Liv culture and the Livonian language, a fishing culture which is evaluated both by locals and tourists. Also, in research about Youth motivation to visit NVBR (Arklina et al., 2020), these were mentioned as hot spots in NVBR by Generation Z and Generation Y (see Figure 6.2), and also when compared with top three places in survey results they were exactly the same: Skanaiskalns and Mazsalaca, Lake Burtnieks, and Vidzeme sea coastline.

In Figure 6.3, the most prospective tourism objects and places by residents of NVBR in 2018 is illustrated. When compared to a population survey conducted in 2008, one new outlook for tourism territory has been named (XI). Experts mentioned that living in NVBR does not restrict residents and their daily life. But being in NVBR for those who work in tourism and hospitality

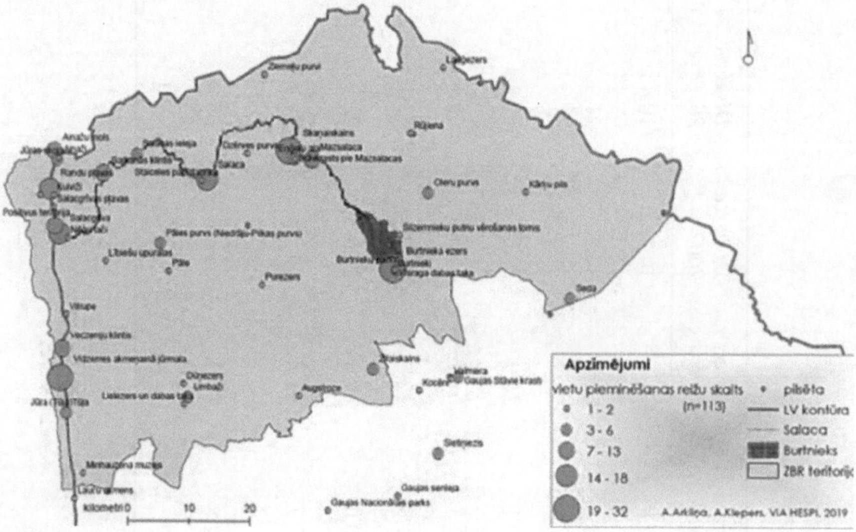

Figure 6.2 The most attractive places for youth in NVBR.

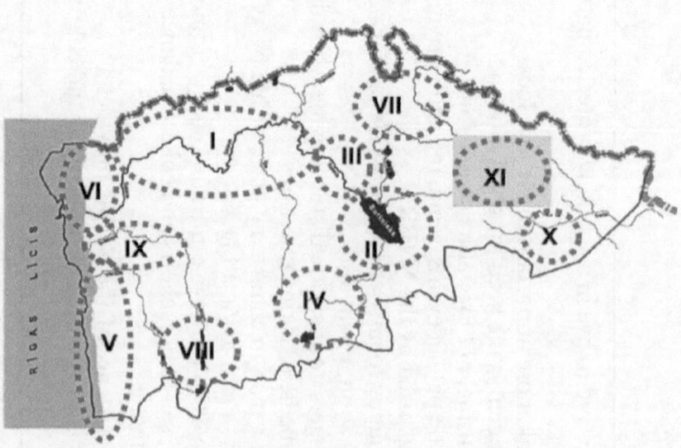

Figure 6.3 Prospective tourism places in NVBR by residents of NVBR, 2018.

Table 6.3 Comparison of preferred and performed activities (during COVID-19 pandemic) for well-being

Type of activities	*What kind of activities would you prefer to enhance your well-being? (%)*	*What have you done/used for your well-being last month? (%)*
Social activities (culture event, party, volunteering)	76.3	30.1
Outdoor activities	59.1	80.6
Education	47.3	50.5
Physical indoor activities	39.8	29.0
Health assistance (consultations)	34.4	11.8
Additional assistance (pets)	17.0	20.4
Assistive technologies (artificial intelligence, apps, smartphones, websites, sensors, other devices)	14.0	23.7
Others	11.0	6.6.

Source: Survey results, 2021.

provides local production and local nature-based services is an added value and they feel proud of that (sense of place). As mentioned in literature studies, CES are relevant for human well-being (Elwell et al., 2020). We asked residents of Latvia how they explore the meaning of well-being. 69.9% of respondents explained that well-being for them means how they feel, 12.9% satisfaction of their needs, 10.8% how they function, and 15.4% other explanations as funny name, new word, balance between emotions, mental, and physical conditions. The most impending obstacles for well-being are their health conditions, low self-esteem, low material conditions, and low socializing options. How satisfied are you with your life now (Spring, 2021 – COVID-19 pandemic obstacles)? More than half of respondents are very satisfied (11.8%) or satisfied (44.1%), partly satisfied 39.8%, and only 4.3% unsatisfied with life.

Table 6.3 shows that the most preferred activities for human well-being are social activities, outdoor activities, and education. The same activities are mentioned as used activities during the COVID-19 pandemic period, but the most used are activities outdoors because it was legally allowed. All three types of most preferred and performed activities correspond to CES.

6.7 Discussion

The concept of biosphere reserves is as living laboratories for sustainable development under diverse ecological, social, and economic contexts, where three main goals are merged: conservation of biodiversity and cultural diversity; economic development that is socio-culturally and environmentally sustainable; and development through research, monitoring, education, and training. As it is mentioned before, NVBR is not managed as joint tourist destination, and there is no direct or obligatory tracking of visitors in tourism attractions, nature trails, accommodations, etc. But due to the high nature values, NVBR area is a rich provider of cultural ecosystem services as recreation and ecotourism, sense of place aesthetic, cultural, social relations, and educational CES and makes an impact on livelihoods in NVBR.

Our study illustrated that in the 10-year period, a new area more inland (see Figure 6.3) number XI covered by swamps and forests was identified as prospective tourism area by locals. There are two main factors that have an impact on the place development: (1) long-term environmental awareness activities and (2) newcomers and active young people in the area with motivation to do things and create public activities.

Since 2016, NVBR administration has been organizing the event NVBR Traveler days, which includes such activities as guided hiking and cycling routes, visits to local businesses, educational, and creative workshops. The Traveler days let a unique opportunity for quiet people to visit neighbors' offerings and see their farmsteads. It provides an opportunity to familiarize themselves and to establish co-operation between those who have not yet engaged in joint activities. Due to the COVID-19 pandemic in 2020, travelers were offered a prepared route that could be taken individually. In 2019, the event was two days long and 296 participants participated in all activities. In 2021, two Traveler days were organized – in spring and summer. In spring, it was campaign "Travel around NVBR," there was a specially prepared information map. Four hundred printed maps were taken, and also it was possible to print them out from online resources. In summer, it was a joint event of both organized and guided tours and offers for individual excursions. In total, 155 participants took part in organized events. In 2021, NVBR participated in the European Day of Parks (organized by EUROPARC Federation) with an activity "Jewels of North Vidzeme Biosphere Reserve parks" where a free-of-charge map with a photo orientation sheet was prepared and the park in area was accessible any time. The map got 1,494 views.

In 2005 Urtans and Soms wrote:

> The "biosphere reserve" becomes a real umbrella for co-operation across the formal municipality and District borders and can bring together partners and stakeholders which traditionally have not been used to communicate, cooperate and compromise. […] connection of the biosphere reserve concept and real activities to the actual biosphere reserve and its zones is not very well developed and will be one of the important matters to be solved during the next years.
>
> (Urtans & Soms, 2005)

Our research findings showed that during the ten-year-period, the amount of residents who are proud of life in the NVBR is increasing. At the same time, as mentioned in the interviews people are self-sufficient and the locals have become younger, and they are busier. Newcomers and younger people have added value for place development if it is additional and provides diversity from other experiences, including cultural issues. This is also supported by talent attraction study in the region (Līviņa & Rozentāle, 2020). Locals are satisfied with their lives in the NVBR, and 61.4% of respondents stated that they are not planning to move from their current place to other.

Due to the changes in administrative management of biosphere reserve and two territorial administrative structural reforms in Latvia, the visibility of NVBR as a united destination is increased and there are still problems with the biosphere reserve concept and the real activities. But due to this status, locals and visitors have acquired ecosystems with which they interact and obtain such high valued and different CES – for visitors to experience, but most importantly, for local residents to impact positively their livelihoods and sustainable development.

6.8 Conclusions

The study confirms that the general concept of biosphere reserves and branding of the biosphere reserve for tourism as a unified brand is still weak. Visitors and tourists do not recognize many popular tourism and recreation sites as an offer of the biosphere reserve area but as individual unique tourism objects and sites. In the local community, residents use primary two types of CEC as participation in public events (cultural events: concerts, creative workshops) and leisure activities in the forest (mainly by picking berries, mushrooms) and along the areas by the water bodies.

According to research findings, it is necessary to improve the name of NVBR through joint development strategy, which includes common communication about NVBR with local residents, development of biosphere reserve partnership program, and ambassadors' program. These initiatives will help to improve both livelihoods for local residents and sustainable development of biosphere reserves and cultural ecosystem services in biosphere reserve. During the COVID-19 pandemic lockdown, tourist specialists stated that the number of tourists increases, most important – from far distances. Promoting NVBR as joint tourism destination will open new products, services, and ways of corporations of local entrepreneurs and society, as well as build on better livelihoods.

Additionally, it is important to take into account for government and local municipalities that people are proud to live in such protected areas as biosphere reserve and do not plan to move to another place, when they are planning services and development strategies for this complex area. Involvement of the population in the development of planning documents, not only as a parish, county, or planning region but also as a biosphere reserve territory is necessary. This aspect will increase a sense of belonging and responsibility in terms of NVBR and economic activity in the area.

As important as above is not only to communicate with locals about NVBR but also to educational work, through seminars, workshops, and events. As results showed, main economic activities are agriculture, forestry, fish processing, and the food industry, from all – only 310 ecological farms.

It needs to be explained how both local society (entrepreneurs) and biosphere reserve area can help each other rather than hindering or restricting.

Acknowledgments

The elaboration of the article was supported by the National research program project "The Significance of Documentary Heritage in Creating Synergies between Research and Society" (Project No. VPP-IZM-2018/1-0022). The survey of well-being was financed by the ERASMUS+ project "Engaged and Entrepreneurial European University as Driver for European Smart and Sustainable Regions."

References

Acharya, P.R., Maraseni, N.T., & Cockfield, G. (2021) Estimating the willingness to pay for regulating and cultural ecosystem services from forested Siwalik landscapes: Perspectives of disaggregated users. *Annals of Forest Science*, 78(51). DOI: 10.1007/s13595-021-01046-3

Almenar, B.J., Elliot, T., Rugani, B., Philippe, B., Gutierrez, N.T., Sonnemann, G., & Geneletti, D. (2021) Nexus between nature-based solutions, ecosystem services and urban challenges. *Land Use Policy*, 100, 104898.

Arklina, A. (2018) *Assessment of Sustainable Development Profile Indicators of the North Vidzeme Biosphere Reserve*. Bachelor Thesis, Vidzeme University of Applied Sciences.

Arklina, A., Grinberga, K., Singh, N., & Livina, A. (2020) Influence of cognitive and emotional advertisements on biosphere reserve image and visitation intention for youth. *Journal of Environmental Management and Tourism*, 6(46). DOI: 10.14505/jemt.v11.6(46).09

Azadi, H., Van Passel, S., & Cools, J. (2021) Rapid economic valuation of ecosystem services in man and biosphere reserves in Africa: A review. *Global Ecology and Conservation*, 28, e01697.

Baro, F., Palomo, I., Zulian, G., Vizcaino, P., Haase, D., & Gomez-Baggethun, E. (2016) Mapping ecosystem service capacity, flow and demand for landscape and urban planning: A case study in the Barcelona metropolitan region. *Land Use Policy*, 57, 405–417.

Beyene, A.D., Mekonnen, A., Hirons, M., Robinson, E.J., Gonfa, T., Gole, T.W., & Demissie, S. (2020) Contribution of non-timber forest products to the livelihood of farmers in coffee growing areas: Evidence from Yayu Coffee Forest Biosphere Reserve. *Journal of Environmental Planning and Management*, 63(9), 1633–1654.

Bhatta, D.L., Chaudrhary, S., Pandit, A., Baral, H., Das, J.P., & Stork, E.N. (2016) Ecosystem service changes and livelihood impacts in the Maguri-Motapung Wetlands of Assam, India. *Land*, 5(15). DOI: 10.3390/land5020015

Bieling, C. (2014) Cultural ecosystem services as revealed through short stories from residents of the Swabian Alb (Germany). *Ecosystem Services*, 8. DOI: 10.1016/j.ecoser.2014.04.002

Bires, Z., & Raj, S. (2019) Determinants of environmental conservation in Lake Tana Biosphere Reserve, Ethiopia. *Heliyon*, 5(7), e01997.

Bires, Z., & Raj, S. (2020) Tourism as a pathway to livelihood diversification: Evidence from biosphere reserves, Ethiopia. *Tourism Management*, 81, 104159.

Bolzonella, C., Lucchetta, M., Teo, G., Boatto, V., & Zanella, A. (2019) Is there a way to rate insecticides that is less detrimental to human and environmental health? *Global Ecology and Conservation*, 20, e00699.

Bradbury, R.B., Butchart, S.H.M., Fisher, B., et al. (2021) The economic consequences of conserving or restoring sites for nature. *Nature Sustainability*. DOI: 10.1038/s41893-021-00692-9

Bridgewater, P., & Babin, D. (2017) UNESCO–MAB Biosphere Reserves already deal with ecosystem services and sustainable development. *Proceedings of the National Academy of Sciences of the United States of America*, 114(22), E4318. DOI: 10.1073/pnas.1702761114

Chakraborty, S., Saha, K.S., & Selim, A.S. (2020) Recreational services in tourism dominated coastal ecosystems: Bringing the non-economic values into focus. *Journal of Outdoor Recreation and Tourism*, 30. DOI: 10.1016/j.jort.2020.100279

Chitakira, M., Torquebiau, E., & Ferguson, W. (2012). Unique combinations of stakeholders in a transfrontier conservation area promote biodiversity-agriculture integration. *Journal of Sustainable Agriculture*, 36(3), 275–295.

Cortinovis, C., & Geneletti, D. (2018) Ecosystem services in urban plans: What is there, and what is still needed for better decisions. *Land Use Policy*, 70, 298–312.

Cruz-Garcia, G., Sachet, E., Blundo-Canto, G., & Quintero, M. (2017) To what extent have the links between ecosystem services and human well-being been researched in Africa, Asia, and Latin America? *Ecosystem Services*, 25, 201–212. DOI: 10.1016/j.ecoser.2017.04.005

CSB. (2021) *Collection of Statistics Tourism in Latvia*. Available at: https://admin.stat.gov.lv/system/files/publication/2021-08/Nr_18_Turisms_Latvija_2021_%2821_00%29_LV_EN.pdf

Dabas aizsardzības pārvalde. (2021) *Ziemeļvidzemes biosfēras rezervāts*. Available at: https://www.daba.gov.lv/lv/ziemelvidzemes-biosferas-rezervats (Accessed 29.06.2021).

Delgado-Aguilar, M.J., Konold, W., & Schmitt, C.B. (2017) Community mapping of ecosystem services in tropical rainforest of Ecuador. *Ecological Indicators*, 73, 460–471.

Druva-Druvaskalne, I. (2008) Sustainable tourism development on the basis of cultural heritage in North Vidzeme Biosphere Reserve, Latvia. In: *WSEAS International Conference on Cultural Heritage And Tourism (CUHT'08)*, Heraklion, Crete Island, Greece, July 22–24, 2008.

Druva-Druvaskalne, I., & Livina, A. (2019) Environmental awareness perception of young people living in the biosphere reserve in Latvia. Society. Integration. Education. In *Proceedings of the International Scientific Conference*, Volume V., May 24–25, 2019, pp. 109–118. DOI: 10.17770/sie2019vol5.3928

Druva-Druvaskalne, I., & Slara, A. (2007) Sustainable tourism development: Lake Burtnieks as destination or part of tourism activities in Latvia. In: Nemeth, A., & David, L. (eds.), *Handbook of Lakes and Reservoirs. A Sustainable Vision of Tourism*. Karoly Robert College, Gyongyos, pp.46–55.

Duan, W., & Wen, Y. (2017) Impacts of protected areas on local livelihoods: Evidence of giant panda biosphere reserves in Sichuan Province, China. *Land Use Policy*, 68, 168–178.

Elwell, L.T., Lopez-Car, D., Gelcich, S., & Gaines, D.S. (2020). The importance of cultural ecosystem services in natural resource-dependent communities: Implications for management. *Ecosystem Services*, 44. DOI: 10.1016/j.ecoser.2020.101123

Epanda, M.A., Fotsing, A.J.M., Bacha, T., Frynta, D., Lens, L., Tchouamo, I.R., & Jef, D. (2019) Linking local people's perception of wildlife and conservation to livelihood and poaching alleviation: A case study of the Dja Biosphere Reserve, Cameroon. *Acta Oecologica*, 97, 42–48.

Eurostat. (2021) *Short-stay Accommodation Offered via Online Collaborative Economy Platforms*. Available at: https://ec.europa.eu/eurostat/statistics-explained/index.php?title=Short-stay_accommodation_offered_via_online_collaborative_economy_platforms

Fletcher, R. (2009). Ecotourism discourse: Challenging the stakeholders theory. *Journal of Ecotourism*, *8*(3), pp. 269–285.

Florian, C., & Hubert, Job. (2019) Community involvement and tourism revenue sharing as contributing factors to the UN Sustainable Development Goals in Jozani–Chwaka Bay National Park and Biosphere Reserve, Zanzibar. *Journal of Sustainable Tourism*. DOI: 10.1080/09669582.2018.1560457

Folke, C., Fabricius, C., Cundill, G., & Schulze, L. (2005) Communities, ecosystems, and livelihood. In: Giampietro, M., Wilbanks, T., & Jianchu, X. (eds.), *Ecosystems and Human Well-being: Multiscale Assessments*. Island Press, Washington, DC, pp. 261–277.

Guerry, A.D., Polasky, S., Lubchenco, J., Chaplin-Kramer, R., Daily, G.C., Griffin, R., … & Vira, B. (2015) Natural capital and ecosystem services informing decisions: From promise to practice. *Proceedings of the National Academy of Sciences*, 112(24), 7348–7355.

He, J., Enomana, H., Dupras, J., Kermagoret, C., & Poder, T. (2019) Measuring recreation benefit loss under climate change with revealed and stated behavior data: The case of Lac Saint-Pierre World Biosphere Reserve (Québec, Canada). *Environmental Management*, 64(6), 746–756. DOI: 10.1007/s00267-019-01219-x

Heinze, A., Bongers, F., Marcial, N.R., Barrios, L.G., & Kuyper, T.W. (2020) The montane multifunctional landscape: How stakeholders in a biosphere reserve derive benefits and address trade-offs in ecosystem service supply. *Ecosystem Services*, 44, 101134.

Hermes, J., Haaren Von, C., Schmuker, D., & Albert, C. (2021) Nature-based recreation in Germany: Insights into volume and economic significance. *Ecological Economics*, 188. DOI: 10.1016/j.ecolecon.2021.107136

Hugé, J., Rochette, A.J., de Béthune, S., Paitan, C.P., Vanderhaegen, K., Vandervelden, T., … & de Bisthoven, L.J. (2020) Ecosystem services assessment tools for African Biosphere Reserves: A review and user-informed classification. *Ecosystem Services*, 42, 101079.

Huluka, A.T., & Wondimagegnhu, B.A. (2019) Determinants of household dietary diversity in the Yayo biosphere reserve of Ethiopia: An empirical analysis using sustainable livelihood framework. *Cogent Food & Agriculture*, 5(1), 1690829.

Ishwaran, N., Persic, A., & Tri, N.H. (2008) Concept and practice: The case of UNESCO biosphere reserves. *International Journal of Environment and Sustainable Development*, 7, 118–131. DOI: 10.1504/IJESD.2008.018358

Kaltenborn, B., Linnell, D.C.J., & Gomez Baggethun, E. (2020) Can cultural ecosystem services contribute to satisfying basic human needs? A case study from the Lofoten archipelago, northern Norway. *Applied Geography*, 120. DOI: 10.1016/j.apgeog.2020.102229

Kermagoret, C., & Dupras, J. (2018) Coupling spatial analysis and economic valuation of ecosystem services to inform the management of an UNESCO World Biosphere Reserve. *PLoS One*, 13. DOI: 10.1371/journal.pone.0205935

Livina, A., & Reddy, M. (2017) Nature park as a resource for nature based tourism. Environment. Technology. Resources, Rezekne, Latvia. In: *Proceedings of the 11th*

International Scientific and Practical Conference, Volume I, pp. 179–183. DOI: 10.17770/etr2017vol1.2590

Līviņa, A., & Rozentāle, S. (2020) Challenge of talent attraction in small and medium urban areas: Case of Valmiera City, Latvia. In: Rehm, M., Saldien, J., & Manca, S. (eds.), *Project and Design Literacy as Cornerstones of Smart Education. Smart Innovation, Systems and Technologies*, vol. 158. Springer, Singapore.

Lyon, A., Hunter-Jones, P., & Warnaby, G. (2017) Are we any closer to sustainable development? Listening to active stakeholder discourses of tourism development in the Waterberg Biosphere Reserve, South Africa. *Tourism Management*, 61, 234–247.

Ma, X., Zhu, J., Zhang, H., Yan, W., & Zhao, C. (2020) Trade-offs and synergies in ecosystem service values of inland lake wetlands in Central Asia under land use/cover change: A case study on Ebinur Lake, China. *Global Ecology and Conservation*, 24, e01253.

Maikhuri, R.K., Negi, V.S., Rawat, L.S., & Pharswan, D.S. (2017) Bioprospecting of medicinal plants in Nanda Devi Biosphere Reserve: Linking conservation with livelihood. *Current Science*, 113, 571–577.

Mayer, M., & Woltering, M. (2018) Assessing and valuing the recreational ecosystem services of Germany's national parks using travel cost models. *Ecosystem Services*, 31. DOI: 10.1016/j.ecoser.2017.12.009

MEA (Millenium Ecosystem Assessment). (2005) *Ecosystems and Human Well-being. Vol 2: Current States and Trends*. Island Press, Washington DC, p. 917.

Misiune, I., Julian, P.J., & Veteikis, D. (2021) Pull and push factors for use of urban green spaces and priorities for their ecosystem services: Case study of Vilnius, Lithuania. *Urban Forestry & Urban Greening*, 58. DOI: 10.1016/j.ufug.2020.126899

Mondino, E., & Beery, T. (2019) Ecotourism as a learning tool for sustainable development. The case of Monviso Transboundary Biosphere Reserve, Italy. *Journal of Ecotourism*, 18(2), 107–121.

Moreno-Llorca, R., Méndez, P.F., Ros-Candeira, A., Alcaraz-Segura, D., Santamaría, L., Ramos-Ridao, Á.F., … & Vaz, A.S. (2020) Evaluating tourist profiles and nature-based experiences in Biosphere Reserves using Flickr: Matches and mismatches between online social surveys and photo content analysis. *Science of the Total Environment*, 737, 140067.

Negev, M., Sagie, H., Orenstein, D.E., Shamir, S.Z., Hassan, Y., Amasha, H., … & Izhaki, I. (2019) Using the ecosystem services framework for defining diverse human-nature relationships in a multi-ethnic biosphere reserve. *Ecosystem Services*, 39, 100989.

Pan, H., Page, J., Cong, C., Barthel, S., & Kalantari, Z. (2021) How ecosystems services drive urban growth: Integrating nature-based solutions. *Anthropocene*, 35. DOI: 10.1016/j.ancene.2021.100297

Pahl-Wostl, C. (2007). The implications of complexity for integrated resources management. *Environmental Modelling & Software*, *22*(5), pp. 561–569.

Plieninger, T., Dijks, S., Oteros-Rozas, E., & Bieling, C. (2013) Assessing, mapping, and quantifying cultural ecosystem services at community level. *Land Use Policy*, 33. DOI: 10.1016/j.landusepol.2012.12.013

PMLP. (2021). *Iedzīvotāju reģistra statistika 2021*. gadā. Available at: https://www.pmlp.gov.lv/lv/iedzivotaju-registra-statistika-2021-gada (Accessed 29.06.2021).

Pool-Stanvliet, R., & Coetzer, K. (2020) The scientific value of UNESCO biosphere reserves. *South African Journal of Science*, 116(1–2), 7432. DOI: 10.17159/sajs.2020/7432. Art. #7432, 4 pages

Pour, M.D., Motiee, N., Barati, A.A., Taheri, F., Azadi, H., Gebrehiwot, K., ... & Witlox, F. (2017) Impacts of the Hara biosphere reserve on livelihood and welfare in Persian Gulf. *Ecological Economics*, 141, 76–86.

Praptiwi, R.A., Maharja, C., Fortnam, M., Chaigneau, T., Evans, L., Garniati, L., & Sugardjito, J. (2021) Tourism-based alternative livelihoods for small island communities transitioning towards a blue economy. *Sustainability*, 13(12), 6655.

Rescia, A.J., Willaarts, B.A., Schmitz, M.F., & Aguilera, P.A. (2010) Changes in land uses and management in two Nature Reserves in Spain: Evaluating the social–ecological resilience of cultural landscapes. *Landscape and Urban Planning*, 98(1), 26–35.

Ribeiro, M.S., Filho, S.B., Costa, L.W., et al. (2018) Can multifunctional livelihoods including recreational ecosystem services (RES) and non-timber forest products (NTFP) maintain biodiverse forests in the Brazilian Amazon? *Ecosystem Services*, 31, part C. DOI: 10.1016/j.ecoser.2018.03.016

Robinson, E.B., Zheng, H., & Peng, W. (2019) Disaggregating livelihood dependence on ecosystem services to inform land management. *Ecosystem Services*, 36. DOI: 10.1016/j.ecoser.2019.100902

Rodríguez-Robayo, K.J., Perevochtchikova, M., Ávila-Foucat, S., & De la Mora Dela Mora, G. (2020) Influence of local context variables on the outcomes of payments for ecosystem services. Evidence from San Antonio del Barrio, Oaxaca, Mexico. *Environment, Development and Sustainability*, 22, 2839–2860.

Rosa, W. (2017). Goal 14. Conserve and Sustainably Use the Oceans, Seas, and Marine Resources for Sustainable Development. *A New Era in Global Health: Nursing and the United Nations 2030 Agenda for Sustainable Development, 359*.

Ruskule, A., Klepers, A., & Veidemane, K. (2018) Mapping and assessment of cultural ecosystem services of Latvian coastal areas. *One Ecosystem*, 3. DOI: 10.3897/oneeco.3.e25499

Schaaf, T. (2009) Mountain Biosphere Reserves – A people centred approach that also links global knowledge. *Sustainable Mountain Development*, 55, 13–15.

Taff, B. D., Benfield, J., Miller, Z. D., D'antonio, A., & Schwartz, F. (2019). The role of tourism impacts on cultural ecosystem services. *Environments*, *6*(4), p. 43.

Thomas, C.C., & Koontz, L. (2021) *2020 National Park Visitor Spending Effects Economic Contributions to Local Communities, States, and the Nation*. Available at: https://www.nps.gov/nature/customcf/NPS_Data_Visualization/docs/NPS_2020_Visitor_Spending_Effects.pdf

Thomas, C.C., Koontz, L., & Cornachione, E. (2018) *2017 National Park Visitor Spending Effects Economic Contributions to Local Communities, States, and the Nation*. Available at: https://www.nps.gov/nature/customcf/NPS_Data_Visualization/docs/NPS_2017_Visitor_Spending_Effects.pdf

UNDP. (2020) *The Next Frontier: Human Development and the Anthropocene Briefing note for countries on the 2020 Human Development Report: Latvia*. Available at: http://hdr.undp.org/sites/default/files/Country-Profiles/LVA.pdf

UNESCO. (1970) *FAQ – Biosphere Reserves*. Available at: http://www.unesco.org/mab/doc/faq/brs.pdf

UNESCO (2002) Ecotourism and sustainable development in biosphere reserves: experiences and prospects; workshop summary report, Quebec City, Canada, May 24-25, 2002. UNESCO. Retrieved from: https://unesdoc.unesco.org/ark:/48223/pf0000127757

United Nations Educational Scientific and Cultural Organization (UNESCO). (2017) *A New Roadmap for the Man and the Biosphere (MAB) Programme and Its World Network of Biosphere Reserves*. UNESCO, Paris.

Urtans, A., & Soms, A. (2005) UNESCO's MAB programme and North Vidzeme Biosphere Reserve in Latvia. In: *Nordic Biosphere Reserves Experiences and Co-operation*. Nordic Council of Ministers, Copenhagen, pp. 85–96.

Vaeliverronen, L., Kruzmetra, Z., Livina, A., & Grinfelde, I. (2017) Engagement of local communities in conservation of cultural heritage in depopulated rural areas in Latvia. *International Journal of Cultural Heritage*, 2, 13–21.

de la Vega-Leinert, A.C., Nolasco, M.A., & Stoll-Kleemann, S. (2012) UNESCO biosphere reserves in an urbanized world. *Environment: Science and Policy for Sustainable Development*, 54(1), 26–37.

Viirret, E., Raatikainen, K.J., Fagerholm, N., Käyhkö, N., & Vihervaara, P. (2019) Ecosystem services at the archipelago sea biosphere reserve in Finland: A visitor perspective. *Sustainability*, 11(2), 421.

Vlami, V., Kokkoris, P.I., Zagaris, S., Kehayis, G., & Dimopoulus, P. (2021) Cultural ecosystem services in the Natura 2000 network: Introducing proxy indicators and conflict risk in Greece. *Land*, 10(4). DOI: 10.3390/land10010004

VRAA. (2021) *Regional Development Indicators Module*. Available at: https://www.vraa.gov.lv/en/regional-development-indicators-module

Willis, C. (2015) The contribution of cultural ecosystem services to understanding the tourism–nature–wellbeing nexus. *Journal of Outdoor Recreation and Tourism*, 10, 38–43. DOI: 10.1016/j.jort.2015.06.002

Wisely, M.S., Alexander, K., Mahlaba, T., & Cassidy, L. (2018) Linking ecosystem services to livelihoods in southern Africa. *Ecosystem Services*, 30. DOI: 10.1016/j.ecoser.2018.03.008

Yadav, P.K., Saha, S., Mishra, A.K., Kapoor, M., Kaneria, M., Kaneria, M., … & Shrestha, U.B. (2019) Yartsagunbu: Transforming people's livelihoods in the Western Himalaya. *Oryx*, 53(2), 247–255.

7 Protecting the Malaga tourism ecosystem

Role of residents

Lidia Caballero-Galeote and Elena Cruz-Ruiz

7.1 Introduction

As indicated by Moore et al. (2006), ecosystem protection is vital for sustainability. Its preservation and balance have become a topic of interest in the aftermath of the pandemic (Floetgen et al., 2021; Jiang and Stylos, 2021). And it has not been only because of their degradation but because of the resetting that the halt in human activity has caused. As indicated by Ozorhon and Ozorhon (2021), crises can be transformed into opportunities and now that the vaccine seems to fill the cities with life again, it is vital to carry out a development plan that allows returning to normal. In this sense, this could be the chance to hit the reset button and build a better normal. To this aim, it will be necessary to reach local consensus on how every one of the stakeholders could contribute. Therefore, nature and human beings are inseparably linked, although at times they appear to be independent entities. This is due to the destruction and loss of ecosystems (Sandeep et al., 2021). They can be disturbed up to a point without loss of function, but if that disturbance exceeds their resilience, they become degraded because of their inability to adapt rapidly to the new state. This degradation increased recently because of the expansion of industry and human demand (Fundación Biodiversidad, 2021). The tourist activity may be detrimental to the protection of our ecosystems. In fact, as indicated by Camarda and Grassini (2003), the negative impact of this sector can make resources dry up (Figure 7.1).

Thus, the continuity of a tourism destination depends to a large extent on the management of its resources (Ruban, 2021). In this sense, as mentioned previously, a crisis can become a moment to reflect on what the destination needs and what it should pursue. There is no doubt that tourism has been and is one of the sectors most affected by the COVID-19 due to its global repercussions (Ranasinghe, 2021). It has been much more impactful in crowded destinations. Although despite some authors' claims that only a few will return to overtourism (Arora and Sharma, 2021), the truth is that the cancellation of restrictions is causing an exponential increase in travel in the case of Malaga (MálagaHoy, 2021). Malaga, despite its efforts, continues to exploit mass

DOI: 10.4324/b23145-10

Figure 7.1 La Misericordia Beach.
Photograph by the Authors.

tourism, so it is of vital importance to carry out studies that not only describe the current situation but also involve the agents that coexist in the city.

Mass tourism is characterized by exceeding the carrying capacity of destinations (Jenkins, 2007), by saturating them at certain times of the year (Torres, 2002), and by the poor relationship between tourists and residents (Gursoy et al., 2010) (Figure 7.2).

Mass tourism was questioned as an unsustainable tourism model in the capital of the Costa del Sol (Diario Sur, 2021; Elconfidencial, 2021; Fájula and Domínguez, 2021). Partly because of the degradation of resources and the agglomeration between tourists and residents in the summer months. The city government, alerted by the path that the tourist activity was taking, began to put their efforts in search of another model of tourism (Figure 7.3).

Besides, as the coronavirus progressed, the destinations emptied to the point of total paralysis of activity, border closures, and home confinement. It

Figure 7.2 Guadalmedina River.

Photograph by the Authors.

Figure 7.3 Pompidou Museum.

Photograph by the Authors.

has been a turning point for the tourism sector, but it has also had other repercussions, the consequences of which could be beneficial in the future. It would reward quality over quantity, and that would also help to preserve the environment. To this end, the creation of several museums, the renovation of the historic center, the pedestrianization of the Alameda (one of its most important streets), the union of the city center with the port, and the renovation of services and mobility on the beaches were promoted.

All these efforts focused on decentralizing tourism from the central zone. It would improve the perception of security and maintain the ecosystem. However, the crisis that has hit the sector in a province whose airport receives 19 million passengers, 75% of which are tourists, has a productive fabric deeply rooted in the service sector. And it has been this dependence that has caused devastating economic effects. However, there is one undeniable fact. The use of resources has decreased so much that the city seems to have recovered another aspect (Figure 7.4).

Residents have proven to play a key role in this whole process. The border closure has turned the resident into a tourist. It has helped not only to save the tourism sector but also to broaden knowledge about their territory. In fact, in some destinations, different policies were implemented to attract the resident to become the new discoverer of the territory. Based on the above, this chapter aims to learn about the role of the residents of Malaga in the preservation and conservation of the ecosystem. The findings obtained can contribute to the city and its governors so that by taking advantage of the opportunity that COVID-19 has brought, this territory can move toward more sustainable tourism that promotes the protection of its ecosystem.

Figure 7.4 Los Montes de Malaga.

Photograph by the Authors.

7.2 Ecosystems and communities

A community is a set of people linked by common characteristics or interests or living in a given territory (RAE, 2021). Following this definition, the academic literature has paid attention not only to the role played by communities in the territories but also to their perceptions. Indeed, as Moges et al. (2018) indicate, this is a top-level issue. Territorial development programs constantly try to include locals not only to involve them in the policies being implemented but also to become an example for other locals and visitors. Hence, community-based local monitoring has emerged as a movement for social change. It is an initiative that helps build relationships between the community and other partners and aims to establish partnerships, create projects, and make decisions (Bliss et al., 2001).

Rasoolimanesh et al. (2020) stated that there is a large body of research on tourism and its importance for sustainable development. However, although many of them talk about ecosystems, there are very few studies where the aim is the resident and their perceptions. The power of communities to manage their resources contributes to both the health and well-being of the resident population. In addition, it showed that residents who participate in the conservation of their environment have a greater sense of belonging (Moore et al., 2006). It is even though ecosystem management has always been in public hands with little citizen participation. However, as stated by Duane (1997), community participation is necessary for good ecosystem management, for the identification of concerns, and for resolving conflicts within the community itself.

If the community does not play an important role, management will fail. As Sahai (2015) indicates that process involves the management of the biophysical environment which would include biotic and abiotic components as well as all human relationships with the economic, cultural, and social environment. Much has been said about human capital as the set of skills possessed by labor power (Goldin, 2016), however, little literature has focused on in-depth research on the natural capital of destinations and less of it all about tourism. It was not until 1997 when Costanza et al. estimated that the economic value of global ecosystem goods and services exceeded the gross domestic product of the entire planet. Ecosystem services would be all the direct and indirect benefits to the community from the function of the ecosystem. When we talk about ecosystem services, we refer, for example, to the production of clean water, soil formation, climate regulation by forests, and pollination. Although humans do not pay enough attention to them, it is essential to conserve ecosystem services because they support our health, economy, and quality of life (CREAF, 2016).

7.2.1 Preservation and protection of ecosystem

The impacts of man's hand on certain areas are transforming the environment at a rate that may present irreversible problems (Daily, 2000). For some decades now, there have been concerns about the proper management of the territory. The central objective is the establishment of priorities that allow the protection of the environment, an issue that is increasingly addressed by advanced communities that want to curb the difficulties involved in the present and in the medium and long term in various areas (Freitas, 2019; Fujino et al., 2017; Mihigo and Cliquet, 2020).

The establishment of environmental policies that improve our society must involve users, understanding what their interests and preferences are, considering their attitudes toward conservation (Martín-López et al., 2007), and having positive repercussions for the sustainability of the territory (Floris et al., 2020; Zhai et al., 2020).

Economic growth will be a priority for all countries in the years following COVID-19. The challenge is to implement measures to protect ecosystems and transition to net-zero emissions (Allan et al., 2020). Global economic growth and the development of a resilient world after COVID-19 will have to take into account a number of concrete recommendations in numerous fields (Ibn-Mohammed et al., 2021), most particularly in the tourism industry (UNWTO, 2020; ours in press turitec) and the aviation sector (Le Quéré et al., 2020; Muhammad et al., 2020). But what really interests us is the great impact that the pandemic has had on issues related to the environment, such as pollution (Wang et al., 2020) and the excessive use of plastics (Prata et al., 2020).

7.2.2 Malaga and its ecosystem

Malaga has become a fashionable city. Not only for its location in southern Europe but also for its tourist offer as the capital of the Costa del Sol. The local government has shown its interest in leaving behind the model of mass tourism to achieve the sustainability of the territory through the search for a quality tourist. That is why in recent times investments in museums, parks, and beaches have offered a renewed image of this territory not only to the tourist but also to the resident. The capital of the Costa del Sol has an area of 154 km^2. It is the second-most populated city in the Autonomous Community of Andalusia with 571,026 inhabitants. Malaga has a forest ecosystem. In addition, reforestation of more than 2,000 hectares of land in the east of the city is currently planned. Despite this, it is one of the cities with the fewest trees in Spain. Its average is less than 28 trees per inhabitant. Most are concentrated in the Serranía de Ronda and the Sierras Tejeda and Almijara. It is important to highlight the mountain range of El Torcal de Antequera. The Urban Environment Observatory (OMAU) has launched a Climate Plan 2050, whose main objective is to build a large green ring with more than 3 million shrubs and trees. In Malaga, there is no desert ecosystem, although it can be found in other nearby regions. Malaga has a freshwater ecosystem in which the rivers of Sierra Bermeja and Tejeda stand out. The main basins are the Mediterranean slopes and those of the Guadiaro (Genal river), Verde, Fuengirola (rivers: Ojén and Alaminas or de las Pasadas), Guadalhorce (rivers: Guadalteba, Turón, Grande and Campanillas),

Guadalmedina, and Vélez. Malaga coastline could be considered the most biodiverse in Europe due to its enclave between the Atlantic and the Mediterranean. It is worth mentioning the more than 100 species of marine or aquatic birds, more than 30 coastal plants, more than 15 different cetaceans, 4 species of sea turtles, 3 species of marine phanerogams, more than 200 species of fish, more than 500 marine invertebrates including mollusks, crustaceans, echinoderms, coelenterates, polychaetes bryozoans, and more than 150 species of algae (Senda Litoral, 2021).

7.3 Methodology

The capital of the Costa del Sol has become a fashionable city, or as *The Times* newspaper wrote: "Malaga is more than a gateway to the coasts." Its location in the south of Spain has contributed to it being the nexus between Africa and Europe. Its climate, more than 320 days of sunshine a year, the opening of the Picasso and Pompidou museums, the construction of a commercial port, and the restoration of its historic center have succeeded in getting the attention of international tourism. In fact, despite the significant loss of tourists, the city has recovered by September 2021, 60% of the tourism that came to the city by air (80% of the city's tourism arrives through its airport). It is due to the increase in public investment, which has helped the city to move toward a cultural destination and leave behind the sun and beach model. This study applies a mixed methodology because despite that most studies focus on the quantitative point of view, as indicated by Ruíz et al. (2019), the use of qualitative methodology can provide a much broader view of the object of study. First, a pilot study was conducted with 25 residents to test the relevance of the questionnaire. In addition, they were provided with a list of ecosystems that residents associate with tourism. After this, it was necessary to add photographs to the final survey to improve understanding of the questions. The questionnaire was composed of ten questions (See Appendix 7.1). First, the resident was asked about the definition of ecosystem through an open-ended question. As reflected in the literature on the link between ecosystem and tourism, the second question asked whether, according to the resident's perception, the ecosystem was important for tourism and vice versa. Taking into account the demonstrated benefits of ecosystem maintenance, the third question is based on the advantages and disadvantages for health, economy, society, culture. The fourth question is focused on analyzing whether residents participate in the maintenance and protection of the ecosystem of their territory. The next question examines whether tourism protects the ecosystem and the seventh question deals with whether the pandemic has contributed to the improvement of the ecosystem or not and in what sense. The last part of the survey deals with the types of ecosystems that exist in the province and which of them, according to their perceptions, are currently protected.

Table 7.1 Protecting the Malaga tourism ecosystem: role of residents

Geographical Area: The City of Malaga
Universe (Residents): 571,026
Sampling error: 4.5%
Procedure: Simple random sample
Reliability: 95%
Fieldwork activities: 2021

Source: Author.

For this purpose, participants were shown a list with a photograph of all the ecosystems present in the province of Malaga and were asked to choose the ones they knew and those that were related to tourist activity. The study was carried out between January and July in the city of Malaga. The survey was sent by mail and WhatsApp thanks to the LimeSurvey application. The data were entered, categorized, coded, and interpreted in NVivo Pro and SPSS. A total of 476 people participated in the study on the role of the resident in the preservation of the Malaga ecosystem (Table 7.1). All respondents gave their consent for the handling, analysis, and publication of the data.

Regarding gender distribution, 51.05% were women and 48.95% were men. Therefore, the sample was representative of the population of Malaga (51.97% women and 48.02% men). The highest number of responses was received from the 25 to 45 age group (29.63% women and 36.05% men). In relation to work activity, 27% of the women are in an employment regulation file due to the pandemic, compared to 15.05% of the men. 35.29% of the respondents stated that their work activity was dependent on the tourism sector, while 7.77% indicated that their work consisted of maintaining the province's ecosystems. 79.41% responded that they had been living in the city for more than 15 years or since birth.

7.4 Findings

This section presents the findings. The first question was related to the ecosystem concept. To illustrate residents' perception of the definition of an ecosystem, the following graph is presented (Figure 7.5).

As can be seen in the graph above, the most repeated terms used to define an ecosystem were divided into negative and positive. As positive, residents define an ecosystem as everything related to beaches, trees, animals, and the environment. It is interesting to note that they relate the ecosystem with the color green. However, residents mention negative aspects of the ecosystem such as pollution, excesses, tourist activity, and lack of respect. It is interesting to mention that they do not personalize the negative effects on the tourists but on the tourism activity. Concerning the lack of respect for the ecosystem,

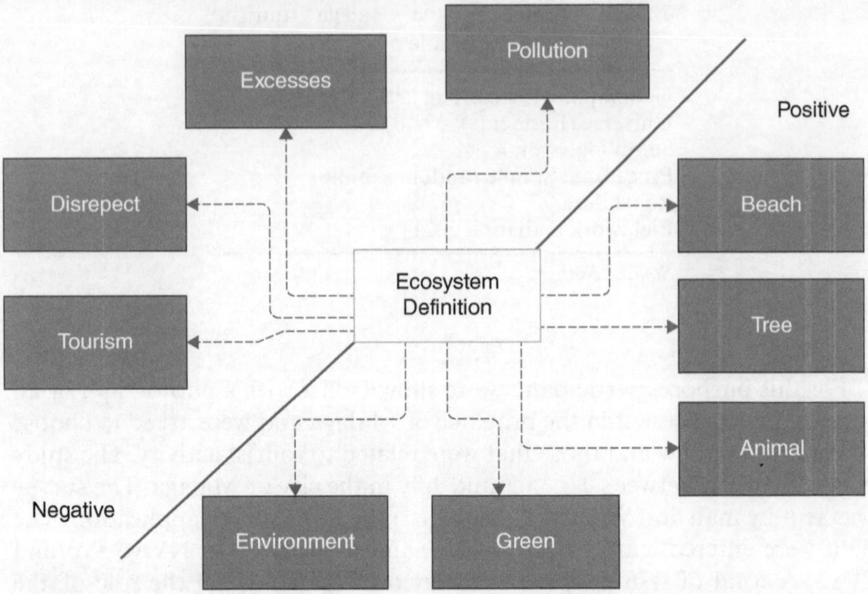

Figure 7.5 Definition of ecosystem according to residents.
Authors from NVivo Pro.

they are talking about themselves. To easily understand the answers, some transcriptions are presented:

Resident case 38:
"Ecosystem is the set of trees and animals"
Resident case 79:
"The beaches are our ecosystem"
Resident case 193:
"Tourism activity is contrary to ecosystems"
Resident case 281:
"Ecosystem is everything with the colour green we can find in the nature"
Resident case 431:
"Ecosystem is a set of systems related to trees, species of our territory. We are not aware of its importance. If we do not respect our own ecosystem, visitors will not respect it."

Following the previous question, participants were asked what type of ecosystem they were experiencing in the city of Malaga. The findings of what ecosystems they think exist in the city of Malaga are shown in Table 7.2.

Regarding the relationship between tourism and ecosystem, the participants concluded through a closed question that the ecosystem is relevant for tourism and tourism for the ecosystem. However, the findings show a difference between genders. While 62.14% of women agreed with both statements, only 47% of men agreed. It is interesting to note that most of the responses indicating "yes" were received by residents under 45 years of age (see Table 7.3).

Table 7.2 Ecosystem in the city of Malaga

Type of ecosystem	*No. of participants experiencing*	*Percentage*
Terrestrial ecosystem	450	94.54%
Forest ecosystem	301	63.24%
Grassland ecosystem	30	6.3%
Desert ecosystem	3	0.63%
Tundra ecosystem	5	1.05%
Freshwater ecosystem	475	99.79%
Marine ecosystem	475	99.79%

Source: Authors from SPSS.

Table 7.3 Importance of the ecosystem for tourism

	Women			*Men*		
	Yes	*No*	*Total*	*Yes*	*No*	*Total*
18 to 24 years	56	0	56	40	30	70
25 to 45 years	68	4	72	32	52	84
46 to 65 years	22	34	56	35	34	59
More than 65	5	44	49	3	17	20
	151	92	92	110	133	133
	62.14%	38%	243	47%	57.08%	233
			100%			100%

Source: Author from SPSS.

Figure 7.6 shows the answers to the question "Do you think that tourism activities help to protect and conserve the ecosystem of your territory?" The findings are very striking. A total of 84% answered "No."

Figure 7.7 shows the findings on the benefits and detriments of the ecosystem for residents. Concerning the benefits, the ecosystem improves their quality of life, allows them to have places to walk and disconnect from their daily lives, and according to their perception, ecosystems are beneficial because they support the sustainability of the planet.

In addition, residents were asked what was the main reason why they should protect the ecosystem. The findings are presented in the following word cloud (Figure 7.8).

In relation to the protection of the ecosystem by the residents, the answers to the question "Do you take actions to protect the ecosystem?" are shown in the following Table 7.4.

As can be observed, most of the respondents affirm that they do not carry out actions to protect their ecosystem. It should be noted that in all age groups the answer was "No."

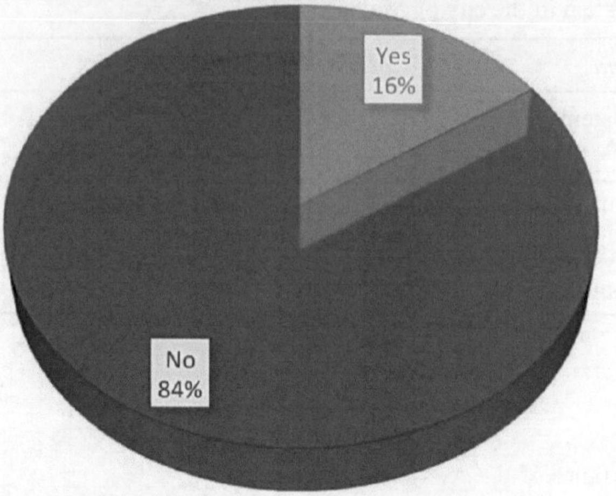

Figure 7.6 Ecosystem protection by residents.
Authors from SPSS.

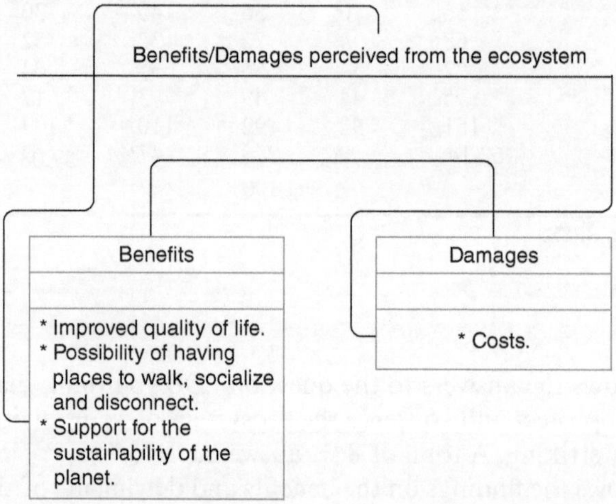

Figure 7.7 Benefits of the ecosystem.
Authors from NVivo Pro.

In relation to the degree of agreement or disagreement with the statement "The pandemic has positively affected the ecosystem of my territory," the findings are shown in Figure 7.9.

As can be seen in Figure 7.9. The findings show that the pandemic has positively affected the ecosystem. The highest number of responses is concentrated in the statements "I agree" and "I strongly agree."

The sustainability of the ecosystem is a highly relevant topic. When participants were asked about who should protect them, the answers provided valuable information (Table 7.4). Women between 18 and 24 years old believe that the visitor/tourist should protect the ecosystem more than themselves. In

Figure 7.8 Why ecosystems are protected?

Authors from NVivo Pro.

Table 7.4 Do the residents take action to protect their ecosystem?

	Women			*Men*		
	Yes	*No*	*Total*	*Yes*	*No*	*Total*
18 to 24 years	16	40	56	20	20	40
25 to 45 years	9	63	72	13	71	84
46 to 65 years	12	44	56	23	35	59
More than 65	20	29	49	10	10	20
	57	35	92	66	136	133
			243			233
			100%			100%

Source: Authors from SPSS.

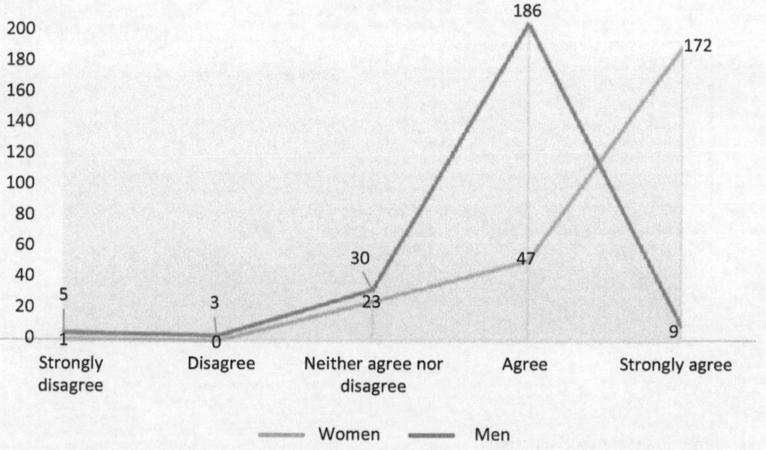

Figure 7.9 Pandemic positively affects ecosystem.

Authors.

this sense, they think that they should protect the ecosystem more than the local or national government. In the case of men, ecosystem protection should be granted by the government. Despite this, they highlight that residents should be concerned about this topic. The figures to the statement "Everyone should be concerned about the ecosystem" are striking.

It has been shown that the policies that achieve the best acceptance are those that take into account the perceptions and opinions of residents. In this sense, residents were asked if they felt listened to by the Authorities about ecosystem conservation. 84.87% stated that they did not feel listened to. Of these, 42.2% are women.

The relationship between tourism activity and ecosystem preservation seems to be linked. Regarding the ecosystems that exist in the province and their relationship with tourism activity, 97% mentioned the 8 shown in the following graph. Then, they were asked to indicate from 1 to 10, where 1 was very little and 10 very much, the level of protection of the ecosystems. The findings show that the ecosystems that residents relate to tourism are the ones that have the most protection.

Table 7.5 Who must protect the ecosystem?

	Women				*Men*			
	18 to 24	*25 to 45*	*46 to 65*	*More than 66*	*18 to 24*	*25 to 45*	*46 to 65*	*More than 66*
Locals	40	32	23	23	62	80	50	9
Local Government	15	30	25	40	68	80	55	19
National Government	36	70	51	48	69	81	54	20
Companies	6	8	1	30	57	72	44	13
Visitors/Tourists	56	68	49	35	69	84	59	18
All the above	16	31	28	49	67	79	53	16

Source: Authors from SPSS.

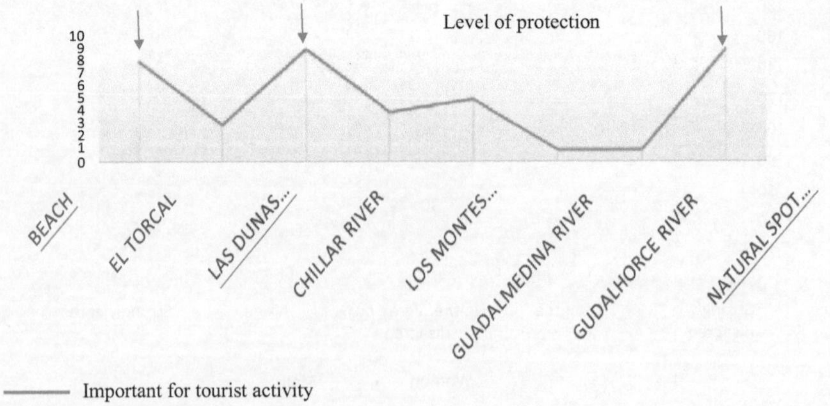

Figure 7.10 Ecosystems most protected by public authorities.

Authors from SPSS.

7.5 Conclusion

The objective of this chapter was to analyze the role of residents in ecosystem management and protection. The most modern definition of ecosystem is the one that describes it as the set of living organisms that share the same habitat. The first definition included only the physical and biological components of the environment; however, in 1935, Arthur George stated that an ecosystem is based on the interactions between the individuals of a community and their environment. Conclusions about the definition and benefits or detriments of an ecosystem have shown that residents do not have a thorough understanding of the subject. In fact, many identify the ecosystem with the beach, with the trees, or with the green color present in nature. They also believe that the ecosystem has negative aspects such as its relationship with tourism, excesses, and lack of respect for the environment.

Although actions are being taken to improve the ecosystem, the interest lies in continuing to create an attractive product for tourists and not for residents. It is not possible to separate tourism activity from the ecosystem. The ecosystem is important because the main economic activity of the city of Malaga, which is tourism, depends on it. Therefore, it is necessary to value the role of the communities and visitors in order to achieve sustainable development that maintains and develops the ecosystem and protects it from aggressions.

Considering the importance of ecosystems for human beings, it is relevant to observe the responses on whether residents take actions to protect the ecosystem. Although younger residents state that they are aware of the relationship between tourism and the ecosystem, they recognize that they do not take actions to help protect it. And this, in a city like Malaga where tourist activity is the main source of income. Perhaps this is because they observe in their daily lives that, although they are important elements, the ecosystem is maintained for the city to be a product that attracts tourists and not for the resident to benefit from it. It is interesting the differences between the age groups so the actions to be carried out to involve the resident in the processes of improvement, management, maintenance, and protection should be different. Younger people are less confident that the government will take action, while older people put more emphasis on the need for politicians to be in charge of protection. The fact that both men and women do not point out that the ecosystem should be protected by all indicates that there is still a lot of work to be done and a lot of training to be provided.

Establishing policies in relation to training is indispensable as this can not only contribute to unite politicians and locals under the same aim but can also increase the sense of belonging and culture. It is necessary that private initiative, especially that related to tourism, joins this and not as part of a purely economic interest but as a cornerstone of the future of both its life as a human being and its business. Therefore, tourism needs the ecosystem to endure and to develop and, for this, as Bulc (2011) indicated, innovation is necessary.

Appendix 7.1: Questionnaire

1 What is an ecosystem?
2 What type of ecosystem exists in your territory?
3 Is ecosystem important to the tourist activity in your territory? And vice versa?
4 Do you think that tourism activities help to protect and conserve the ecosystem of your territory?
5 According to your perceptions, what are the benefits and detriments of the ecosystem in your life?
6 What is the main reason to protect the ecosystem? Do you take actions to protect the ecosystem?
7 What is your level of agreement with this statement? "The pandemic has positively affected the ecosystem of my territory"
8 Has the pandemic affected the ecosystem?
9 Who should be in charge of ecosystem protection?
10 What of these ecosystems are the most protected by authorities and which of them are important for tourism activity?

References

Allan, J., Donovan, C., Ekins, P., Gambhir, A., Hepburn, C., Reay, D., Robins, N., Shuckburgh, E. & Zenghelis, S. (2020). A net-zero emissions economic recovery from COVID-19. COP26 Universities Network. Available at: https://orca.cardiff.ac.uk/id/eprint/132131/1/COP26%20Universities%20Network%20Briefing%20-%20Economic%20Recovery%20from%20COVID-19.pdf.

Arora, S. and Sharma, A. (2021), Covid-19 Impact on Overtourism: Diversion from Mass Tourism to Alternative Tourism, Sharma, A. and Hassan, A. (Ed.) *Overtourism as Destination Risk (Tourism Security-Safety and Post Conflict Destinations)*, Emerald Publishing Limited, Bingley, pp. 275–283. https://doi.org/10.1108/978-1-83909-706-520211018

Bliss, J., Aplet, G., Hartzell, C., Harwood, P., Jahnige, P., Kittredge, D., Lewandowski, S. and Soscia, M.L., 2001. Community-based ecosystem monitoring. *Journal of Sustainable Forestry*, *12*(3-4), pp.143–167.

Bulc, V., 2011. Innovation ecosystem and tourism. *Academica Turistica-Tourism and Innovation Journal*, *1*(1), 27–34.

Camarda, D. and Grassini, L., 2003. Environmental impacts of tourism Bari: CIHEAM, p. 263–270 Available at: https://tamug-ir.tdl.org/bitstream/handle/1969.3/29338/04001977.pdf?sequence=l.

Costanza, R., d'Arge, R., De Groot, R., Farber, S., Grasso, M., Hannon, B., … & Van Den Belt, M., 1997. The value of the world's ecosystem services and natural capital. *Nature*, *387*(6630), pp.253–260.

CREAF. 2016. ¿Qué son los servicios ecosistémicos? Available at: <http://blog.creaf.cat/es/conocimiento/que-son-los-servicios-ecosistemicos/> [Accessed 2 August 2021].

Daily, G.C., 2000. Management objectives for the protection of ecosystem services. *Environmental Science & Policy*, *3*(6), pp. 333–339. Doi: 10.1016/S1462-9011(00)00102-7

Diario Sur, 2021. Andalucía descarta la «turismofobia» y aboga por distribuir la presión turística. [online]. Available at: <https://www.diariosur.es/turismo/andalucia-descarta-turismofobia-20170804000648-ntvo.html> [Accessed 1 August 2021].

Duane, T.P., 1997. Community participation in ecosystem management. *Ecology LQ*, *24*, p. 771.

Elconfidencial.com, 2021. De Málaga a 'Malaguf': el turismo 'low cost' aterriza en la capital de la Costa del Sol. [online]. Available at: <https://www.elconfidencial.com/espana/andalucia/2021-06-28/turismo-malaga-policia-peleas-magaluf-ruido-vecinos_3155816/> [Accessed 1 August 2021].

Fájula, D. and Domínguez, J., 2021. La turismofobia brota en Málaga. [online] ELMUNDO. Available at: <https://www.elmundo.es/andalucia/2017/08/12/598df439468aebb41a8b45cf.html> [Accessed 2 August 2021].

Floetgen, R.J., Strauss, J., Weking, J., Hein, A., Urmetzer, F., Böhm, M. and Krcmar, H., 2021. Introducing platform ecosystem resilience: Leveraging mobility platforms and their ecosystems for the new normal during COVID-19. *European Journal of Information Systems*, *30*(3), pp.304–321.

Floris, M., Gazale, V., Isola, F., Leccis, F., Pinna, S. and Pira, C., 2020. The contribution of ecosystem services in developing effective and sustainable management practices in marine protected areas. The case study of "Isola dell'Asinara". *Sustainability*, *12*(3), p. 1108.

Freitas, F.L., 2019. Brazilian public protection regulations and the preservation of ecosystem services and biodiversity. Doctoral dissertation, KTH Royal Institute of Technology.

Fujino, M., Kuriyama, K. and Yoshida, K., 2017. An evaluation of the natural environment ecosystem preservation policies in Japan. *Journal of Forest Economics*, *29*, pp.62–67.

Fundación Biodiversidad, 2021. Available at: <https://fundacion-biodiversidad.es/sites/default/files/PDF_ordenados/Anejos/Anejo_02.pdf> [Accessed 2 August 2021].

Goldin, C. (2016). "Human Capital." In *Handbook of Cliometrics*, ed. Claude Diebolt and Michael Haupert, 55–86. Heidelberg, Germany: Springer Verlag.

Gursoy, D., Chi, C.G. and Dyer, P., 2010. Locals' attitudes toward mass and alternative tourism: The case of Sunshine Coast, Australia. *Journal of Travel Research*, *49*(3), pp.381–394.

Ibn-Mohammed, T., Mustapha, K.B., Godsell, J., Adamu, Z., Babatunde, K.A., Akintade, D.D., … and Koh, S.C.L. (2021). A critical analysis of the impacts of COVID-19 on the global economy and ecosystems and opportunities for circular economy strategies. *Resources, Conservation and Recycling*, *164*, p. 105169.

Jenkins, C.L., 2007. 'Mass Tourism' is an out-dated concept – A misnomer? *Tourism Recreation Research*, *32*(1), pp. 112–113.

Jiang, Y. and Stylos, N., 2021. Triggers of consumers' enhanced digital engagement and the role of digital technologies in transforming the retail ecosystem during COVID-19 pandemic. *Technological Forecasting and Social Change*, *172*, p. 121029.

Le Quéré, C., Jackson, R.B., Jones, M.W., Smith, A.J., Abernethy, S., Andrew, R.M., … and Peters, G.P., 2020. Temporary reduction in daily global CO_2 emissions during the COVID-19 forced confinement. *Nature Climate Change*, *10*(7), pp. 647–653.

MálagaHoy, 2021. Available at: <https://www.malagahoy.es/malaga/aeropuerto-malaga-operaciones-fin-semana_0_1597041716.html> [Accessed 1 August 2021].

Martín-López, B., Montes, C. and Benayas, J., 2007. The non-economic motives behind the willingness to pay for biodiversity conservation. *Biological Conservation*, *139*(1-2), pp.67–82.

Ntirumenyerwa Mihigo, B. P., & Cliquet, A., 2020. Payment for ecosystem services in the Congo Basin: Filling the gap between law and sustainability for an optimal preservation of ecosystem services. In Mauerhofer, V., Rupo, D. & Tarquinio, L (eds.) *Sustainability and Law*. Springer, Cham.

Moges, A., Beyene, A., Triest, L., Ambelu, A. and Kelbessa, E., 2018. Imbalance of ecosystem services of wetlands and the perception of the local community towards their restoration and management in Jimma Highlands, Southwestern Ethiopia. *Wetlands*, *38*(6), pp. 1081–1095.

Moore, M., Townsend, M. and Oldroyd, J., 2006. Linking human and ecosystem health: The benefits of community involvement in conservation groups. *EcoHealth*, *3*(4), pp. 255–261.

Muhammad, S., Long, X. and Salman, M., 2020. COVID-19 pandemic and environmental pollution: A blessing in disguise? *Science of the Total Environment*, *728*, 138820.

Ozorhon, G. and Ozorhon, I.F., 2021. Crisis or opportunity: Looking at the past for the resilience of settlements. *Structural Studies, Repairs and Maintenance of Heritage Architecture, XVII & Earthquake Resistant Engineering Structures, XIII*, p. 337.

Prata, J.C., Silva, A.L., Walker, T.R., Duarte, A.C. and Rocha-Santos, T., 2020. COVID-19 pandemic repercussions on the use and management of plastics. *Environmental Science & Technology*, *54*(13), pp.7760–7765.

RAE, 2021. Available at: <https://dle.rae.es/comunidad≥ [Accessed 31 July 2021].

Ranasinghe, R., 2021. After Corona (COVID-19) impacts on global poverty and recovery of tourism based service economies: An appraisal. *International Journal of Tourism and Hospitality*, *1*(1), pp. 52–64.

Rasoolimanesh, S. M., Ramakrishna, S., Hall, C. M., Esfandiar, K. and Seyfi, S., 2020. A systematic scoping review of sustainable tourism indicators in relation to the sustainable development goals. *Journal of Sustainable Tourism*, pp. 1–21. https://doi.org/10.1080/09669582.2020.1775621

Ruban, D. A., 2021. Natural Resources of Tourism: Towards Sustainable Exploitation on a Regional Scale. *Sustainability*, *13*(12), pp. 1–5

Ruiz, E.C., de la Cruz, E.R.R. and Vázquez, F.J.C., 2019. Sustainable tourism and residents' perception towards the brand: The case of Malaga (Spain). *Sustainability*, *11*. Doi:10.3390/su11010292.

Sahai, R. 2015. Community Participation in Environmental Management: Role of Women. In *Int. Conf. Recent Res. Dev. Environ. Soc. Sci. Humanit*. New Delhi: India (pp. 173–180).

Sandeep, P., Reddy, G.O., Jegankumar, R. and Kumar, K.A., 2021. Modeling and assessment of land degradation vulnerability in semi-arid ecosystem of Southern India using temporal satellite data, AHP and GIS. *Environmental Modeling & Assessment*, *26*(2), pp. 143–154.

Senda Litoral, 2021. Available at: <http://www.sendalitoral.es/es/6219/flora-fauna-marina> [Accessed 16 July 2021].

Torres, R., 2002. Cancun's tourism development from a Fordist spectrum of analysis. *Tourist Studies*, *2*(1), pp.87–116.

UNWTO, 2020. *Impact assessment of the COVID-19 outbreak on international tourism*. United Nation World Tourism Organization, Madrid, Spain.

Wang, P., Chen, K., Zhu, S., Wang, P. and Zhang, H., 2020. Severe air pollution events not avoided by reduced anthropogenic activities during COVID-19 outbreak. *Resources, Conservation and Recycling*, *158*, 104814.

Zhai, T., Wang, J., Fang, Y., Qin, Y., Huang, L. and Chen, Y., 2020. Assessing ecological risks caused by human activities in rapid urbanization coastal areas: Towards an integrated approach to determining key areas of terrestrial-oceanic ecosystems preservation and restoration. *Science of the Total Environment*, *708*, 135153.

Part IV

Ecosystems Services and Sustainable Tourism Development

8 Ecosystems Services for Urban Destination Development

Potential for Sustainability

Michael Johansson and Jan Henrik Nilsson

8.1 Introduction

The COVID-19 pandemic has fundamentally changed travel patterns to and within urban regions, and destinations have therefore seen a dramatic shift from over-tourism to undertourism (Månsson & Eksell, 2021). Today we have identified a mainstreaming of ethical concerns around consumption of urban space due to urbanization, in what reflects an increasing anxiety with, and accompanying sense of responsibility for, the challenges of sustainable urban planning. In that context, ecosystem services can act as a strategic tool for understanding the turn towards sustainable and conscience consumption of destinations. This paper provides knowledge that is relevant for ethical urban decision-making. Tourism can no longer be seen as a trivial or exceptional activity. Instead, tourism has become fundamentally important for many societies, especially for local employment, economy, and culture (Nilsson, 2018). Tourism, as a whole, has grown rapidly in the latest decades. Tourism has become a global phenomenon, where large parts of the world have become both generating regions and destinations (Cohen & Cohen, 2019). As a consequence of the increased consumption of travel, tourism is estimated to account for between five and eight per cent of global greenhouse gas emissions (IPCC, 2014). Transport, mainly aviation and car traffic, causes 75 per cent of tourism-generated greenhouse gas emissions, accommodation 20 per cent, and activities the remaining five per cent (Scott, Hall & Gössling, 2012). Climate change is global, while the effects of climate change are regional and very diverse. Urban areas are particularly vulnerable to the effects of climate change, through heat waves and excessive precipitation. Many cities are located in coastal areas facing the risk of rising sea levels, for example the Swedish cities Malmö and Helsingborg.

Despite significant research on the relationship between tourism and climate change (Gössling, 2011), we still know little about the relationship between climate change, ecosystem services, and urban tourism. For example, in connection with the IPCC report Scott, Hall and Gössling (2016:15) mention that "very little research has examined the potential climate change effect on urban tourism." In general, sustainable urban tourism is an under-researched field. Sustainable tourism research has traditionally had a

DOI: 10.4324/b23145-12

strong focus on rural tourism, particularly in relation to nature tourism, i.e. ecotourism. The lack of research on the relationship between urban tourism and sustainability is striking, bearing in mind that urban tourism has been growing very fast up until the pandemic, both for business and for leisure.

The development of low-cost flights and digital booking systems in recent decades have systematically reduced travel prices and transaction costs. New business models have also greatly increased the range of accommodation opportunities and accessibility of urban tourist services in transport, food, and local experiences (Nilsson, 2020). As urban tourism has grown, larger parts of cities have been affected, i.e. they have become parts of the destination. Traditionally, urban tourism was concentrated in central cities, where most of the visitor attractions and amenities are usually located (Williams, 2009). To prevent urban degradation and negative impacts from urban tourism flows, it is important to relate to the global sustainability goals, such as goal 11 (sustainable cities and communities): *Make cities and human settlements inclusive, safe, resilient and sustainable*. In order to do that, this chapter addresses both opportunities and challenges in urban planning by implementing the concept of ecosystem services in the context of urban tourism development. Studies of urban tourism have not yet approached the concept and value of ecosystem services, nor have we so far seen any significant interest in green infrastructure as part of tourism planning. By green infrastructure, we mean the spatial structures where ecosystem services actually *take place*. Urban tourism offers a variety of ways in which various forms of green infrastructure may enhance, compliment, support, or improve the tourist experience – and vice versa (Terkenli et al., 2019). By focusing on the green infrastructure, this chapter seeks to emphasize the link between the creation of ecosystem services and urban spatial planning, a fundamental factor behind attractive environments that may create both ecosystem services and experience value for visitors.

The focus of this chapter is on the dynamics between tourism-relevant ecosystem services in urban environments and the development of urban green infrastructure. The aim is to provide an overview of this dynamic, identify value-creating functions in the form of urban ecosystem services, and discuss their importance for sustainable destination development. The chapter, based on a literature review and two Swedish case studies provides a basic overview of the state-of-art knowledge and a review of central concepts and definitions. Thereby, it provides a theoretical platform for continued research and development in the subject area. Furthermore, this chapter investigates how urban ecosystem services can be utilized to improve experience value in urban tourism and how existing ecosystem services could be considered in sustainable urban development. The results will support practitioners in future work on sustainable urban planning and place development with ecosystem services as important sustainable indicators. Findings will affect national policymaking and urban planning agencies for tourism, regional tourism organizations, local stakeholders in municipalities, and partner organizations.

8.2 Literature Review

8.2.1 Urban Sustainable Tourism

Tourism, as an economic activity, is broadly understood as a generator of employment and income, and that can also promote social development. Tourism is one of the world's fastest-growing industries and a major source of income for many countries. Urban tourism stands out from other types of tourism, in that people travel to places with an already high population density and that they spend less time spent there than at other destinations (Edwards et al., 2008). Sustainable urban tourism refers to policies and practices by destinations management organizations and tourist businesses, aiming to minimize negative environmental impact on destinations or regions. Negative impacts to a destination include damage to the natural urban environment and overconsumption of space due to over-tourism. In addition to being important destinations for both business and leisure tourism, cities have a significant role in the overall tourist system through their central position in transport networks (Spirou, 2010). As visitors' and residents' desires and needs intersect, urban destinations face different, and sometimes contradictory, demands for services and facilities. The visitor's consumption of urban space can therefore compromise its balance, possibly reducing urban quality of life (Johansson & Nilsson, 2021; La Rocca, 2014). The challenges connected to over-tourism in destinations may require a radical transition of tourist activities. However, this issue needs to be further addressed by all relevant destination stakeholders (Nevens et al., 2013). According to Seraphin et al. (2018), the issue of over-tourism and the survival of a destination is linked to the conflict between human and natural capital.

Besides being a policy area, sustainable tourism is a growing research field. A significant part of previous research on sustainable tourism discusses ways in which destinations and tourism businesses work with changing their internal activities in more sustainable directions. Hence, a substantial part has dealt with various measures to reduce the industry's climate impact (mitigation). Transport has gained a lot of research interest since it accounts for most of tourism's greenhouse gas emissions. Because of its rapid global growth, aviation has been a particularly important sector to study. Knowledge of its policies, business models, and development trend is essential for understanding its dynamic (Gössling & Upham, 2009). The applied perspective on carbon management aims to understand ways to reduce the ecological footprint of tourism (Gössling, 2011). This issue concerns all geographical scales from the European Union to single destinations and businesses. At destinations, an important concern is to form networks to coordinate sustainable practices, develop knowledge and best practice, and promote sustainable alternatives to visitors. A small part of the research deals with how tourism activities and destinations adapt locally to ongoing climate change (adaptation). In many cases, tourist activities need to adjust to a changing climate, and changing patterns of seasonality. It can also be a matter of changing the location of activities, such as developing skiing areas at higher altitudes to ensure sufficient amounts of snow (Scott, Hall & Gössling, 2012). So far there are only very limited research done on the role of ecosystem-based adaptation to climate change in relation to tourism and tourist destination. The examples used are mainly from Pacific islands facing the consequences of rising levels (Nalau & Becken, 2018). There is definitely room for developing similar research in other geographical contexts. A fundamental idea behind this chapter is that local and regional climate adaptation creates opportunities to promote different types of ecosystem services, services where ecosystems create values for local destinations, also in urban contexts.

8.2.2 Ecosystem Services

Ecosystem services are a socio-ecological concept (Berkes & Folke, 1998). Hence, ecosystem services are defined as "the benefits that humans directly or indirectly receive from nature's ecosystems" (Costanza et al., 1997). Everard (2017) uses the metaphor of a cascade to illustrate their meanings, a "cascade runs from ecosystems, through the functions they perform, to the ecosystem services (benefits to people) flowing from them, leading to the values people accord them in how they are used…" (ibid, p. 9). The fact that human beings depend on the structures, processes, and functions of ecosystems for survival and well-being is of fundamental importance. The concept of ecosystem services gained interest in the 1990s (Costanza et al., 1997; Daily et al., 2009). The concept of ecosystem services was broadened in an interdisciplinary direction by approaching scientific fields such as ecological economics (Arrow et al., 1995; Costanza & Daly, 1992; Johansson, 1994). In an early attempt to facilitate the understanding of the actual value of nature for humankind, De Groot et al. (2002) developed a classification system of the ecosystems' relationship between its function as goods (products) and services.

Urban ecosystem services are the ecosystem services that exist in urban environments (Bolund & Hunhammar, 1999; Keane et al., 2014). Urban ecosystem services were later linked in urban planning to green infrastructure (Escobedo et al., 2018). The concept of ecosystem services had an impact outside the scientific community through the Millennium Ecosystem Assessment (MEA, 2005) in the mid-2000s. Today, much of the research related to ecosystem services relates to the need to inform decision-makers about ecosystem services, sustainable development, and its significance for the urban environment (Childers et al., 2013; Sitas et al., 2014).

The scientific literature talks about four categories of ecosystem services: regulatory, provisioning, cultural, and supporting. Regulatory services means that ecosystem cleans soil and water from pollution. Provisioning services encompass production of food and raw material. Cultural services refer to how people use nature and ecosystem for experiences and recreation. Supporting services are essential for the delivery of the other ecosystem services (Everard, 2017; MEA, 2005). Of these categories, regulatory ecosystem services are important for urban climate adaptation by mitigating the effects of climate change in the form of increased urban heat islands and stronger and more frequent rainfall. The value creation of cultural ecosystem services takes the form of health-promoting recreation and rest, aesthetic experiences, and spaces for different types of outdoor activities. Tourism activities are largely recreational and thereby dependent on cultural ecosystem services (Kulczyk et al., 2014). Cultural ecosystem services benefit both residents and visitors. Thus, a combination of ecosystem services is relevant for urban destination development. In this context, the green infrastructure has a dual significance.

Ecosystem services take place in green infrastructure, and the green infrastructure plays a vital part in creating sustainable tourist experiences. Green infrastructure is defined as ecologically functional networks of habitats and structures in the form of green urban spaces (Gómez-Baggethun et al., 2010). Urban green infrastructures exist in many geographical scales, from large parks and urban forests to allotment areas and street plantations. The development and preservation of green infrastructure have climate-adapting functions in urban areas. Cultural ecosystem services used and experienced by residents and visitors take place in the green infrastructure. It is the place where outdoor activities, such as walking, running, and water sports, are possible in urban areas. Studies that link the relation between green infrastructure with cultural ecosystem services, destination development, and attraction of urban tourism are scarce (Johansson & Nilsson, 2021). Existing research only weakly captures the cultural value of ecosystem services together with green infrastructure in a tourism-related perspective. A scientific gap also appears to exist in addressing tourism issues related to already existing natural-specific ecosystem services in urban areas, particularly in city centres (Haase et al., 2014). Loss of urban ecosystems involves long-term economic costs and can affect many other social and cultural values (Gómez-Baggethun & Barton, 2013).

8.2.3 Valuation of Ecosystem Services

The concept of ecosystem services has a clear pedagogical value since it addresses the link between ecosystems and human welfare. Since policy decisions are often evaluated through cost-benefit assessments, an economic analysis can help make ecosystem service research operational. There are difficulties in valuation of urban ecosystem services in monetary terms. This is mainly because ecosystem services are normally not bought or sold in an open market situation. The economic value of ecosystem services is often not apparent, although it can constitute a parameter in decisions about investment and development in urban planning (Gómez-Baggethun et al., 2010). Furthermore, valuing urban green space in destinations in a city is often based on a whole chain of events that can be difficult to separate from each other. For example, the sea's ability to purify water from environmental toxins in turn means good water quality, which creates good conditions for swimming, which many visitors value as part of attractive destination development. Hence, water treatment can be seen as a direct ecosystem service, and good water quality as an indirect ecosystem service, and the opportunity for swimming can be seen as a specific value creation aspect (Naturvårdsverket, 2021). Another example is that vegetation in the urban environment contributes to natural noise reduction, which in turn positively affects our health (Bolund & Hunhammar, 1999). The scientific literature mentions three types of values in urban ecosystem services: biophysical (or ecological), economic, and socio-cultural. These three categories are linked to different methods of calculating values (Gómez-Baggethun & Barton, 2013).

The biophysical values of ecosystem services come from the natural ability to strengthen the supporting and regulating ecosystem services, and their contribution to biodiversity. The biophysical values can be evaluated by using traditional scientific metrics, as in carbon sinks, water and air filtration, or noise reduction. Cultural ecosystem services, however, are difficult to evaluate in biophysical terms. A direct biophysical value provided by green infrastructure is its impact on local climate conditions, as a means to improve adaptation to the effects of climate change. For example, the city of Berlin has made a thorough investigation of how forests, parks, and open land have positive effects on local summer heat waves. Even small green structures in the form of tree lines in streets and micro parks proved to have positive effects locally (Berlin, 2016). In this case, the biophysical value of ecosystem services is measured as differences in temperature. This may in turn have effects on local living conditions and health.

However, most often economic values are discussed in terms of ecosystem services, making it possible to avoid direct costs and expenses for example emissions or temperature regulation in cities. The green urban infrastructure can also create direct values, for example in urban cultivation of food or access to biofuels or improved water supply. Collecting and harvesting urban grass can be used as raw material for biogas production in biogas plants, together with a fraction from pre-treatment of source-sorted food waste. Johansson et al. (2020) show that the city of Helsingborg's harvest from 370 hectares of municipal grassland can supply 250 biogas cars with fuel, with an annual mileage of about 1500 km each. The study also shows that with access to more sources of urban green biomass, the potential could be tripled. Biogas can also be converted and used for electricity production.

The social and cultural values of ecosystem services are considered as non-material values and therefore difficult to directly link to specific economic values. Socio-cultural values are quite simply difficult to measure and therefore difficult to include in the decision-making process: "articulation of social and cultural values in decision-making processes may require, in most cases, some sort of deliberative process, use of locally defined metrics, and valuation methods based on qualitative description and narration" (Gómez-Baggethun & Barton, 2013). Ecosystem services are often public goods, which means that many people can enjoy them without affecting other peoples' enjoyment. A natural urban area can have different values depending on how, and by whom, it is used. It can be important for wildlife, it can serve as inspiration for residents or visitors, and it can serve as a playground or area for outdoor life and recreation (Everard, 2017). Tourist experiences also represent a set of potential values to be gained from developing ecosystem services. There are a few recent publications discussing the assessment of cultural ecosystem services in relation to tourism (Church, Coles & Fish, 2017; Ram & Kay Smith, 2019). These are mainly discussing nature-based tourism. In our knowledge, no similar assessments have been made in urban contexts.

8.2.4 Cultural Ecosystem Services Take Place in Green Urban Infrastructure

In order to operationalize studies of ecosystem services, the spatial structures in which they take place must be studied, i.e. the urban green infrastructure. Green infrastructure is defined as local or regional ecological functional networks of habitats and ecological structures. There, landscape elements are designed, used, and managed in such a way that biodiversity is preserved. Coherent recreational areas and green urban structures are very important for gaining value of urban ecosystem services (Beatley, 2000; Fouchier, 1995), and the loss of important urban ecosystem services is often the direct result of unsustainable urban planning (Beatley, 1995, 2000).

It is important for society that ecosystem services are promoted throughout the landscape (Naturvårdsverket, 2020; Persson et al., 2018). A coherent green infrastructure in urban areas and destinations is a given prerequisite for ecosystems to be able to deliver ecosystem services at all. At the same time, the boundaries between different urban ecosystems and their services are often diffuse (Moll & Petit, 1994). The city as a destination can be defined as a large ecosystem or as a network composed of several individual ecosystems (Rebele, 1994).

> The concept of ecological infrastructure captures the role that water and vegetation in or near built environment play in delivering ecosystem services at different spatial scales. It includes all green and blue spaces that may be found in urban and peri-urban areas, including parks, cemeteries, gardens and yards, urban allotments, urban forests, single trees, green roofs, wetlands, streams, rivers, lakes and ponds.
>
> (Gómez-Baggethun & Barton, 2013)

Urban green infrastructure thus encompasses a wide range of environmental functions and urban values.

Cultural ecosystem services are central to developing attractive environments for both visitors and local population, which requires extensive development of green infrastructure in urban environments. Environments that are primarily developed with the well-being of the local population in mind can therefore also be developed to be valuable also to visitors. An attractive living urban environment for the local population increases the city's attractiveness and will also be visible to visitors. As a result, a doubled value creation process can take place. In an urban planning process, value creation for the tourism industry could therefore be put forward as a potentially important gain from cultural ecosystem services.

8.3 Methodology

The empirical studies presented in the chapter focus on two local cases in the south of Sweden: the cities of Malmö and Helsingborg. Hence, the objective is to present local and practical cases on the relation between ecosystem services value, urban place, and destination development in a Swedish context. The case studies focus on how urban planning projects, primarily aimed at mitigating GHG emissions and adapting to climate change, can be extended to develop places where experience values for both residents and visitors are created alongside other kinds of ecosystem services. The case studies are chosen to represent different types of destinations, facing different challenges and with contrasting motives for identifying ecosystem services value to sustainable urban tourism. They illustrate how a broad range of green infrastructure can be reflected in destinations. They are also chosen because they are well-known best practices in a Swedish context, well documented, and because the authors have in-depth knowledge about them. The case studies are based on a combination of document analysis, interviews, and on-site observations.

8.4 The Case of Malmö

Malmö emerged as a trading city during the late middle ages. The goods from the ships through Öresund between Sweden and Denmark were then rowed to the shallow shore of Malmö. The fact that Malmö finally got a port during the second half of the 18th century marks the city's rebirth. This was later followed by large warehouse buildings, railways, silos, slaughterhouses, and shipyards. In the 1970s crisis, manufacturing production largely disappeared from the city. Kockum's ship production was closed down in 1987. The city went through a long period of decline; new strategies were therefore needed to invigorate Malmö. The symbolically most important parts of the transformation of Malmö from an industrial city to a sustainable city of knowledge took place in Kockum's old shipyard area, known as the sustainable Western Harbour (Västra Hamnen).

8.4.1 Destination: Western Harbour

The Western Harbour in Malmö covers an area of 217 hectares (2.17 km^2). The area began to be built in the year 2000, initially with two large exploitation projects as flagships: the Swedish national housing fair "Bo-01" and the skyscraper Turning Torso. During the same period, the bridge between Malmö and Copenhagen was built, and Malmö got a university located in Western Harbour. This created a sense of a new urban dynamic; the population started to grow again. Later investments would have been more difficult to implement without a bridge and university. Western Harbour was originally planned as a development area, where housing and workplaces would be mixed. Socially and ecologically sustainable housing development was in focus of the planning process, with large elements of both green and blue infrastructure. There are several parks where ecosystem services are linked to various recreation activities. Public parks are supplemented with tree plantations, green plots, and roof plantations. There are also old docks and quays, canals, and ponds for vegetation and animals. The blue infrastructure creates attractive environments with increasing biodiversity contributing to both purification and delay of stormwater. However, the Öresund (the straits between Sweden and Denmark) with beaches and quays is the main marine tourist attraction in the Western Harbour. In the area, walking, cycling, and public transport are prioritized. Car traffic is strictly regulated, not to take up too much space that could be used for other purposes. Public transport benefits from the short distance to the city centre. The area is also centrally located in the region; it takes only 20 minutes with the railway from Malmö Central Station to Copenhagen Airport.

The Western Harbour has spatial connections to green and blue infrastructure in two different directions. The area is a strategic part of a cohesive route along the water from the city centre, via the Western Harbour to the touristic beach area Ribersborg towards the West. Ribersborg's main attraction besides the beach itself is the old bathhouse for outdoor winter bathing, a wooden structure dating back to 1898. The city's main parks, Slottsparken, Kungsparken, and Pildammsparken, form an extensive green belt stretching from the sea towards the southern housing districts. Malmö canal, the former moat, plays an important role as green and blue infrastructure in its path around Malmö's older parts. It creates proximity to water and wildlife and is important for activities such as fishing, canoeing, and rowing. The canal and parks are connected to attractions like the Casino, the City Library, the Art Gallery, the Opera, and Malmö Football Stadium. An interesting aspect of Malmö's active work with sustainable urban development is that it has given the city great international recognition, through different kinds of awards and in media reports focusing on Malmö's assets as a sustainable destination. The Western Harbour with its exciting architecture, innovative ecological solutions, and central location has received the most international attention, but other projects where green infrastructure is a key part of urban development, mainly Augustenborg and Hyllie, have attracted study visits from politicians, architects, and urban developers.

8.4.2 Challenges and Development Potential for Ecosystem Services in Malmö

Even though Malmö in many ways has been successful in managing sustainability and planning of green urban infrastructure, the city has several challenges. Malmö is a growing and is a densely populated and compact city in a flat landscape close to the sea. There are few unused areas in the central parts of the city that can be used for developing new green urban areas. For that reason, there is a need to be innovative and look for small-scale solutions.

In Malmö, there are a couple of interesting examples of how to develop green infrastructure in street environments, where you simultaneously develop ecosystem services and create attractive places for both residents and visitors. In the densely populated district of Möllevången, a well-known "creative" area, a tree-planting project is currently underway. Drought-resistant and heat-accepting trees are planted along several streets. They will provide shade for streets and pavements in summer; in this way, the effect of urban heat waves can be reduced. Tree plantations also provide a cost-effective contribution for taking care of rainwater during heavy rains, which are frequent in the Swedish summer. This effect is enhanced if tree plantings are combined with so-called rain gardens and flower plantations along the pavements where rainwater is led from streets and the drainpipes of properties. In this way, the pressure on the city's water and sewage system is reduced in the event of heavy rain (Malmö stad, 2017). This type of activity thus provides space for different types of ecosystem services. In addition to the climate-adapting elements, the plantations increase the biological diversity by creating habitat for birds and insects and providing a more calming environment for humans.

8.5 The Case of Helsingborg

Helsingborg is one of Sweden's oldest cities. The distance to Helsingör in Denmark is only about 2 kilometres. Being at this narrow passage called Öresund between the Atlantic and the Baltic Sea has always been of great importance to Helsingborg. However, it was not until the growth of industry and large-scale trade in the 19th century that it became one of Sweden's largest cities. At present, Helsingborg has many challenges. Helsingborg plans to significantly grow in the coming decades by urbanization. According to the 2017 city plan, the population is expected to increase from today's almost 150,000 to 200,000 by 2050. The growth is primarily intended to take place through densification processes, which places great demand on future spatial land and water use. Both Helsingborg and Malmö are located in one of the most densely populated and highly exploited parts of Southern Sweden. Only 10 per cent of Helsingborg's area consists of forest, wetlands, meadows, beaches, and other open lands. The surrounding landscape is heavily dominated by arable land with several smaller forests and streams in between. However, the landscape is fragmented and there is a shortage of continuous green urban infrastructure within the municipality. The City of Helsingborg's local environmental and climate work is extensive; it's recently ranked among the most ambitious cities in Sweden. Helsingborg is, like Malmö, a member of ICLEI (Local governments for sustainability).

Every city, place, or destination has its special and unique potential of ecosystem services in place development. The visual impression of a certain place affects our general experience of a city as a tourist destination. In Helsingborg, the identity-creating green urban infrastructure is largely located along a central natural geological elevation, called "Landborgen." The "Landborgen trail" is a publicly planned urban experience and recreational trail, following the coast of Öresund from Sofiero in the northern parts of the city to Råå and Raus in the south.

8.5.1 Destination: Landborgspromenaden – An Urban Recreational Trail

Landborgen is a so-called abrasion slope, it extends along the Öresund coast and rises between 20 and 40 meters above sea level. It has its steepest part approximately at the Kärnan tower in central Helsingborg. Along the trail, Landborgen consists of many public meeting places, parks, and valleys such as Pålsjö city forest, Jordbodalen, and Ramlösa. In the central parts of Helsingborg, several of the older parks are located close to Landborgen. Along the trail, there are many interesting places to visit. Many of these cultural attractions (including recreation and tourism destinations) along the Landborgen trail also serve as important supportive, regulatory, and cultural ecosystem services. Landborgen thus offers, together with nature reserves such as Pålsjö forest and Rååns valley, a calmer and quieter side of Helsingborg. The green urban infrastructure along Landborgen can therefore strengthen surrounding ecosystems, regulate local temperatures, increase property values, sequester carbon dioxide, capture local emissions, and reduce the risk of flooding. Thus, Landborgen is a green urban infrastructure that is cost-effective, resilient, and provides many public benefits.

8.5.2 Destination: Sofiero castle

Sofiero is a castle with a 15-hectare arranged park garden in Helsingborg. The garden and the park took the current form after the Swedish Crown Prince Gustav Adolf moved into Sofiero in 1905. The Crown Princess Margareta formed and created the world-famous rhododendron ravine and spectacular flower garden. At his death in 1973, Gustav VI Adolf donated Sofiero castle and the surroundings to Helsingborg city, which has run the facilities since then. Exhibitions and other events are often held in the surrounding gardens. Sofiero has, among other things, northern Europe's largest collection of rhododendrons, about 10,000 rhododendron bushes of 300 different varieties. Sofiero overlooks the Öresund towards Denmark and consists of a steep ravine towards water. Hence, Sofiero is one of Sweden's most-visited public gardens. In 2010, Sofiero was named Europe's most beautiful park.

8.5.3 Destination: Pålsjö City Forest

Pålsjö forest as a destination is a 70-hectare urban recreational area located in the northern part of Helsingborg, which since 2016 is a part of a nature reserve. Pålsjö forest consists mostly of beech and oak. The forest is a natural part of the regional hiking trail Skåneleden and the forest an essential part of the Landborgen trail. Adjacent to Pålsjö forest, a 100-hectare area with natural meadows spreads out. In the forest, there are several running, and walking trails equipped with electric lights. Pålsjö forest offers opportunities for everyday recreation and exercise activities for both residents and visitors. The forest also offers possibilities for spontaneous activities, power walking, dog walking, outdoor activities, and sports training. It is an advantage that everyday recreation can take place close to urban areas, and is accessible for residents and visitors. Hornsten (2000) concludes that the relationship with the forest is still close, but public use is changing from harvesting towards the purely recreational. Recreation is therefore seen as one of the most valuable forest ecosystem services.

8.6 Findings

The effects of a changing climate pose major challenges for long-term sustainable and local environmental management. Other urban challenges may be that intensive land use and in some cases overexploitation of urban environments create barrier effects and fragmentation of natural habitats. In this section, some possible solutions to such challenges are investigated. Based on the cases presented above, we discuss the role of green infrastructure in relation to urban planning, and as a means to create ecosystem-based adaptation to the effects of climate change. Both these perspectives are directly or indirectly relevant for tourism and destination development.

8.6.1 Urban Planning

It is a major challenge to create enough space for green infrastructure in urban areas, there is strong competition between different sorts of land use. In planning green infrastructure, it is important that green areas are spatially connected to one another. It is very valuable if the green infrastructure can form long coherent elements in the landscape, for example in forests or along river valleys. In doing this, the spatial structure of other urban development needs to be considered. A well-known example is the Danish finger plan. The expansion of Copenhagen is built along five fingers, containing housing districts and transport routes, stretching from the city centre outwards. In between the fingers, green wedges provide land for agriculture and recreational purposes (Cervero, 1998). Other advantages of this kind of spatial structure are that paths and routes are easily recognizable to the public. They are also necessary for plants' and animals' possibilities to migrate between areas, and thereby to sustain local biodiversity.

Apart from developing viable spatial structures of green infrastructure, it is also an advantage if different sorts of features and attractions are located close to one another. Many people like to combine nature-based and cultural experiences on the excursions. Cultural and nature-based amenities can, together, draw more visitors, create advantages of coordination, and add extra value to visitors. This also creates possibilities to coordinate cultural preservation and natural conservation. The importance of the castle Sofiero in the northern part of Landborgspromenaden can't be overestimated. It is an attraction in its own right, but it also encourages extensive walks in the neighbouring area. It is also a node for local traffic, by car, bicycle, and public transport. The cold bath house on the Ribersborg beach in Malmö has a similar function, as a landmark and attraction for the brave. In Pålsjö forest, they have managed to locate different forms of outdoor activities in specific places; this kind of variety increases public appeal. Increasing numbers of visitors to a natural area may have negative impact on the wildlife. This can be solved if activities and human presence is concentrated along specific paths, leaving other areas as wild zones. Wild areas may also make a significant local contributor to climate change mitigation acting as a local carbon sink.

It can be shown in the cases that proximity and good accessibility to green infrastructure with public transportation, by walking and biking, play an important role for peoples' possibilities for outdoor recreation. Green urban infrastructure destinations often have facilities such as toilets and parking spaces, which are particularly important for people who have difficulties to move around by themselves, for example, older adults or families with small children. Tourism and other organized outdoor life make use of the traditional Swedish right of public access. This means that everyone is allowed to visit most land, except arable land and private gardens, for activities as long as they behave appropriately and don't destroy plants or disturb the wildlife. (Naturvårdsverket, 2011). In nature close to city centres, the right of public access is important. It enables many people to take part in nature activities close to home. This has been very important during the recent pandemic, when many people have avoided meeting family and friends indoors.

8.6.2 Ecosystem-Based Solutions to Adaptation

Conscious policies for climate adaptation reduce vulnerability in local and regional urban planning. Building hard infrastructure, for example advanced drainage systems, has been the most common way to improve local climate adaptation. This chapter has shown that knowledge-raising measures in green infrastructure and ecosystem services can lead to improvement of climate adaption. Vegetation in cities is vital for functioning ecosystems, which improves the ability to withstand local drought or flooding. To preserve and develop biodiversity and functioning ecosystems in green infrastructure is therefore important in order to strengthen society's resilience to climate change. Resilient ecosystems are a key factor in long-term sustainable climate adaptation.

There are good conditions for ecosystem-based adaptation to create additional benefits for society. In an already built-up area, for example, the area for handling large amounts of water in the event of heavy rain may be limited. One solution is to develop multifunctional surfaces, places on low ground that can handle large amounts of water temporarily, but the rest of the time is used for other purposes, such as football pitches, parks, and playgrounds. There are a couple of examples of this in Western Harbour, and the Möllevången district, in Malmö. They are seen in many sizes and scales, from neighbourhood parks, and street trees, to small rain gardens. Apart from their adaptive functions, they can improve biodiversity and have positive local climate effects. Moreover, they make the place more pleasant for locals and visitors. The potentials in this area are not yet fully explored. However, what is evident from the Malmö case is that the city's profile as a sustainable city has attracted large numbers of study visits. The profile has also had positive effects on city marketing through considerable media visibility.

Based on this study, it seems to be very clear that ecosystem services play an important role for cities' possibilities to adapt to the effects of climate change. The development of green infrastructure is also vital for creating space for recreation and leisure activities. Urban aesthetic and experience values can also be enhanced along with the development of green infrastructure. In particular, ecosystem-based solutions can increase both biophysical value in the forms of biodiversity and climate adaptation, and recreational values for locals and tourists. Thus, by including the value ecosystem services bring to recreation and tourism as a factor in planning and developing green infrastructure cities, investments in sustainable solutions may prove to be more profitable than if only direct economic values were part of calculations. This way both the local environment and the tourist may profit. This argument is quite significant from a political perspective too.

8.7 Conclusions

The character and development of urban tourism have gained a lot of interest in recent years, first because of its rapid growth ending in "over-tourism," later due to the COVID-19 crisis, which laid many cities empty from tourists. These recent problems have put focus on the long-term sustainability of tourist cities. This chapter has investigated the dynamics between ecosystem services in urban environments and the development of urban green infrastructure relevant to destination management. The aim has been to provide an overview of this dynamic, identify value-creating functions in the form of urban ecosystem services, and discuss their importance for sustainable destination development. Ecosystem services, taking place in green infrastructure, are viewed as a link between sustainable urban planning, and the development of attractive environments that may create both recreation and experience value for locals and visitors. There is a shortage of research about the relationship between urban tourism and ecosystem services. In this chapter, we address this research gap.

In the review section, literature from three related research areas was discussed: urban sustainable tourism, ecosystem services, and valuation of ecosystem services. This section ends with a focus on the role of green infrastructure as the spatial structure where ecosystem services take place. Therefore, green infrastructure was in focus in the two case studies, of the Swedish cities Helsingborg and Malmö. The case studies were based on qualitative methods: document analysis, interviews, and on-site observations. Our findings show that the development of green infrastructure, in various forms and scales, may have significant impact in the areas of urban planning, development of ecosystem-based solutions to adaptation to the effects of climate change, and as arenas for recreation and tourism. By including the value ecosystem services bring to recreation and tourism as a factor in planning and developing green infrastructure, the possibilities of improving both experience values and ecological values increase. Based on the findings of this study, we would like to point at the following management relevant recommendations:

- Green infrastructure that is already in place needs to be carefully conserved and managed. Further developments could preferably be linked with other sights of natural or cultural interest.
- It is important to include ecosystem services in urban planning processes, to safeguard environmental sustainability.
- Take advantage of possibilities to develop means of adaptation in ecosystem services. Thereby, traditional physical infrastructure can be supplemented.
- Use ecosystem services strategically to increase experience values, and urban attraction, for locals and visitors.

Acknowledgement

The research behind this publication was supported by Formas (Swedish Research Council for Sustainable Development), project 2018-02238.

References

Arrow, K., Bolin, B., Costanza, R., Dasgupta, P., Folke, C., Holling, C.S., Jansson, B.O., Levin, S., Maler, K.-G., Perrings, C., & Pimentel, D. (1995). Economic growth, carrying capacity, and the environment. *Science* Vol. 268: 520–521.

Beatley, T. (1995). The many meanings of sustainability. *Journal of Planning Literature* Vol. 9(4): 339–349.

Beatley, T. (2000). *Green Urbanism: Learning from European Cities*. Island Press, Washington, DC.

Berkes, F. & Folke, C. (1998). *Linking Social and Ecological Systems: Management Practices and Social Mechanisms for Building Resilience*. Cambridge University Press, Cambridge: UK

Berlin. (2016). *Stadtentwicklungsplan. Klimaanpassung in der Wachsenden Stadt*. Senatsverwaltung für Stadtentwicklung und Umwelt.

Bolund, P. & Hunhammar, S. (1999). Ecosystem services in urban areas. *Ecological Economics* Vol. 29: 293–301.

Cervero, R. (1998). *The Transit Metropolis: A Global Inquiry*. Island Press. Washington DC: USA.

Childers, D.L., Pickett, S.T.A., Grove, J.M., Ogden, L., & Whitmer, A. (2013). Advancing urban sustainability theory and action: Challenges and opportunities. *Landscape and Urban Planning* Vol. 125: 320–328.

Church, A., Coles, T., & Fish, R. (2017). Tourism in sub-global assessments of ecosystem services. *Journal of Sustainable Tourism* Vol. 25: 1529–1546.

Cohen, S. & Cohen, E. (2019). New directions in the sociology of tourism. *Current Issues in Tourism* Vol. 22(2): 153–172.

Costanza, R. & Daly, H. (1992). Natural capital and sustainable development. *Conservation Biology* Vol. 6: 37–46.

Costanza, R., d'Arge, R., de Groot, R., Farber, S., Grasso, M., Hannon, B., Limburg, K., Naeem, S., O'Neill, R., Paruelo, J., Raskin, R., Sutton, P. & van den Belt, M. (1997). The value of the world's ecosystem services and natural capital. *Nature* Vol. 387(15): 253–260.

Daily, G., Polasky, S., Goldstein, J., Kareiva, P.M., Mooney, H.A., Pejchar, L., Ricketts, T.H., Salzman, J., & Shallenberger, R. (2009). Ecosystem services in decision-making: time to deliver. *Frontiers in Ecology and the Environment* Vol. 7(1): 21–28.

De Groot, R.S., Wilson, M.A., & Boumans, R.M.J. (2002). A typology for the classification, description and valuation of ecosystem functions, goods and services. *Ecological Economics* Vol. 41: 393–408.

Edwards, D., Griffin, T., & Hayllar, B. (2008). Urban tourism research: Developing an agenda. *Annals of Tourism Research* Vol. 354: 1032–1052.

Escobedo, F.J., Giannico, V., Jim, C.Y., Sanesi, G., & Lafortezza, R. (2018). Urban forests, ecosystem services, green infrastructure and nature-based solutions: Nexus or evolving metaphors? *Urban Forestry & Urban Greening* Vol. 37: 3–12.

Everard, M. (2017). *Ecosystem Services: Key Issues*, (1st ed.). Routledge. London: UK. https://doi.org/10.4324/9781315531816

Fouchier, V. (1995). *The Ecological Paradoxes of the Density-Nature Dialectics. The Case of Paris' Region*. Paper presented at SGB/NUREC/PRO/RMNO colloquium 26–27 October, in Dordrecht, The Netherlands, 1995.

Gómez-Baggethun, E. & Barton, D.N. (2013). Classifying and valuing ecosystem services for urban planning. *Ecological Economics* Vol. 86: 235–245.

Gómez-Baggethun, E., De Groot, R., Lomas, P.L., & Montes, C. (2010). The history of ecosystem services in economic theory and practise: From early notions to market payment schemes. *Ecological Economics* Vol. 69: 1209–1218.

Gössling, S. (2011). *Carbon Management in Tourism: Mitigating the Impacts on Climate Change* (1st ed.). Routledge. London: UK. https://doi.org/10.4324/9780203861523

Gössling, S. & Upham, P. (2009) *Climate Change and Aviation: Issues, Challenges and Solutions*. Earthscan from Routledge. New York: USA.

Haase, D., Larondelle, N., Andersson, E., Artmann, M., Borgström, S., Breuste, J., & Kabisch, N. (2014). A quantitative review of urban ecosystem service assessments: Concepts, models, and implementation. *Ambio* Vol. 43(4): 413–433.

Hornsten, L. 2000. *Forest Recreation in Sweden – Implications for Society and Forestry*. Doctoral thesis, Swedish University of Agricultural Sciences, Uppsala.

IPCC. 2014. Summary for policymakers. In: C.B. Field, V.R. Barros, D.J. Dokken, K.J. Mach, M.D. Mastrandrea, T.E. Bilir, M. Chatterjee, K.L. Ebi, Y.O. Estrada, R.C. Genova, B. Girma, E.S. Kissel, A.N. Levy, S. MacCracken, P.R. Mastrandrea, & L.L. White (eds.), *Climate Change 2014: Impacts, Adaptation, and Vulnerability. Part A: Global and Sectoral Aspects*. Contribution of Working Group II to the Fifth Assessment Report of the Intergovernmental Panel on Climate Change. Cambridge University Press, Cambridge, UK and New York, USA, pp. 1–32.

Johansson, O. (1994). *Investing in natural capital: the ecological economics approach to sustainability*. Island Press. Washington DC: USA.

Johansson, M. & Nilsson, J.-H. (2021). *Hållbar urban turism Värdeskapande kulturella ekosystemtjänster i den gröna infrastrukturen*. Lund University, Rapport.

Johansson, M., Bramryd, T., Svensson, S.-E., Törner, L., Narvelo, W., Syde, N. & Blom, A. (2020). *Biogas Potential from Urban Grass Areas – Feasibility Study with the City of Helsingborg as a Case*. Sveriges Lantbruksuniversitet, p. 10.

Keane, Å., Stenkula, U., Wijkmark, J., Johansson, E., Philipson, K., & Louise Hård af Segerstad L. (2014). *Ekosystemtjänster i stadsplanering – en vägledning*. C/O City.

Kulczyk, S., Wozniak, E., Kowalczyk, M., & Derek, M. (2014). Ecosystem services in tourism and recreation. Revisiting the classification problem. *Economics and Environment* Vol. 4(51): 84–93.

La Rocca, R.A. (2014). The role of tourism in planning the smart city. *TeMA – Journal of Land Use, Mobility and Environment* Vol. 73: 269–283.

Malmö stad. (2017). *Skyfallsplan för Malmö*.

Månsson, M. & Eksell, J. (2021). *A Communicative Approach to Resilience in Urban Regions*. Presentation to Atlas SIG Meeting Urban Tourism, Netherlands.

MEA. (2005). *Ecosystems and human well-being – Synthesis*. Washington, DC. & TEEB (2010) *Mainstreaming the Economics of Nature: A Synthesis of the Approach, Conclusions and Recommendations of TEEB*.

Moll, G. & Petit, J. (1994). The urban ecosystem: Putting nature back in the picture. *Urban Forests* Vol 14: 8–15.

Nalau, J. & Becken, S. (2018). *Ecosystem-based Adaptation to Climate Change: Review of Concepts*. Griffith Institute for Tourism, Research Report No. 15.

Naturvårdsverket. (2011). *Right of Public Access – A Unique Opportunity*.

Naturvårdsverket. (2020). *Global utvärdering av biologisk mångfald och ekosystemtjänster*. Sammanfattning för beslutsfattare. Rapport 6917.

Naturvårdsverket. (2021). *Naturbaserade lösningar – ett verktyg för klimatanpassning och andra samhällsutmaningar*. Rapport 6974.

Nevens, F., Frantzeskaki, N., Gorissen, L., & Loorbach, D. (2013). Urban transition labs: Co-creating transformative action for sustainable cities. *Journal of Cleaner Production* Vol. 50: 111–122.

Nilsson, J.H. (2018). Gränsregional urban utveckling och hållbarhetens gränser – Destinationsutveckling i Greater Copenhagen. *Ymer*, årgång 138.

Nilsson, J.H. (2020). Conceptualizing and contextualizing overtourism: The dynamics of accellerating urban tourism. *International Journal of Tourism Cities* Vol. 6: 657–671. DOI: 10.1108/IJTC-08-2019-0117.

Persson, G., Wikberger, C., & Amorim, J.H. (2018). *Klimatanpassa nordiska städer med grön infrastruktur*. Report SMHI, 2018.

Ram, Y. & Kay Smith, M. (2019). An assessment of visited landscapes using a cultural ecosystem services framework. *Tourism Geographies*. DOI: 10.1080/14616688.2018.1522545

Rebele, F. (1994). Urban ecology and special features of urban ecosystems. *Global Ecology and Biogeography Letters* Vol. 4: 173–187.

Scott, D., Hall, C.M., & Gössling, S. (2012). *Tourism and Climate Change: Impacts, Adaptation and Mitigation* (1st ed.). Routledge, London: UK. https://doi.org/10.4324/9780203127490

Scott, D., Hall, C.M., & Gössling, S. (2016). A review of the IPCC fifth assessment and implications for tourism sector climate resilience and decarbonization. *Journal of Sustainable Tourism* Vol. 24: 8–30.

Seraphin, H., Sheeran, P., & Pilato, M. (2018). Over-tourism and the fall of Venice as a destination. *Journal of Destination Marketing & Management* Vol. 9: 374–376.

Sitas, N., Prozesky, H.E., Esler, K.J., & Reyers, B. (2014). Exploring the gap between ecosystem service research and management in development planning. *Sustainability* Vol. 6: 3802–3824. DOI: 10.3390/su6063802

Spirou, C. (2010). *Urban Tourism and Urban Change: Cities in a Global Economy* (1st ed.). Routledge, London: UK. https://doi.org/10.4324/9780203835807

Terkenli, T., Bell, S., Živojinović, I., Tomićević, J., Panagopoulos, T., Straupe, I., Tosković, O., Kristianova, K., Straigyte, L., & O'Brien, L. (2019). Tourist uses of urban green infrastructure in Europe: A cross-cultural study. *Poster*. DOI: 10.26226/morressier.5d5fdb2bea7c83e515cbf8c3

Williams, S. (2009). *Tourism Geography: A New Synthesis* (2nd ed.), Routledge, London: UK. https://doi.org/10.4324/9780203877555

9 Next-Practice Platforms as an Enabler of Sustainable Tourism Destination Management

Perspectives from an Asian Chain Hotel's Sustainability Journey

Zeynep Gulen Hashmi and Munaza Kazmi

9.1 Introduction

The COVID-19 pandemic has highlighted the magnitude and scope of tourism's global importance and also its vulnerability. The unprecedented social challenges brought by the pandemic present opportunities for the hardest-hit tourism industry to shift its focus toward sustainability (Gretzel et al., 2020). OECD work on managing tourism development for sustainable and inclusive recovery, for instance, looks at destinations that have introduced initiatives supporting the green transition and sustainable tourism development as part of the COVID-19 response and recovery actions (OECD, 2020). Destinations are increasingly focusing on balancing economic development with protection of the environment and social uplifting of communities (Kumar et al., 2012; Fyall and Garrod, 2019), since they are unarguably the areas where tourism's main impacts on the ecosystems services are felt most powerfully (Wall and Mathieson, 2006). Destinations are also considering how their efforts to develop within sustainability principles may be supported by tourism businesses, which, in a fragmented tourism industry, pursue aggressive growth strategies with competing demands and strong profit maximization motive.

The tourism industry directly interacts with ecosystems and produces both beneficial and detrimental environmental and socio-cultural impacts. Unfortunately, many tourism businesses are not fully conscious of the extent of their dependence and impact on ecosystems and the possible consequences. However, there is potential for the private sector to play a key part in tackling current sustainability issues such as poverty eradication, hunger, climate change, healthy lives, and well-being. In developing country contexts such as Pakistan where the concept of sustainable tourism lags behind due to rapid economic and regional development concerns (Ullah et al., 2021), tourism businesses that are leading in sustainability have a bigger role to leverage their sustainability expertise and contribute toward sustainable destination management.

There have already been numerous examples of successful businesses that are helping society by providing healthy lives and well-being, education, and food security (Lloyd, 2015). Such corporate behaviors are aligned with Porter

DOI: 10.4324/b23145-13

and Kramer's (2011) Creating Shared Value (CSV) concept, which may be considered as light at the end of the tunnel. The CSV concept is particularly helpful as its focus is on identifying and expanding the links between economic and social progress, which is much needed in sustainable tourism destination management. The concept of CSV caters to the pressing need for the tourism industry to see ecosystem change and sustainability as a source of business risk and opportunity.

Goodwin (2015) asserted that tourism destinations can create shared value by helping rural communities, with hotels investing in and improving their ecosystem. He further went on to suggest that shared value would not exist without collaboration between tourism businesses, destination authorities, and the local communities. Indeed, central to the concept of CSV is the creation of next-practice platforms, which are innovations that lead to next practices (Nidumolu et al., 2009). These innovations would require becoming a network of alliances either across the value chain or in partnership with other tourism businesses, or in collaboration with policymakers.

A considerable amount of research has been carried out in determining how to apply sustainability principles to the management of tourist destinations; however, a big portion of this research and academic literature has been devoted to lists of indicators that may be used to assess the extent of sustainability of a tourist destination (Asmelash and Kumar, 2019). Moreover, although most of the literature on sustainability of destinations has followed a supply-based perspective, none of them has really emphasized the potential and opportunities of shared value creation and the role of next-practice platforms in creating shared value for tourism businesses and the society at the same time.

Thus, this case study aims to contribute to the literature by investigating the concept of CSV from both a supply-based perspective as well as a sustainable destination management perspective. The case study not only sheds light on Islamabad Serena Hotel's (ISH) shared value creation activities as a best practice that strives to serve the common good but also assesses its potential to further create next-practice platforms as an enabler of its contribution to sustainable destination management. In this regard, our study pursues the following two research objectives:

1 Understand how Islamabad Serena Hotel engages in shared value creation as a leading hotel in sustainability.
2 Explore various perspectives of the hotel's various stakeholders about the hotel's potential for creating next-practice platforms for sustainable destination management.

In preparing this case, we first conduct a review of the literature on the concepts of sustainable tourism destination management, creating shared value and next-practice platforms. Then, these concepts are considered in the current roles and sustainability strategies of tourism businesses in sustainable destination management. Third, the study explores, through the lens of CSV theory, the current sustainability concern, organizational perspective, and value creation of Islamabad Serena Hotel as a luxury chain hotel leading in sustainability. Finally, the chapter explores the extent to which the hotel is ready to create next-practice platforms for sustainable destination management by identifying the main opportunities and challenges for it to do so. Based on the findings, theoretical and practical implications are concurrently discussed, and new areas for future research are unveiled in a recovering and rebuilding tourism context.

9.2 Literature Review

9.2.1 Sustainable Destination Management and Hotel Companies: The Business Opportunity

As the COVID-19 pandemic revealed limitations and flaws in mass tourism models, environmental and social concerns are increasingly dominating new business strategies. Destinations around the world are slowly beginning to reopen while protecting local communities and their economies. With a growing awareness among businesses, governments, and consumers for the need to not only prioritize profit but also people and the planet, companies are increasingly being held accountable for ecosystem services degradation, climate change, or unsustainable practices in general (Porter and Kramer, 2011). Considering that only 55% of tourism businesses implemented some form of sustainability strategy, compared to 70% of consumer packaged good industries in 2020 (Euromonitor, 2020), it appears that tourism and hospitality businesses have a lot to cover in terms of their sustainability strategies and contribution toward sustainable tourism destination management.

The World Tourism Organization defines sustainable tourism as 'tourism which leads to management of all resources in such a way that economic, social and aesthetic needs can be filled while maintaining cultural integrity, essential processes, biological diversity and life support systems' (UNWTO Report, 2018). Sustainable tourism has its roots in the concept of sustainable development (Hardy and Beeton, 2001). The Brundtland Report produced the most widely used definition of sustainable development as 'meeting the needs of the present without compromising the ability of future generations to meet their own needs' (World Commission on Environment and Development, 1987:43). In this regard, a destination can be considered sustainable if an appropriate balance is achieved and maintained in the economic, socio-cultural, and environmental aspects of tourism destination management (Edgell, 2012), which highlights the triple-bottom-line approach.

In its broadest meaning, every place for a holiday, every place to visit may be considered a destination. However, based on a number of common definitions found, a tourism destination may be intended as 'a locality that offers the tourist the opportunity of exploiting a variety of attractions and services' (Machiavelli, 2001), usually based on geographical criteria. Sustainable destination management allows us to minimize environmental impacts and maximize socio-economic advantages of tourist destinations (Vidishcheva and Bryukhanova, 2017).

Tourist destinations are increasingly being called upon to tackle social, cultural, economic, and environmental challenges. The European Commission, for instance, has developed the European Tourism Indicators System (ETIS) for Sustainable Management at Destination Level, intended to be used by policymakers, tourism businesses, and other stakeholders, to supervise, manage, measure, and enhance their sustainability performances. Another initiative is the Global Sustainable Tourism Council (GSTC) Industry Criteria. Considered as the global baseline standards for sustainability in tourism and travel, the GSTC criteria are used for policy-making for businesses, government agencies and other organization types, education and awareness-raising, measurement and evaluation, and as a basis for certification. While these initiatives are valuable, in that they encourage sustainable destination management, they do not take into consideration the business opportunities and sustainable competitive advantage from the point of view of hotel businesses; they rather focus on risk mitigation and internal efficiencies such as cost savings to increase the bottom line.

Promoting competitiveness through cost reduction is rooted in the resource-based view of enterprise, which argues that business will engage in sustainability actions to gain a competitive advantage that others cannot quickly imitate (Hart, 1995). Cost and risk reduction support eco-efficiency, in the sense that reduced use of resources would potentially result in quarterly earnings growth and reduction in exposure to liabilities and other potential losses (Hart and Milstein, 2003). However, as Brønn and Vidaver-Cohen (2009) assert, it is actually more desirable for businesses to have less regulation in order to have more freedom in decision-making to meet market and social factors. Some large hotel chains have already increased awareness of the business opportunities inherent in environmental and social challenges, thus have gone far enough to go beyond compliance and engage in shared value creation strategies.

9.3 Enriching the Creating Shared Value Strategy through Next-practice Platforms

The shared value proposition focuses on the scale of impact and degree of innovation that companies can bring toward societal challenges where governments and traditional NGOs have often lacked. According to Porter and Kramer (2011), businesses acting as businesses, not as charitable, philanthropic givers, are arguably the most powerful force to address pressing society's issues. They argue that adopting the CSV lens makes companies see social and environmental issues not as disconnected and externally imposed but as real opportunities and strategic targets for business decisions. According to the authors, CSV is about creating social and business values simultaneously. While the social values refer to the positive improvements in the social challenges targeted by the business, and social outcomes or social changes that are to be achieved, the business values are the actual economic benefits to the businesses (Porter et al., 2012). In this respect, CSV is a more recent development of Triple Bottom Line (TBL), which measures the multidimensional business contributions to sustainability (Elkington, 1997).

It must be highlighted here that solving social problems such as sustainability challenges requires innovative practices (Pfitzer et al., 2013), collective impact (Kania and Kramer, 2011), and changes to the traditional CSR mindset. Critics of CSV claim that CSV addresses the crisis of capitalism poorly, and they further state that CSV is unauthentic and doesn't accord for the differences between economic and societal issues (Crane et al., 2014). Although the concept of CSV is a state-of-the-art contribution toward linking businesses to society at large, and thus a progressive move from the conflicting views of shareholder value management (Friedman, 1970) and stakeholder value management (Freeman, 1984), it is limited to those issues that prioritize the 'business case' for sustainability and thus economic value for business (Dyllick and Muff, 2013).

There are also differences between CSR and CSV, which puts the concept of CSV in a more favorable position for a win-win value creation tool. While CSR focuses on '*doing good*', CSV focuses on '*doing good by doing new with others*'. CSV is about embedding sustainability and CSR into a company's business model or portfolio. In other words, CSV embraces the concepts of CSR, sustainability, and the stakeholder theory. Both the developed and developing countries are benefiting from the shared value concept as more and more companies are keying into social change. While the focus in developed countries is more on education, income inequalities, environment, and sustainability, the CSV focus by companies in developing markets is expansion of the value chain, raising the standard of living in the communities in destinations, empowerment, entrepreneurship, financing, and capacity-building (Kramer, 2012).

Porter and Kramer (2011) state that companies can create shared value in three levels: reconceiving products and markets, redefining productivity in the value chain, and building supportive industry clusters in the destinations where the company is operating. In order to re-conceive products and markets, a business needs to re-assess its relationships with consumers, and then incorporate shared value into the process of goods exchange. Rebranding of a more socially conscious corporate mission, addressing unmet needs of consumers, and strong advertisement can be great tools to utilize in this level. As Porter et al. (2012) posit, this level is about market share, revenue growth, and profitability that arise from the environmental and social or economic development benefits delivered by the business's products and services.

As for redefining productivity in the value chain, this can be done through resource use, energy use and logistics, location, procurement, sub-contracting, and distribution. This level focuses on gaining internal efficiencies that improve quality, productivity, cost, and resource access through environmental improvements that are associated with better resource utilization, better supplier competencies, and employee skills development. The Intercontinental Hotel Group (IHG) is a relevant example of a multinational organization that has been working on improving productivity at this level of CSV. The hotel group launched its *Green Engage* program – an online sustainability tool – in 2009 with the objective of testing alternative solutions for reducing water consumption, energy use, and waste disposal in pilot hotels. This provided the hotel group with energy savings of up to 25% as well as substantial reductions in the hotel's operating costs (IHG, 2012).

Creating shared value from enabling cluster development originates from improving the external environment for the business through community investments and strengthening local institutions, local suppliers, and local infrastructure in ways that also enhance business productivity. In fact, without a supporting cluster, productivity suffers. The success of every company is affected by its ecosystem. Clusters help increase the likelihood of survival during a business' infancy through reduction of costs related to the start-up phase. In terms of tourism cluster development, global hotel groups can certainly create shared value, for instance, by financing training and development to offer sophisticated and skillful workforce for the industry. This would simultaneously decrease the hotels' human resources expenditures in recruiting, training, and development activities. Open and transparent markets are a key aspect of cluster building in both developing and developed countries. According to Porter and Kramer (2011), companies can support cluster development in their destinations by identifying shortcomings in areas such as suppliers, distribution channels, and educational institutions (Porter and Kramer, 2011).

Nidumolu et al. (2009) posit that by treating sustainability as a goal today, early movers can develop competencies that rivals will be hard-pressed to match and that these companies will stand out among the crowd because sustainability will always be an integral part of development. In their study regarding the sustainability initiatives of 30 corporations, Nidumolu et al. (2009) revealed that sustainability is a key driver of innovation. Although the concept of next-practice platforms has been coined by Nidumolu back in 2009, the concept has not been widely used, except in the innovation literature. In this study, we will refer to next-practice platforms as innovative cross-sectorial or sector-specific collaborative business model platforms. These platforms are usually created at the level of the interaction of companies with their environment (including research institutes), allowing the formation of clusters at local level. Kramer and Pfitzer (2016) assert that companies must sometimes team up with governments, NGOs, and even rivals to capture the economic benefits of social progress. In this regard, next-practice platforms enable value creation beyond a company's own value chain, leading to a value ecosystem in itself.

9.4 Methodology

9.4.1 Research Design

Qualitative approach was used as a general methodology for this study. Qualitative studies are rich in description of phenomena and enable the researcher to understand phenomena from different angles (Creswell, 2012).

As a qualitative research method, this study conducts a case study, and probes into the 'why and how' of relations or interactions between factors and events in the selected case (Yin, 2003). In case studies, the researchers experience the case's activities as they occur in their context and situation. Single case studies can provide important insights into novel or under-explored contexts and draw practitioners' attention to the presence of an unusual phenomenon (Daly et al., 1992). This study takes Islamabad Serena Hotel (ISH) as the focus case of a best practice in the field of sustainability.

9.4.2 Context

Serena Hotels, operated under the name Tourism Promotion Services Pakistan (TPS-P), has been at the forefront of social change since its foundation. The company is renowned for bringing positive change for all its stakeholders and communities it operates in, which is clearly stated in its corporate-level mission of '*To be an agent of positive change for the Stakeholders and Community by pursuing an ethical and sustainable business*' (www.akdn.org). Operating 36 hotels, resorts, safari lodges, palaces, and forts located in Southern Asia, East Africa, and Mozambique, Serena Hotels not only contributes to economic development of the places where its hotels are located but also improves the quality of life of its locations through various health initiatives, educational, and cultural projects and sustainability-related collaborations.

The flagship hotel of the company is Islamabad Serena Hotel (ISH), which is a five-star deluxe hotel and a member of the 'Leading Hotels of the World'. The 387-room hotel stands in six acres of landscaped gardens in the center of Islamabad, with magnificent views of the Margalla Hills and Rawal Lake. The hotel stands out among its competitors because of its architecture, quality of service, and guest satisfaction. The hotel's mission statement originates from a holistic societal concern than the industry's traditional focus on guests only:

> We are the leading luxury city-resort hotel where our highest mission is to promote tourism and fulfill the unexpressed wishes of people with warmth, love, and inspiration by creating a sense of belonging, a place where dreams are fulfilled, and an environment in which growth and spirituality are one.

ISH has an environmental policy '*We must conserve rather than exploit*' since 2007. The hotel is also compliant with several important international standards such as ISO 9001 for Quality Management System, ISO 14001 for Environmental Management System, ISO 45001 for Occupational Health and Safety Management System, and ISO 22000 for Food Safety Management System, and ISO 50001 for Energy Management System. However, ISH's primary sustainability focus is on delivering cultural and educational hospitality (Hashmi and Muff, 2015), which centers around engaging with external stakeholders in the form of various sectorial/cross-sectorial collaborations.

In addition to its traditional CSR projects on issues of poverty, health, and education, ISH actively engages in stakeholder dialogues to contribute positively in these areas. For instance, the hotel organizes an internal '*I live Green*' competition to motivate and engage its employees on environmentally friendly initiatives. Similarly, its 'Karighar' CSR program provides entrepreneurship training to various segments of women for a sustainable growth and quality of life. Under its Social and Environmental Education Development (SEED) Program, which was initiated in 2009, the hotel creates awareness for its internal stakeholders regarding the importance of using organic substances for growing vegetables and fruits. The hotel also promotes food sustainability through the use of herbs and spices from its in-house herb garden. The SEED program also helps enhance the guest experiences at ISH, contributing to educational hospitality.

Under the auspices of its Cultural Diplomacy program, ISH sponsors an annual craft festival aimed at economic empowerment of craftsmen. ISH also holds various food festivals, thus creating awareness about balanced diet, different aspects of healthcare, as well as prevention of diseases such as cancer. Under its Adventure Diplomacy, the hotel contributes to promoting healthy pursuits, and under its Sports Diplomacy, it encourages various stakeholder groups to participate in sports promoting different games as well as Pakistan.

ISH has further become the first company in Pakistan to have entered into collaboration with the International Finance Corporation (IFC) to work for the worldwide Economic Dividends for Gender Equality (EDGE) certificate. ISH is the first company in Pakistan and second hotel internationally to receive the EDGE Gender Certification. In 2018, SHI signed a Memorandum of Understanding (MoU) with UNDP to establish a strategic collaboration to achieve the Sustainable Development Goals (SDGs). This collaboration is designed to help build and share solutions toward sustainable development of tourism in terms of biodiversity protection, wildlife conservation, and promoting responsible management of natural resources and sustainable livelihoods.

In 2019, in partnership with Telenor Pakistan, ISH launched a revolutionary, first of its kind digital platform called 'S Tel', which is a complimentary service for Islamabad Serena Hotel guests, allowing them to enjoy free international and local calls along with 4G internet and hotspot options, all packed in a dedicated smartphone placed in their rooms. In 2019, ISH also partnered with Ecole Hoteliere de Lausanne (EHL) of Switzerland to develop capacity-building of senior and mid-level executives of Pakistan's tourism industry. For skills development of frontline employees, the hotel is presently collaborating with National Vocational and Technical Training Commission (NAVTTC) of Pakistan on an apprenticeship program. Finally, in 2021, ISH initiated a strategic partnership with NUST Business School (NBS) to foster gender equality through Serena Management Trainee Program. The collaborative platform offers an opportunity to female graduates of NBS for hands-on experiential learning and training in the field of hospitality management. More recently, the hotel committed to improving and preserving the environment through natural resource management and started collaborating with National Productivity Organization (NPO) and Asian Productivity Organization (APO) to develop a case study on energy efficiency and conservation with the support of Japanese experts.

ISH was an ideal case for our research considering the abundance of sustainability initiatives the hotel has been leading in collaboration with various stakeholder groups. Although the hotel has a past record of creating next-practice platforms, perceptions of the hotel's various stakeholders can differ regarding ISH's potential for next-practice platforms at the destination management level.

9.4.3 Data Collection and Interviews

Our data sources include both secondary data and primary data in the form of semi-structured interviews. While secondary data entails information on the official website of ISH, media coverage, press releases, social campaigns, and different corporate documents such as the hotel's code of conduct, environmental management system reports, and CSR report, our semi-structured interviews probe into 'how' ISH engages in shared value creation actions and behaviors.

Semi-structured interviews were conducted by the authors during June and July 2021. In order to tap a wide range of different perspectives, the sample included ISH's internal stakeholders consisting of senior corporate managers and external stakeholders consisting of high-ranking representatives from tourism governance bodies in Islamabad and general managers of other chain hotels that are considered key players in the Islamabad hotel market. Two of the interviewees (external stakeholders) officially engaged in destination policy-making at the federal level (Pakistan Tourism Development Corporation, Sustainable Tourism Foundation Pakistan). The GM and three corporate directors of ISH (internal stakeholders) were selected in accordance with the years of their professional experience with Serena Hotels (operated under Tourism Promotion Services). The two GMs of other five-star chain hotels, three executives from NGOs (external stakeholders), and the hotel's two sustainability experts participated in the research based on their largely acknowledged expertise.

Each interview was recorded with the interviewee's permission and lasted 60–75 minutes. The researchers had no existing relationships with the interviewees but personal connections were used to contact the first interviewees (Aguinis and Solarino, 2019). The primary data gathering process started with the corporate directors working at ISH, in order to form an initial understanding of the core concepts regarding shared value creation and next-practice platforms, as perceived by senior management. Interviewees were selected in a non-probability, information-oriented manner that aims at maximizing information based on expectations about the content (Flyvberg, 2011). This was followed with face-to-face interviews with sustainability experts working as employees at the hotel. Finally, purposive sampling was conducted with the hotel's various external stakeholders, to uncover the various perspectives of potential decision-makers regarding CSV and sustainable destination management. A snowball sampling approach was also harnessed, asking interviewees to identify other potential interviewees (Gobo, 2005), which helped to easily reach out to the external stakeholders for the interviews.

A total of 13 people were interviewed, which is in line with the number suggested by McCracken (1988). Interviews of other senior managers and GM of the hotel were convened by the Corporate Director of Sales, which assured a high degree of participation. Participation was voluntary and participants were informed of the anonymity and informed consent requirements of this study. The interview guide consisted of seven questions that explored the drivers, benefits, and challenges of adopting a CSV approach, the hotel's main business and societal concerns as well as the types of value it creates, and its potential to create next-practice platforms to better contribute to sustainable destination management. The interview questions were adapted to make the interviewee feel easy and to get a convenient flow. To guarantee the consistency and reliability of the interview findings, the discussions were based on a carefully developed interview protocol with semi-structured questions. To enhance the validity in the data collection, triangulation was ensured (Denzin, 2012; Houghton et al., 2013) through the presence of two interviewers in the data collection process.

9.4.4 Data Analysis

For high quality and internal validity of this case study, three data sources were triangulated (Yin, 2009): semi-structured interviews, a thorough literature review, and corporate documents (either web-based or hard copies provided by the respondents themselves). The company website and materials were used primarily to collect and supplement the information about the case, which was also compared with interview transcripts to ensure the information accuracy. Two researchers were engaged in the transcription process. The interviews were analyzed following a content analysis approach to organize and analyze data systematically. According to Krippendorf (1980), content analysis is a systematic and objective way of describing and quantifying phenomena. To identify and select themes, interviews were transcribed in full, based on a deductive approach (Elo and Kyngäs, 2008). Theme by theme, using several coding methods, the answers were codified. Thus, based on the codes, categories per theme were built up (Saldana, 2009). For each stakeholder group, clear examples of what the interviewees said as well as the emphasis used were noted for a rich analysis. To respect the anonymity of the interviewees who might have been easily identifiable from their quotes, data were presented in a dispersed rather than listed manner.

9.5 Findings

Interviews started by discussing ISH's major role in contributing to sustainable destination management. Several interviewees spoke of preserving the destination, including the need for dealing with overtourism and the need to become more engaged with destination consultative planning. A second group, which consisted of the majority of interviewees, spoke more about destination branding, including the hotel's role in effective marketing and promotion of tourism, attendance at international tourism fairs and sports events, prioritization of high value-added segments, and deployment of an authentic digital-marketing strategy. A third group articulated on visitor management, including management of city's various attractions, development, and implementation of service quality standards for small-to-medium-sized hotels.

Interviewees were then asked about their opinion regarding the hotel's various motives for contributing to society with a positive impact. Eight interviewees, all of whom were internal stakeholders of the hotel, mentioned awareness of social responsibility to promote livelihoods, including capacity-building to promote employment and livelihood through traditional arts, crafts, and culture. A second group, which consisted of ISH's external stakeholders, talked about the parent company's altruistic values, influencing the hotel's strategy, while a third group consisting of both internal and external stakeholders emphasized the business case for sustainability.

Interviewees were asked what kind of benefits ISH receives from contributing to sustainability challenges. Five respondents saw an opportunity for increased profits while four mentioned legitimization for good corporate citizenship. The rest of the respondents were all external stakeholders of the hotel and talked about improved relationships with government and third parties and business growth, as benefits ISH could derive from positive contribution to society.

Discussion of the main challenges of making a positive contribution to society and creating economic value simultaneously was next. With this question, it became clear that there was confusion between the concepts of CSR and CSV as three interviewees could not distinguish between them without an explanation. Most respondents saw challenges regarding development of collaborative partnerships and public–private partnerships, mentioning divergent interests of the public sector and private sector. A few others mentioned a potential lack of focus in meeting guests' needs and catching up with latest market trends, and two interviewees saw lack of an integrated approach for proper enforcement of standards and certifications to measure positive societal impacts.

The type of value creation the hotel engages in was also discussed. More than half of the interviewees said that ISH creates both social value and business value at the same time. Those that had concerns over the hotel's social value creation justified their answers with the business case for sustainability. There was a strong sense that the residents of Islamabad were not perceived as a key stakeholder in destination management.

The final two questions asked the interviewees to reflect on the future opportunities and challenges of creating next-practice platforms as a way of creating shared value for sustainable destination management. For this, the concept was first explained to each interviewee. Although the concept seemed to sound a bit too aspirational for the interviewees at first, nine respondents mentioned cultural and environmental protection as the biggest avenue for exploring next-practice platforms. They mentioned that ISH could still cover a lot in this domain, building upon its current expertise of establishing these platforms in the social domain. They mentioned that environmental stewardship looked more promising compared to other domains of sustainable destination management. Two respondents mentioned development of services and facilities including management of tourist attractions while one respondent highlighted promotion of livelihoods and another articulated on the local products and suppliers as other potential avenues.

Regarding the challenges involved in the creation of next-practice platforms, governance changes were cited the most. The interviewees drew attention to the short-termism of private sector versus long-termism of public sector in the case of public-private partnerships in destination management. While the hotel's internal stakeholders and some external stakeholders focused more on their frustration related to the public sector, the external stakeholders mostly mentioned limited interest in planning for the longer term as a challenge. Deficiency of the tourism services infrastructure, lack of specific standards and lack of reliable research data to inform innovative business model partnerships were also mentioned but to a lesser extent.

9.6 Discussion

One of the objectives of this case study was to understand how a luxury Asian hotel – ISH – engages in shared value creation as a leading hotel in sustainability. This required understanding the business concerns (external vs. internal) and organizational perspective (inside-out vs. outside-in) of the hotel's various stakeholders as well as their perception of the type of business value (social vs. economic) the hotel generated as a sustainable and successful business. Based on our findings, we argue that ISH is clearly inspired by external concerns in its ecosystem, which, in the case of sustainable destination management, relates to destination branding of Islamabad and Pakistan primarily. The hotel addresses societal issues such as improvement of livelihoods through sports and culture, and capacity-building for the hospitality industry in Pakistan, as voiced by the majority of the interviewees. Finally, through its state-of-the-art corporate mission, ISH aims to contribute to the larger purpose of promoting tourism with a sense of belonging, individual growth, and spirituality. This is indeed in line with Dyllick and Muff (2013)'s statement that truly sustainable companies have external concerns rather than internal concerns, which are limited to profit maximization.

As more than half of the interviewees mentioned, awareness of the social responsibility to promote livelihoods is the primary concern for the hotel to engage in sustainable destination management.

This finding confirms the common CSV focus in the developing country context, which relates to raising the standard of living of communities in destinations. However, this is in contrast with Friedman's (1970) argument that if businesses act as good agents in the interests of shareholders, the overall welfare would be enhanced. As findings further suggest, ISH's strong concern for the uplifting of communities appears to be embedded in its parent company's moral values and principles. This is consistent with the theory that one of the three frames that explains pro-sustainability behavior in hospitality enterprises is through values rather than formalized plans (Matten and Moon, 2008).

ISH's current environmental initiatives also show its concern for environmental stewardship as the hotel has started green technology investments to decrease its negative impacts on the environment while achieving resource efficiency gains, mainly through water and energy. This has led ISH to engage in various sustainable practices such as energy monitoring and reduction, water use reduction, and sustainable sourcing. Complying with voluntary third-party accreditation criteria such as ISO 14001 and ISO 5000, the hotel is successfully improving productivity in its value chain while respecting the environment, thus creating shared value (Porter and Kramer, 2011).

Regarding ISH's organizational perspective, we can easily say that ISH possesses an inside-out perspective as the hotel is heavily focused on embedding sustainability throughout the organization in compliance with various environmental management systems, accreditation criteria, and employee trainings. This finding is consistent with the arguments of various authors and researchers who posit that businesses will need to embed sustainability in their strategies and operations, governance and management processes as well as organizational culture and governance systems, rather than simply manage risks and opportunities externally (Laszlo and Zhexembayeva, 2011; Belz and Peattie, 2012; Eccles et al., 2012).

Focusing on the type of value created is another dimension to understand ISH's engagement in shared value creation, which relates to the output of its business operations and strategies. Perceptions of ISH's various stakeholders show that the hotel is engaged in simultaneous creation of both shareholder value and social value by managing new opportunities for increased profits, seeking legitimization for good corporate citizenship, improving stakeholder relationships, and solving societal challenges at the same time. This is in alignment with Porter and Kramer's (2011) CSV concept that highlights the need for business to go beyond business value creation and create social value at the same time. At this point, differentiating between CSV and CSR would be necessary as various CSR projects still constitute a substantial part of the ISH's sustainability initiatives for the society. This supports Deale's (2013) assertion that hotels aim to increase efficiencies throughout the value chain with separate CSR activities in the form of donations of goods, services, and more recently volunteer hours. Yet, these CSR activities only help redistribute existing business value rather than create new business value (Porter and Kramer, 2011).

As one can see from ISH's various social and environmental initiatives as well as the interviews, the hotel harnesses a triple bottom line business sustainability across different bottom lines (social value, environmental protection, profit). Rather than maximizing shareholder value or social profits, the management team focuses on what they perceive as a balanced value for business and society. The hotel seeks differentiation in terms of products and services by organizing various cultural, sports, and adventure events under its various diplomacies and innovative digital platform offerings to its guests such as S Tel. Thus, value is created not just as a side effect of the hotel business activities but also as the result of deliberately defined goals and programs addressed at specific sustainability issues. As the majority of interviewees mentioned, the hotel sees economic value creation as an outcome of running a responsible and meaningful business. This is in accordance with the concept of Triple Bottom Line, which measures the multi-dimensional business contributions to sustainability (Elkington, 1997), a foundation on which the concept of CSV rests.

The other objective of this study was to explore various perspectives of the hotel's various stakeholders about the hotel's potential for creating next-practice platforms in sustainable destination management. Based on interview findings, it is promising to see that the hotel is progressing in its sustainability journey by expanding its focus from organizational transformation toward systems building and cluster development, creating shared value through next-practice platforms. According to the interview findings, potential opportunities for establishing next-practice platforms were found to be related to cultural and environmental protection, services and facilities, local products and suppliers, and promoting livelihoods. These are indeed aligned with the GSTC Sustainable Tourism Industry Criteria (GSTC, 2019) and are also listed in the 12 Aims for Sustainable Tourism (UNWTO and UNEP, 2012). ISH's proactive engagement in societal challenges such as education and gender equality is explicitly reflected in the various cross-sectorial collaborations the hotel creates with educational institutions such as NAVTTC, EHL, and NUST university. This supports the findings of Nidumolu et al. (2009), who state that sustainability can lead to interesting next-practice platforms that originate from innovations.

Among the stakeholders interviewed, there seems to be agreement that deficiency of the tourism services infrastructure, governance changes, lack of specific standards, and lack of reliable research data for decision-making constitute the main challenges of establishing next-practice platforms and implementing CSV. These are consistent with academic research, which shows that tourism companies often find it difficult to incorporate sustainability in their operations due to a lack of specific tools, standards, and methods in the industry (Bagur-Femenías et al., 2015).

The takeaway message from this discussion is that creation of next-practice platforms for sustainable destination management is challenging and multi-dimensional, which implies that suggestions and challenges presented herein may be only local and incomplete, considering the unique features of the industry (fragmented, seasonality, mainly SMEs, overdependence, productivity). Thus, a CSV perspective requires an understanding of decision-making and power relations that exist between the engaged stakeholders. The conclusion to this study will now look at specific implications of CSV and next-practice platforms for academia and tourism businesses.

9.7 Conclusions

On the academic side, this case study closes a significant research gap since it is first to showcase a best practice hospitality business leading in sustainability in the context of Pakistan, where the tourism industry is in its infancy. Although the CSV approach is being embraced and applied by most global multinational corporations, there is no known study in Pakistan. Therefore, a study of the current CSR practices is recommended on whether the CSV concept is already being practiced by indigenous companies in Pakistan. To the best of our knowledge, this study is also the first study to conduct empirical research in the tourism sector, incorporating the theory of CSV and the concept of next-practice platforms in sustainable destination management. This case study adds to the existing CSV literature by exploring how CSV is perceived and implemented in the eyes of various stakeholders of a luxury chain hotel in Pakistan.

Islamabad Serena Hotel (ISH) is a leading hotel in sustainability, which is part of the success of Islamabad as a tourist destination. The luxury hotel has many awards won for its considerate and proactive initiatives for the betterment of its stakeholders and society at large. Thus, ISH is often referred to as an international model with Asian values. The authors believe that there is scope in sharing examples of best practices in tourism and hospitality, which is the worst-hit industry by the COVID-19 pandemic. It is thus essential for hospitality businesses to realize that they need to work in tandem with other organizations in order to create shared value, and move the sustainable destination management agenda forward in the post-pandemic era we are currently in. The study contributes to raising managers' awareness of sustainable tourism practices and improving their networking and collaboration potential for establishing next-practice platforms, new sustainable alliances, and implementing joint sustainability initiatives with DMOs in sustainable destination management. In this regard, the study's findings encourage networking and inter-firm collaboration across different sub-sectors of the tourism industry.

This study also has a few limitations: for instance: (i) limited identification and inclusion of all key stakeholders such as local residents, tour operators, hotel guests, and suppliers and (ii) the limited interview questions that did not probe into the CSV and next-practice platforms concepts in detail. Although some of these limitations lie beyond the scope of this study, others emerged either due to limited time available of, or a lack of access to, the participants. Additionally, some limitations emerged during the data-analysis stage. Although we could argue that CSV and next-practice platforms could present a good leap for hotel companies toward sustainable destination management, it is extremely difficult to make recommendations on the application of CSV in this case study. This is probably because of the complex nature of various stakeholder groups and their power relations that exist in destination management (Fyall, 2011).

Although interview findings emphasize opportunities relating to creation of next-practice platforms in environmental stewardship, management of tourist attractions and promotion of livelihoods, destination management organizations (DMOs) might not be competent to engage with the hotel companies as public–private partnerships are usually seen to slow down the process of innovation remarkably (Vaidyanathan and Scott, 2012). Similarly, interview findings show that external stakeholders, including destination management organization representatives, are challenged by the concepts of CSV and next-practice platforms. They perceive benefits of ISH's potential sustainability leadership in sustainable destination management to be oriented toward improved relationships with government and third parties as well as business growth rather than simultaneous creation of social and economic value. Although these interviewees mentioned a desire to preserve the destination, their emphasis transformed to the need for business growth and destination growth for the capital city of Islamabad eventually. This justifies how tourism companies as well as DMOs are often in a position of embracing growth to compensate for lagging behind in establishing innovative collaborations and partnerships with each other (Pechlaner et al., 2012).

Future research could focus on more case studies of best practices to understand how the CSV strategies of hospitality businesses may contribute to a tourism destination's sustainability. Understanding the demand side of CSV and next-practice platforms is especially important as consumers' opinion often impacts how brands are perceived by stakeholders (Bies, 2000). This is also true for perceptions about a specific tourism destination. Next, a longitudinal study in this area of research could possibly investigate the long-term benefits of next-practice platforms as it may establish its positive (or negative) effects on tourism stakeholders in a destination. In this regard, a quantitative method to objectively measure benefits of next-practice platforms with a wider sample of stakeholders constitutes an important next step to solidify our understanding and application of next-practice platforms in creating shared value. Finally, additional research avenues could focus on potential misperceptions of practitioners regarding CSV, given that three interviewees confused CSR activities with CSV referring to donations and charities as part of a CSV strategy. There remains a need for a more coherent narrative about CSV and the role of next-practice platforms in achieving CSV, compiled by academia (Dembek et al., 2016).

The benefit of this case study is that it brings forward perceptions of certain stakeholders into sustainable destination management, allowing for an understanding of the challenges that ISH might be facing in its sustainability journey ahead. This is useful for future preparedness of all parties in managing complexities of prospective initiatives toward sustainable destination management. ISH is highly appreciated for its willingness to participate in this study, its transparency, and for providing the opportunity for academia and practitioners to learn from its sustainability journey.

References

Aguinis, H. and Solarino, A.M. (2019). Transparency and replicability in qualitative research: The case of interviews with elite informants. *Strategic Management Journal*, 40(8), pp. 1291–1315.

Asmelash, A.G. and Kumar, S. (2019). Assessing progress of tourism sustainability: Developing and validating sustainability indicators. *Tourism Management*, 71, pp. 67–83.

Bagur-Femenías, L., Martí, J. and Rocafort, A. (2015). Impact of sustainable management policies on tourism companies' performance: The case of the metropolitan region of Madrid. *Current Issues in Tourism*, 18(4), pp. 376–390.

Belz, F.M. and Peattie, K. (2012). *Sustainability Marketing* (2nd ed.). Chichester, England: Wiley.

Bies, R.J. (2000). Interactional (in) justice: The scared and the profane. In J. Greenberg and R. Cropanzano (Eds.), *Advances in Organizational Justice*, pp. 89–118. Stanford, CA: Stanford University Press.

Brønn, P.S. and Vidaver-Cohen, D. (2009). Corporate motives for social initiative: Legitimacy, sustainability, or the bottom line?. *Journal of Business Ethics*, 87(1), pp. 91–109.

Brundtland, G.H. (1987). *Our common future: Report of the world commission on environment and development*. Geneva, UN-Dokument A/42/427. http://www.un-documents

Crane, A., Palazzo, G., Spence, L.J. and Matten, D. (2014). Contesting the value of creating shared value. *California Management Review*, 56(2), pp. 130–153.

Creswell, J.W. (2012). *Educational Research: Planning, Conducting, and Evaluating Quantitative and Qualitative Research* (4th ed.). Boston, MA: Pearson.

Daly J., McDonald, I., and Willis, E. (1992). Why don't you ask them? A qualitative research framework for investigating the diagnosis of cardiac normality. In: J. Daly, I. McDonald, and E. Willis (Eds.), *Researching Health Care: Designs, Dilemmas, Disciplines*, pp. 189–206. London: Routledge.

Deale, C.S. (2013). Sustainability education: Focusing on hospitality, tourism, and travel. *Journal of Sustainability Education*, 4(2), pp. 17–25.

Dembek, K., Singh, P. and Bhakoo, V. (2016). A theoretical concept or a management buzzword? *Journal of Business Ethics*, 137(2), pp. 231–267.

Denzin, N.K. (2012). Triangulation 2.0. *Journal of Mixed Methods Research*, 6(2), pp. 80–88.

Dyllick, T. and Muff, K. (2013). *Clarifying the meaning of sustainable business: Introducing a typology from business-as-usual to true business sustainability*. Retrieved from http://ssrn.com/abstract=2368735 (accessed 29 June 2021).

Eccles, R.G., Miller Perkins, K. and Serafeim, G. (2012). How to become a sustainable company. *MIT Sloan Management Review*, 53(4), pp. 43–50.

Edgell, S. (2012). *The Sociology of Work: Continuity and Change in Paid and Unpaid Work*. London, SAGE.

Elkington, J. (1997). *Cannibals with Forks: The Triple Bottom Line of 21st Century Business*. Oxford: Capstone

Elo, S. and Kyngäs, H. (2008). The qualitative content analysis process. *Journal of Advanced Nursing*, 62(1), pp. 107–115.

Euromonitor International. (2020). *Ferrero & Related Parties in Packaged Food (World)*. Passport. https://www-warc-com.dbgw.lis.curtin.edu.au/content/article/euromonitor/ferrero-and-related-parties-food/90830

Flyvberg, B. (2011). Case study. In N.K. Denzin and Y.S. Lincoln (Eds.), *The Sage Handbook of Qualitative Research* (4th ed.), pp. 301–316. Thousand Oaks, CA: Sage.

Freeman, R.E. (1984). *Strategic Management: A Stakeholder Approach*. Boston, MA: Pitman

Friedman, M. (1970). A Friedman doctrine: The social responsibility of business is to increase its profits. *New York Times Magazine*, 13(33).

Fyall, A. (2011). Destination management: Challenges and opportunities. In Y. Wangand A. Pizam (Eds.), *Destination Marketing and Management: Theories and Applications*, pp. 340–358. Wallingford, Oxfordshire: CABI.

Fyall, A. and Garrod, B. (2019). Destination management: A perspective article. *Tourism Review*, 75, pp. 165–169.

Gobo, G. (2005). The renaissance of qualitative methods. *Forum Qualitative Sozialforschung/Forum: Qualitative Social Research*, 6(3), pp. 1–12`.

Goodwin, C. (2015). Professional vision. In *Aufmerksamkeit*, pp. 387–425. Springer VS: Wiesbaden.

Gretzel, U., Fuchs, M., Baggio, R., Hoepken, W., Law, R., Neidhardt, J., Pesonen, J., Zanker, M. and Xiang, Z. (2020). E-tourism beyond COVID-19: A call for transformative research. *Information Technology & Tourism*, 22, pp. 187–203.

Hardy, A.L. and Beeton, R.J. (2001). Sustainable tourism or maintainable tourism: Managing resources for more than average outcomes. *Journal of Sustainable Tourism*, 9(3), pp. 168–192.

Hart, S.L. (1995). A natural-resource-based view of the firm. *Academy of Management Review*, 20(4), pp. 986–1014.

Hart, S.L. and Milstein, M.B. (2003). Creating sustainable value. *Academy of Management Perspectives*, 17(2), pp. 56–67.

Hashmi, Z.G. and Muff, K. (2015). Evolving towards truly sustainable hotels through a 'well-being' lens: The S-WELL sustainability grid. In M.A. Gardetti and A.L. Torres (Eds.), *Sustainability in Hospitality: How Innovative Hotels are Transforming the Industry*, pp. 117–135. Sheffield: Greenleaf Publishing.

Houghton, C., Casey, D., Shaw, D. and Murphy, K. (2013). Rigour in qualitative case-study research. *Nurse Researcher*, 20(4), pp. 12–17.

IHG. (2012). *Intercontinental hotel group: CSR report IHG green engage*. Retrieved July 15, 2021, from http://www.ihgplc.com/index.asp?pageid=742

Kania, J. and Kramer, M. (2011). *Collective Impact*, pp. 36–41. Beijing, China: FSG. Leland Stanford Jr. University, USA.

Kramer, M. (2012). *Shared value: How corporations profit from solving social problems*. Retrieved July 12, 2021, from http://www.guardian.co.uk/sustainable-business/shared-value-how-corporationsprofit-social-problems?intcmp=122

Kramer, M.R. and Pfitzer, M.W. (2016). The ecosystem of shared value. *Harvard Business Review* 94(10), pp. 80–89.

Krippendorf, K. (1980). *Content Analysis. An Introduction to Its Methodology*. Beverly Hills, CA: Sage.

Kumar, V., Rahman, Z., Kazmi, A.A., and Goyal, P. (2012). Evolution of sustainability as marketing strategy: Beginning of new era. *Procedia Social Behavioral Science*, 37, pp. 482–489.

Laszlo, C. and Zhexembayeva, N. (2011). *Embedded Sustainability: The Next Big Competitive Advantage*. Stanford, CA: Stanford Business Books.

Lloyd, R.A. (2015). A brief history of globalised markets: Implications for managers. *Journal for Global Business Education*, 14, pp. 5–11.

Machiavelli, H.C. (2001). A man of 'His' time: R.B.J. Walker and The Prince. *Millennium*, 30(1), pp. 1–18.

Matten, D. and Moon, J. (2008). "Implicit" and "explicit" CSR: A conceptual framework for a comparative understanding of corporate social responsibility. *Academy of management Review*, *33*(2), pp. 404–424.

McCracken, G. (1988). *The Long Interview*. Vol. 13. Newbury Park, California, USA: Sage.

Nidumolu, R., Prahalad, C.K. and Rangaswami, M.R. (2009). Why sustainability is now the key driver of innovation. *Harvard Business Review*, 87(9), pp. 56–64.

OECD. (2020). *Building resilience to the Covid-19 pandemic: The role of centres of government*. http://www.oecd.org/coronavirus/policy-responses/building-resilience-to-the-covid-19-pandemic-the-role-of-centres-of-government-883d2961/ (accessed on 5 October 2020).

Pechlaner, H., Volgger, M. and Herntrey, M. (2012). Destination management organizations as interface between destination governance and corporate governance. *Anatolia, An International Journal of Tourism and Hospitality Research*, 23(2), pp. 151–168.

Pfitzer, M., Bockstette, V. and Stamp, M., (2013). Innovating for shared value. *Harvard Business Review*, 91(9), pp. 100–107.

Porter, M., Bernard, M., Chaturvedi, R.S., Hill, A., Maddox, C. and Schrimpf, M. (2012). Tennessee Music Cluster. *Microeconomics of Competitiveness. Harvard Business School*. Available at https://www.isc.hbs.edu/Documents/resources/courses/moc-course-at-harvard/pdf/student-projects/MOC_Final%20Paper_TN%20Music%20Cluster.pdf

Porter, M. and Kramer, M.R. (2011). Creating shared value. *Harvard Business Review*, 89(1/2), pp. 62–77.

Saldana, J. (2009). *The Coding Manual for Qualitative Researchers*. London: SAGE.

The Global Sustainable Tourism Council (GSTC) (2019). Available online: https://www.gstcouncil.org/ (accessed on 12 July 2021).

Ullah, Z., Naveed, R.T., Rehman, A.U., Ahmad, N., Scholz, M., Adnan, M., and Han, H. (2021). Towards the development of sustainable tourism in Pakistan: A study of the role of tour operators. *Sustainability*, 13, 4902.

United Nations World Tourism Organisation (UNWTO). (2018). *Tourism and the Sustainable Development Goals – Good Practices in the Americas*; UNWTO: Madrid, Spain.

UNWTO-UNEP. (2012). *Tourism in the green economy: Background report*. Retrieved June 29, 2021, from http://sdt.unwto.org/en/content/publications-1

Vaidyanathan, L. and Scott, M. (2012). Creating shared value in India: The future for inclusive growth. *VIKILPA*, 37(2), pp. 108–113.

Vidishcheva, E.V. and Bryukhanova, G.D. (2017). Analyses of the sustainable tourism development factors: The example of Sochi-city. *Journal of Advocacy, Research and Education*, 4, pp. 172–180.

Wall, G. and Mathieson, A. (2006). *Tourism: Change, Impacts, and Opportunities*. London: UK: Pearson Education

Yin, R.K. (2003). Designing case studies. *Qualitative Research Methods*, 5, pp. 359–386.

Yin, R.K. (2009). *Case study research: Design and methods* (Vol. 5). USA: SAGE Thousand Oaks, California

10 Ethical Consumption of Ecosystem Services in Tourism

A Jurisprudence Approach

Bitok Kipkosgei

10.1 Introduction

Consumers are becoming increasingly familiar with and reacting to systemic inequality in society. The way consumers buy sustainable products is closely related to their status, experience, responsibility, fashion, and income (Esposti et al., 2021). In the Ethical Consumer Market Report 2020, 27% of customers in the UK purchased a post-COVID-19 Fair Trade Certificate, and this trend seems to continue increasing. Fairtrade and organic have outperformed the market and are prepared to gain from the after-pandemic for food buying.

This chapter highlights the importance of understanding ethical consumption from the tourist and general tourism industry perspectives. It also reviews the evolution of ecotourism in mainstream tourism from an ethical focal point. While supporting the work by Francis (2012), this chapter seeks to assess whether the increase in ecotourism is demand-led or hedonistic and self-actualizing. In the demand-led perspective, the review bases consumer desires on ethical behavior. Meanwhile, hedonistic and self-actualizing consumer desires are related to nature or adventurous experiences.

In the recent past, interest in ecosystem services has gained momentum as an environmental policy decision-making tool (Gómez-Betancur et al., 2022). However, there has been limited scientific literature that focuses on the relationship between tourism and ecosystem services. Based on this realization, there is a need to apply the jurisprudence approach to harmonize the tourism discourse and the ecosystem services to comprehend both consumer behavior and the roles of tourists and the tourism industry. As suggested by Taekema (2021), the three types of approaches; philosophical, legal doctrinal, and social-scientific approaches triangulated to the rule of law. This chapter uses triangulation of three aspects of the jurisprudence approach to provide more understanding of the connection between ethical consumption and ecosystem services in tourism.

10.2 Purpose of the Chapter

This study aims at describing the phenomenon of ethical consumption and restoration of ecosystem services from the general behaviors exhibited by

DOI: 10.4324/b23145-14

consumers, and more particularly the tourist and the tourism industry. In its findings, the chapter focuses on and contributes to the ethical tourism literature by relating the implication of ethical consumption to ecosystem services in tourism.

10.3 Concept of Ethical Consumption

When making tourism-related choices, ethics is an essential element (University of Oulu, 2020). A consumer must decide where to go, how to get there, and what type of entertainment to enjoy. Acting and reacting are how ethical consumption is demonstrated. Tourists can actively choose or refuse specific services, thereby affecting social, political, and environmental change. Customer activism is crucial in the issue of responsible and sustainable tourism, as tourism is increasingly becoming a key area for combating climate change and lowering carbon emissions. This chapter intends to contribute to the discussions on ethical consumption by bringing in a new perspective of relooking at consumer behavior through the jurisprudence lens to uncover more ethical behaviors, especially among tourists.

10.3.1 Consumer Behavior and Ethical Decisions

In the past decade, many of the world's largest companies have embraced discussions on ethics, responsibility, and proper practices. Business leaders anticipate a future of ethical consideration and responsible business activities. In the present times, laws are not only passed but stakeholder rights are regularly enshrined in the company's values and principles. Government regulation and supervision of e-commerce are crucial in stimulating online ethical consumption (An et al., 2021). Through the work of civil society, ecological conservation not only brings changes to the market exchange framework but also moral markets, and makes consumers agents of change.

The theory of planned behavior and intuitive theory is demonstrated in the way in which intuitive judgments affect consumer behavior (Zollo et al., 2018). Ethical consumers are willing and prepared to pay for conservation. The dimensions of consumption are environmental, social, and economic, and have a positive significant impact on the decision to engage in ethical consumer behavior (Tomsa et al., 2021). The common good, virtue, utilitarianism, rights, and fairness or justice are the five approaches to ethical standards. There exists a gap between the good intentions of consumers and their actual behavior, especially when ethically minded consumers opt for unethical brands (Rasool et al., 2020). This happens because the consumer is not only considering sustainability when deciding to buy. Some of the factors that influence consumer decisions include price, habits, values, availability and convenience, and social norms. Other factors include the desire to make a difference, peer pressure, and emotional appeal. Consumers communicate status to themselves and others using their consumption patterns. The framework for ethical decision-making recognizes ethical challenges, seeks facts, determines alternative action, and reflects on decisions that have been implemented. The connection between biodiversity, human health, and climate change heralds a new era for innovators to adopt green solutions necessary for healthy living and a sustainable future (Ren et al., 2021). Eco-friendly tourist behavior is stimulated and influenced by environmental awareness, perception of environmental value, consumer perceptions of efficiency, and environmental attitudes.

The ethically minded consumer behavior scale conceptualizes ethical consumer behavior as a series of consumer choices related to corporate social responsibility and the environment (Sudbury and Kohlbacher, 2016). More and more new tools are being developed to help consumers track their ethical behavior and brand behavior. The top five most recent ethical calculators serve various goals: (i) Mastercard allows consumers to estimate the carbon footprint of their purchases; (ii) Watershed analyses the climate impact of remote and hybrid work to determine how working from home can reduce or increase an organization's carbon emissions; (iii) Klarna assists shoppers to trace the environmental effect of their purchases; (iv) The Dubai Tourist Board requires all hotels to use a new carbon calculator to measure and report their monthly carbon emissions, and (v) AdGreen hopes to help advertising agencies monitor carbon emissions in advertisement production. Another collaborative effort launched in 2020 is Tourism Declares, an initiative designed to help people in the travel sector develop plans to reduce carbon emissions as they recover from the pandemic (Berg, 2021). More than 200 companies, from travel agencies to tour operators, have committed to publishing a climate protection plan within 12 months of incorporation.

According to Generation Z and Millennials, businesses should invest in society and should seek solutions to make society better (Digital Marketing Institute, 2019). Millennials want to participate in activities such as voluntary labor or donations to charity efforts at a firm. As more and more firms start to realize the influence their socially and ecologically oriented activities impress on their consumers, they are more likely to start their environmental programs. Other organizations are likely to mimic such environmentally oriented innovations. Our ideas, attitudes, and opinions are affected by fundamental changes in our life. Astute marketers will need to adjust their policies and strategies to reflect a long-term recovery of social marketing concepts and a more responsible business orientation.

10.3.2 Ethical Consumption in Tourism

The ethical impacts of global consumption have engendered initiatives aimed at creating fair trade and sustainable products. The Global Code of Ethics for Tourism is a comprehensive set of principles intended to assist major participants in tourism development and is a vital frame of reference for responsible and sustainable tourism (The World Tourism Organization, 2020). It strives to help governments, the travel industry, communities, and tourists maximize the sector's benefits while reducing its potentially detrimental impact on the environment, cultural heritage, and societies around the world. The Code recognizes the function of the World Committee on Tourism Ethics, to which stakeholders may submit questions about the document's application and interpretation, even though it is not legally binding.

Calls for more accountability in tourism have elicited no persuasive responses from either the tourism sector or the general public (Weeden and Boluk, 2017). Ethical consumption puts a lot of pressure on travelers to make a lot of judgments at a time when hedonic incentives are threatening to take precedence over other considerations. Tensions arise, and concessions are made, as is to be expected. It can be hard for consumers to gather all the information needed to make fully informed decisions, and even supposedly ethical consumers do not always make ethics the top priority in their purchases (Bain, 2021). This chapter offers new insight into the motivations for ethical consumption among tourists and their influence on the restoration of ecosystem services. It looks at how consumers navigate the responsible tourism industry and gives businesses and destinations a better knowledge of the issues they face in encouraging ethical purchases.

Ecotourism is seen as a way to maintain the natural environment by generating revenue, educating people about ethics, and involving local populations in environmental preservation (Wardana et al., 2021). Simultaneously, conservation and development will sustainably take place. Thus, environmental preservation, conservation, economic development, revenue, education, and community involvement are all facets of ecotourism development. Rural tourism, agrotourism, adventure tourism, ecotourism, homestays, and sustainable livelihoods are just a few of the steps that several state governments have done to encourage sustainable and responsible tourism (Ministry of Tourism, 2021). The tourism industry has also taken voluntary steps to become more sustainable, particularly in terms of energy use, material procurement, and environmentally friendly practices. The fundamental issue in promoting sustainability is to integrate it into the mainstream of tourism development and to build methods and procedures to track government, destination, and industry adoption of sustainable projects.

Gram-Hanssen (2021) developed practice theories centered on three discussions: how to conceptualize variation in practice and performance, how different consumption approaches have conceptualized ethics, and general understanding to conceptualize ethics of consumption within theory and practice. Social researchers working with consumer patterns must focus on changing behaviors, especially given the increasing importance of sustainability and consumer responsibility in public discussions (Esposti et al., 2021). The emerging tendencies in sustainable consumption indicate that the social aspect has taken root in recent times, alongside the focus on the environmental impact of sustainability. It is important to acknowledge that there are additional ethical problems such as cost and that ethical and sustainable products are not necessarily a feasible option for all consumers. Consumption of environmentally friendly food items, tourism, trade, and apparel has been given a dominant focus. There are three components of sustainability in nature;

environmental, social, and economic aspects, which have a substantial beneficial effect on consumers' ethical conduct (Tomsa et al., 2021). Alsaad et al. (2022) demonstrated that moral certainty associated with ethical consumption is also strongly connected to religiosity and moral intensity. What makes customers buy or not buy ethical items remains a puzzle to researchers.

While the desire for ethical tourism is expanding in theory and practice, relatively little is known about what motivates tourists to undertake ethical tourism. Teng, Ma, and Jing (2021) used the theory of planned behavior to investigate the impact of personal variables, positive social influence, destination image, and service quality on Chinese visitors' behavioral intentions. The goal of the study was to figure out what motivates Chinese tourists to engage in ethical tourism. The findings suggest that awareness of ethical tourism, attitudes toward ethical tourism, family and friend influences, and the destination's local environment are the important elements that motivate Chinese travelers to engage in ethical tourism. Furthermore, service customization, service assistance, and previous experiences with ethical tourism are all major factors of visitors' happiness with ethical tourism.

The importance of ethical and sustainable consumption in the tourism industry has been highlighted in a previous study. Using the mindful consumption theory, Jirojkul et al. (2021) conducted an exploratory factor analysis of survey data from 400 tourists who visited a town in Krabi, Thailand. According to their findings, four variables make up a mindful mentality, and six factors make up mindful behavior. Mindful mentality consists of the value of community, the value of natural resources, personal values, and the value of tourism to the community, which is a mindset arising from the experiences gained in community tourism. It creates an awareness of the value of tourism that increases economic value, cultural preservation, and building community self-esteem. Meanwhile, mindful behavior consists of (1) behavioral control, or the control of which behaviors should or should not be performed, (2) behavior modification, (3) thought modification, (4) perception of correct thinking, (5) value-building behavior, and (6) decisions about travel and activity selection. These factors demonstrate that the development of a growth mindset leads to the development of positive and ethical behavior.

Past research indicates how tourists react to unethical occurrences and the implications of their actions. After unethical occurrences, Liu, Zhang, and Yao (2021) developed a process model for evaluating how tourist power in social media affects tourism market regulation. Their study demonstrates that tourists' perceptions of tourism market regulation are influenced by the perceived severity of an unethical incident, culpability attribution, and ambiguity tolerance. "Tourists" behavioral responses to unethical activity follow a predictable pattern: market re-configuration, cognition, tourist market control, emotion, and behavior. As a result, the purpose of this chapter is to raise awareness and contribute more to the understanding of market re-configuration following unethical acts.

Destinations are rethinking mass tourism in the aftermath of the coronavirus pandemic, focusing on more sustainable segments and looking for more responsible visitors. This necessitates gathering appropriate data to identify the types of tourists who visit the location while also keeping track of changes in tourist behavior and attitudes. Dias et al. (2021) sought to address this issue by developing a metric to assess tourist responsibility. Civic responsibility and charitable responsibility are two dimensions revealed by their research.

10.4 Concept of Ecosystem Services

Practically all the ecosystems of the earth have changed drastically through human action. As agriculture, urban areas, and infrastructure expand, the conversion of ecosystems is expected to continue. Many of these services are being exploited in unsustainable ways between now and 2050, as people rely on ecosystems (Obiwulu et al., 2021). This has damaging repercussions for human well-being. Ecologists and professionals should carefully examine the potential of ecosystem services to contribute to a variety of approaches in the conservation of nature and improvement of the environment. The connection between humans and nature has a two-way effect on resource use as well as society through human welfare and well-being. While ecosystem services valuation literature has expanded over recent years, it still has few applications in urban environments (Croci et al., 2021). Some of the most common ecosystem services in urban areas include local climate regulation, management of air quality, storage, and carbon sequestration. Besides, art and design as well as aesthetic enjoyment and desire for culture contribute to ecosystem services. The ecosystem service framework should be viewed with a biocentric lens, where ecosystems serve humans, and humans contribute to the health of ecosystems and live symbiotically as part of them. This chapter is designed to address research gaps, including how human values determine how we value nature, the strengths of ecosystem services to support human well-being, as well as identifying effective ways of managing the ecosystem services.

10.4.1 People's Values toward Nature

Values are the results of scientific descriptive evaluations of the relationship between the human being and the ecosystem (Stålhammar, 2021). Values, which impact science and institutions, are viewed as basic ideas and moral concepts of right and wrong. A key objective of value studies is to understand why or how the minds of people converge with the goals of sustainability and use tools to evaluate present conditions in order to integrate change and transformation. For sociological, environmental, and economic reasons, nature's value may change for various peoples. In recent times, relational values are recognized as the third class of natural values. It is believed to be among other recognized segments of axiology such as instrumental and intrinsic values (Stålhammar and Thorén, 2019). It expresses traits like caring, social bonding, connection, and spiritual meaning in relationships between humans and the environment. The most common human values are compassion and self-transcendence, with the feminine gender being the most religious and pro-conservation. Advancement in age connects more with conservation and adaptation to change. Mannarini et al. (2021) conducted research to examine the link between fundamental human and communal values. Their study confirms that communities with low stability and cohesiveness have strong self-transcendence and value for conservation.

It has been proven that a person's link to nature leads to positive attitudes toward the protection of the environment (Oh et al., 2021). Family values are substantially related to a person's relationship with nature's cognitive, emotional, and experiential components. Various elements of the link with nature can be explained by family values, social standards, and natural experiences. The restoration of habitats typically depends on ecocentric ideals and attitudes. Attitudes better than fundamental principles predict behavioral intentions, such as readiness to pay. A variety of instruments are available to detect, describe, quantify, and value ecosystem services (Delpy et al., 2021). These tools cover field surveys, interviews, public involvement, remote sensing, and geographic information system mapping. Some of these tools focus on quantifying ecosystem services biophysically while others analyze social benefits or monetary worth. They are utilized in diverse settings such as conservation and restoration planning, policymaking, environmental impact evaluation, with deliberate participation by local communities. The designers of the built environment have access to these tools, and some can be employed in an urban setting for ecosystem-based adaptation, natural, or regenerative design. Ecology growth, land degradation, erosion, landslides, and deforestation are potential drivers for ecosystem service pricing. Transformation of land-use cover and evaluation of ecosystem services plays key roles in vegetation restoration. The land uses cover is classified as bareland, farmland, grasslands, forest, plantation, settlements, tree-growing, water-body, and woodlands (Shiferaw et al., 2021). Land sharing versus land sparing raises the argument on how to effectively combine wildlife conservation and agricultural production. It involves dividing the land for intensive agricultural productivity and biodiversity conservation.

Economic valuations of ecosystem services supplied by natural environments may be significant instruments for conservation subject to six caveats (Buckley and Chauvenet, 2021). Firstly, they naturally underestimate ecosystems and the people of low-income countries. Secondly, present standards are inadequate and should consider mental health arising from monetary implications to employers, insurers, governments, and companies. Thirdly, geographically and politically they apply at various scales: for some, global or cross-border but for most, local. Fourth, they are powerful for scarce resources in high demand such that one user can stop others from accessing them or that it is possible to stop someone from using them. Fifth, their political strength relies on how the costs and benefits are distributed, for instance, health supersedes conservation. Finally, they depend on human institutions such as carbon pricing.

10.4.2 Strengths of Ecosystem Services to Support Human Well-Being

Ecosystem functions are outcomes of various interactions between ecological and social processes. Three main themes prevail: biological diversity, conservation, and management (Wang, Zhang, and Cui, 2021). In the past five decades, humans have been changing these ecosystems faster and more widely than ever before, so that the ever-increasing demand for food, fresh water, wood, fiber, and fuel can be met (Obiwulu et al., 2021). People feel the impact of the ecosystems to provide services geared toward human prosperity and stability. The provision of ecosystem services linearly and explicitly flows from nature to humans, with no human input or feedback. Human activities determine the benefits of ecosystem services through various mechanisms, such as the management, mobilization, distribution, occupation, and value-addition on ecosystem services. These mechanisms are influenced by people's decisions based on rules, assets, values, and spatial context, as described in the entire ecosystem service cascade.

The impact of COVID-19 has demonstrated a new interrelationship between humans and food systems and the scale, scope, and severity of the potential consequences of encroaching on natural ecosystems. A quarter of the global deaths relate to economic decisions that affect the environment (Paula and Willetts, 2021). Poor environmental decisions are related to exposure to chemicals, micro-plastics, poor air quality, poor water quality, and over-nutrition. Some other decisions that negatively affect the environment include agricultural run-off, zoonotic diseases, and misdirected economic development agenda. Cooperation between the environment and health sectors to form a joint agenda and link policies and plans is essential to promote green recovery.

Wildlife has significant impact on human well-being (Methorst et al., 2020). The impacts range from potentially fatal interactions to beneficial contributions. The non-physical contributions of wildlife to humans relate to health, well-being, identity, and spirituality. Wild animals have a negative effect on humans, such as mostly feeling insecure or harmed by mammals or reptiles. Over the past decade, there are several wildlife publications that have had a positive impact on humans. These include impacts such as good mental health and positive emotions. Nevertheless, future studies should focus on implementing multi-effect approaches to understand wildlife-human conflict. They can do so by jointly assessing the impacts of wildlife on the well-being of the people. These assessments of wildlife for humans have ramifications on actionable scientific results necessary in shaping human and wildlife coexistence, as well as the enhancement of human well-being and health.

Nature is vital in maintaining human well-being, especially health (Methorst et al., 2020). Nature provides pollination of nutrient-rich crops, purification of drinking water, flood protection and weather safety, and health benefits. A key but challenging frontier of research is to clarify how nature promotes physical activity, because it has many benefits for physical and mental health, especially in densely populated urban areas, where opportunities for contact with nature are becoming less and less. Cities naturally promote sports activities and provide information for urban greening and broader health assessments (Marselle et al., 2021). Four pathway domains connect biodiversity to human health: (i) reduction of harm (for example, drug supply and reduction of air exposure); (ii) restoration of capacity (for example, restoring concentration and reducing stress); (iii) ability development (for example, promotion of physical activity, detached experience); and (iv) cause harm (for example, dangerous wild animals, zoonotic diseases, allergens). A better comprehension of these approaches can improve biodiversity conservation as a way of promoting human and natural health.

People have the responsibility of utilizing natural capital for economic development and general well-being. But they need to connect with the ecosystems and natural capital much more sustainably. Factors such as population growth trends have exacerbated the impact of these pressures on human well-being. Human well-being must always be a consideration, to be compared to any other competing political objective. Tropical island countries enjoy enormous ecosystem service benefits despite being densely populated (Sarathchandra et al., 2021). These benefits concern human well-being in unique ways and the ecosystems draw a lot from proper management of resource-rich tropical lands.

10.4.3 Management of Ecosystem Services

On March 1, 2019, The United Nations General Assembly proclaimed that 2021–2030 is the Decade on Ecosystem Restoration (UNEP/FAO Factsheet, 2020). This proclamation is aligned with the Decade of Action for the Sustainable Development Goals. The UN system shares the need to act and the commitment to mainstream biodiversity and natural solutions via collective action. The United Nations Environment Assembly resolved that all ecosystems be conserved and restored. Climate change awareness and other environmental issues in many nations have reached new levels. Protecting the rights and building on rural community knowledge is crucial to the success and protection of a substantial share of biodiversity across the world.

Ecosystem management establishes stakeholder relationships and networks in a very unique and complex manner (Solomonsz et al., 2021). This transaction landscape includes governments, the tourism industry, scientific research, nature conservation organizations, civil society, and international decision-making bodies. Stakeholder groups have varied perceptions of the significance of ecosystem services. It is expected of the ecosystem services framework not only to reflect the power relationships that mediate the flows of ecosystem services but also to integrate the analysis of ecological interactions between stakeholder interactions and ecosystem services (Felipe-Lucia et al., 2015). The most powerful are the stakeholders who manage (but do not use) the key ecosystem properties and services, to intermediate regulatory and final services users. Conversely, the remaining non-excludable ecosystem services can only be accessed by unauthorized actors. These include water quality, freshwater supply, and some cultural services. However, the critical factors that determine the state of ecosystem services are land stewardship, governance, and access rights. It is necessary to analyze the role of the actors and their linkages to promote equitable access for ecosystem services.

Variations in stakeholder types that benefit from and manage ecosystem services are occasioned by deliberate choices, for example, preferences for local benefits (Vallet et al., 2019). Most stakeholders manage water quality and agricultural production. There are stark differences in the identity of the actors managing or benefiting from ecosystem services. It is the local stakeholders and the business sector that are believed to benefit more from ecosystem services, and public and non-governmental organizations are most involved in managing ecosystem services. Fairer governance of ecosystem services may need to involve more stakeholders in decision-making. There is also a need to examine the factors that determine stakeholder roles and power distribution. In particular, it is necessary to understand how some critical aspects underpin inequalities in different social and cultural contexts and these include entitlements, rights, foundations, as well as spatial configurations.

According to Urcuqui-Bustamante et al. (2021), the degree of freedom of choice that stakeholders share varies based on stakeholder participation. The most essential and often underestimated is trust between key actors and institutions because it leads to successful collaboration on ecosystem services initiatives. It explains why program managers and government agencies should know and develop ecosystem services design and implementation for effective natural resource management.

10.5 Jurisprudence Approach and Ethical Consumption of Ecosystem Services

The concept of jurisprudence has given way to the numerous laws, rules, and legislation through which justice can be understood in society. Concepts are representational states of mind (Himma, 2015). To properly analyze each concept may require knowledge about the content of the representation and the nature of the relevant state. Particularly, the relevant state could be dispositions and beliefs of various kinds. Conceptual jurisprudence deals with explaining the central concepts of our legal practice, including their interrelationships. Scholarly writing engages three aspects of jurisprudence, which comprises natural law, analytical jurisprudence, and normative jurisprudence (Market Business News, 2021). As an ethical theory, natural law helps in understanding the laws of economics from scientific and rational perspectives, and how economies should work (The Investopedia Team, 2021). It determines the extent to which economic analysis dictates (or prohibits) public order or business behavior. Self-interest is one of the three natural laws of economics that relates to how people perform their daily duties to fulfill their needs. The law of competition, which is the second natural law of economics, concerns the competition that compels people to create safer and quality products and services. Meanwhile, the law of supply and demand ensures that sufficient goods and services are produced in a market economy at the lowest possible price to satisfy demand. It should be noted that the majority of economic questions are not answered by philosophers of natural law. A comparison of costs and benefits suggests that ecological sustainability is not worth pursuing for welfare reasons from a utilitarian perspective, given that the benefits do not exceed the costs (Zagonari, 2020). If ecological sustainability is achieved from an egalitarian perspective for ethical reasons, it is worth striving to achieve it even with a growing world population. Concerning feasibility, achieving goals relates to having realistic parameter values for a given sustainability paradigm and solutions classified according to technology, preference, and ethics. Empirically, the sustainability paradigms being inconsistent are resolved through theoretical disputes on absolute rights. Thus, ecological sustainability as an ethical issue needs to be treated based on the impact on the sustainable development of tourism.

Analytical jurisprudence is often seen as being co-extensive with an explanation of abstract concepts or linguistic analysis (Twining, 2005). The main goal of legal philosophy is to critically analyze the assumptions and preconditions of legal discourse. Analytical jurisprudence has often been criticized, largely for its association with linguistic analysis and legal positivism. It is believed that law enforcement could reduce the logic of the will, whereby any human activity could be considered permissible or prohibited (Banović, 2021). Consumption is one of the mechanisms that allow people to differentiate themselves, particularly through dietary practices, culture, and exercise. To many social scientists, consumption is not primarily viewed as an individual act but rather a social activity (Toti and Moulins, 2016). During the recent decades, researchers have pointed out the growing importance of ethics in consumption. For example, in decision-making processes, purchasing decisions, and people's behavior are increasingly influenced by ethical considerations. For these reasons, ethics and morals are now integrated into the strategies of organizations through the concept of sustainability and implemented through corporate social responsibility initiatives. Social responsibility combines social and environmental concerns with the economic objectives of organizations. The growing concern among companies about corporate social responsibility can perhaps be explained by ethical consumer decision-making. Most of the consumer decisions we make daily can be ethical in content. In general, our ethical concerns stem from personal beliefs and social influences, not from the law (Dias, 2013). The question then arises as to whether public institutions are entitled to prescribe certain dimensions of ethical behavior and whether consumer law is the appropriate mechanism to achieve such purposes. As observed by Rulia et al. (2021), the consumption model has been largely influenced by the emergence of the obligation to protect the environment. Of international concern at present is the pursuit of organic products and sustainable consumption. There is a renewed drive for green consumption globally as consumers make a significant contribution to achieving sustainable development goals. People decide between green and environmentally friendly services and products. However, the evolution of consumer preferences and behavior has been largely ignored in research studies, especially pursuits for sustainable tourism development in developing countries.

Normative jurisprudence deals with evaluative legal theories. It deals with the aim or purpose of the law as well as the moral or political theories constituting the law. To revitalize normative legal scholarship, normative jurisprudence provides criticism to positive law and proposes reforms in line with the stipulated legalistic ideals and moral values (West, 2011). This jurisprudence is detailed on the three great traditions of jurisprudence that in the past have provided the philosophical basis of normative science. These include natural law, legal positivism, and critical jurisprudence. Normative ethical approaches create ethical standards for marketers to know what is right and wrong and based on observations or data, describe what seems to happen in morally charged situations. About ethical purity, it is difficult to assess intent because it requires an understanding of the internal motivation behind the actions or policies of an organization. The ethics of virtue requires people to hold the highest ethical standards or character justifiable to a relevant moral community. In the 20th century, consumer sociological research largely focused on sustainability and connection to daily and routine consumption (Gram-Hanssen, 2021). The normative ethical approach comes as an alternative away from individualistic psychological or identity-communicative approaches toward sustainable consumption. However, there is a tendency to ignore variations in consumer practices, given the shift toward a practice-based approach. Specifically, the discourse is yet to explain the variations on ethical concerns as well as questions on sustainability. There seems to have been neglect of important questions, and as to whether or not ethics have a role in changing practices in a sustainable direction. This chapter offers detailed theoretical underpinning and practice on conceptualizing variations in ethical practices and how consumers and businesses among other stakeholders can strive for sustainable consumption.

10.6 Methodology

This chapter adopts the methodology of triangulation, which implies using natural law, analytical jurisprudence, and normative jurisprudence to better understand how ethical consumption affects the restoration of ecosystem services for sustainable tourism development. Taekema (2021) conducted a study that focused on methodology and conceptual claims using various articles on the rule of law as provided for by legal philosophers. The author used three specific forms of triangulation to the rule of law: legal doctrinal, philosophical, and social-scientific. Arriving at the same results using different methodological tools is believed to increase support for those results and better justify the conclusions. Triangulation does not always serve to confirm results, but it can serve to complement different findings. While the methods augment each other, the approach can serve to discover aspects that the other study fields cannot or only partially explain. The chapter adopts the qualitative design and justifies the conceptual claims of ethical consumption through recourse to social and economic practices in the restoration of ecosystem services for sustainable tourism development. The analysis is implemented through a comprehensive review of the existing academic literature, as well as theoretical case studies that support the arguments presented.

10.7 Findings

Although there is no precise definition of environmentally friendly consumption, various factors may help to identify the environmental dimension of long-term consumption (Tomsa et al., 2021). Recycling, packaging, raw materials, energy, production, and climate are all tied to these elements. We need to drastically change how we consume and produce goods and resources. Tourism's contribution of 10% of global gross domestic product represents a significant responsibility in production and consumption. Consumers are becoming increasingly familiar with and reacting to systemic inequality in society. The way consumers buy sustainable products is closely related to their status, experience, responsibility, and fashion (Esposti et al., 2021). Consumption patterns are mainly affected by people's income. In the Ethical Consumer Market Report 2020, 27% of customers in the UK purchased a post-COVID-19 Fair Trade Certificate, and this trend seems to continue increasing. Fairtrade and organic have outperformed the market and are prepared to gain from the after-pandemic for food buying. The information particularly on duration, compensation, and their environmental footprint should be transparent, comparable, and harmonized.

Tourism's detrimental socio-cultural, economic, and environmental effects are nevertheless a source of concern. As a result of this concern, several programs have been launched to raise awareness about tourism activities as well as the influence of tourism on the environment and society in general (Mokoena, 2019). Sustainability in tourism encompasses all of the terms used to describe ethical tourist practices, such as responsible tourism, ecotourism, and sustainable tourism (Kenya Tourism Board, 2016). Berg (2021) predicts an increased interest in trips focused on time outdoors and going to less-visited places, movements that facilitate sustainable practices and help reduce the burden of over-tourism. Sustainable consumption and production require both consumer demand and supply from commerce and industry to manage natural resources more sustainably and use them more efficiently. It will require profound changes in the way that society and our livelihoods function. Some visitors' behaviors, including how they choose tourism products and trip places, have already begun to change as a result of this awareness. This surge of knowledge may compel tourist corporations to do business in new and profitable ways in the future while minimizing tourism's harmful effects.

There are varied ethical issues associated with the restoration of ecosystem services. Improved openness in the application of the ecosystem services idea might help in exploiting its capabilities without succumbing to potential disadvantages due to ethical concerns. Many researchers have a somewhat skeptical view of ecosystem services' payments or even deny the usefulness of this conservation tool (Jax et al., 2013; Kaiser et al., 2021). Some writers have criticized the fact that payments for ecosystem services might reduce the complexity of ecosystems, owing to marketing and marketing trends, typically linked with this instrument. The mere concentration on monetary values underscores the existing variety of values. Some scholars acknowledge that payments for environmental services are a tool of neoliberal and market conservation that promotes privatization and perhaps perpetuates current imbalances. The phrases environmental and green consumption can be used interchangeably and sustainable market consumption cannot be achieved without addressing the ethical concerns of the target market. Marketers must become aware of the meaning of these phrases.

Some human values have been found to portray people's value for nature. It has been proven that a person's link to nature leads to positive attitudes toward the protection of the environment (Oh et al., 2021). Family values are substantially related to a person's relationship with nature's cognitive, emotional, and experiential components. Various elements of the link with nature can be explained by family values, social standards, and natural experiences. The restoration of habitats typically depends on ecocentric ideals and attitudes (Matzek and Wilson, 2021). Attitudes better than fundamental principles predict behavioral intentions, such as readiness to pay.

Ecosystem services present enormous strengths for supporting human well-being. People have the responsibility of utilizing natural capital for economic development and general well-being. But they need to connect with the ecosystems and natural capital much more sustainably. Factors such as population growth trends have exacerbated the impact of these pressures on human well-being. Human well-being must always be a consideration, to be compared to any other competing political objective. Ecosystem service assessments increasingly support conservation initiatives (Schröter et al., 2020). Policy guidance and execution are necessary for a variety of situations, including urban, rural, and natural spaces. Tropical Island countries enjoy enormous ecosystem service benefits despite being densely populated (Sarathchandra et al., 2021). These benefits concern human well-being in unique ways and the ecosystems draw a lot from proper management of resource-rich tropical lands.

There are stark differences in the identity of the actors managing or benefiting from ecosystem services. Policymakers and researchers have shown a dependable interest in responsibility in tourism (Saarinen, 2021). Variations in the types of stakeholders that benefit from and manage ecosystem services are occasioned by deliberate choices, for example, preferences for local benefits (Vallet et al., 2019). Most stakeholders manage water quality and agricultural production. It is the business sector and local stakeholders that are believed to benefit more from ecosystem services and public and non-governmental organizations are most involved in managing ecosystem services. Fairer governance of ecosystem services may need to involve more stakeholders in decision-making. There is also a need to examine the factors that determine stakeholder roles and power distribution. In particular, it is necessary to understand how some critical aspects underpin inequalities in different social and cultural contexts and these include entitlements, rights, foundations, as well as spatial configurations.

This chapter analyses the linkage between ethical consumption and restoration of ecosystem services using the three aspects of the jurisprudence approach. Empirically, the sustainability paradigms being inconsistent are resolved through theoretical disputes on absolute rights (Zagonari, 2020). Thus, ecological sustainability as an ethical issue needs to be treated based on the impact on the sustainable development of tourism. Most of the consumer decisions we make daily can be ethical in content. In general, our ethical concerns stem from personal beliefs and social influences, not from the law (Dias, 2013). The question then arises as to whether public institutions are entitled to prescribe certain dimensions of ethical behavior and whether consumer law is the appropriate mechanism to achieve such purposes. There seems to have been neglect of important questions, and as to whether or not ethics have a role in changing practices in a sustainable direction. Tourism and hospitality are intimately linked with rules and regulations. The rules help to guarantee that tourists and industry stakeholders work in a decent and equal environment.

10.8 Conclusion and Recommendations

This chapter provides an in-depth review of several influential contributions to the study of ethical consumption of ecosystem services for sustainable tourism development. It concludes that ethical consumption can no longer be ignored as companies and consumers pursue long-term success in creating fair trade and sustainable tourism products. The description of ethical consumption indicates that the social aspect is taking root alongside the focus on the environmental impact of sustainability. This chapter recommends that consumption of environmentally friendly food items, tourism, trade, and apparel should be given a dominant focus. With this focus in mind, public and private institutions should positively promote religiosity and moral intensity as they have a strong connection to ethical consumption.

Tourists need to be helped to make ethical decisions. In the future, laws need not only be passed but stakeholder rights ought to regularly be enshrined in the company's values and principles. Eco-friendly tourist behavior is stimulated and influenced by environmental awareness, perception of environmental value, consumer perceptions of efficiency, and environmental attitudes. Successful factors in improving ecotourism performance are strategic management priorities to build quality tourism experiences and sustainable tourism. This chapter proposes that civil society, among other stakeholders, needs to sustain the advocacy for ecological conservation, moral markets, and making consumers agents of change.

Restoration of ecosystem services is confronted by ethical issues that require consideration both in theory and practice. The ecosystem services concept has many dimensions of value and ethical challenges that need to be clarified. This chapter recommends that policy guidance and execution are necessary for a variety of situations. Improved openness in the application of the ethical consumption idea, especially environmental and green consumption, might help in exploiting the capabilities of ecosystem services without succumbing to potential disadvantages due to ethical concerns.

Human dependency on nature may contribute to negative consequences on the national and global economy, including human values for nature. Large-scale urbanization is affecting biodiversity hotspots and agricultural lands, which in turn will lead to degradation of ecosystem services and food security in human settlements. This chapter makes a recommendation that environmental restoration should be improved, mainstreamed, and scaled up in two ways: create an alternative view of the human perspective of ecological restoration and formulate a shared agreement for the planning and transformation as well as monitoring the progress of social goals.

Ecosystem functions are the result of complex interactions between social and ecological processes and have the strength to support human well-being. People feel the impact of the ecosystems to provide services geared toward human prosperity and stability. This chapter proposes the clarification of how nature promotes physical activity, especially in densely populated urban areas where opportunities for contact with nature are decreasing.

Several actors establish and decide on the management of ecosystem services. The new macroeconomic world is influenced by an integrated set of markets and governments. This chapter advocates for effective land-use planning for ecosystem services and conservation decision-making. It promotes consultation among the various stakeholders as to the extent of identifying conservation actions and where and how protected areas should maximize ecosystem services.

The jurisprudence approach has been adopted in this chapter to explain more about the linkage between ethical consumption and restoration of ecosystem services. The concept of jurisprudence gives way to the numerous laws, rules, and legislations through which justice can be understood better in society. The concept of ecosystem services has been brought forth by ecologists and economists to guide economic thinking among decision-makers in areas of biodiversity conservation. Nevertheless, it fails to change how we treat nature because it fundamentally lacks practical fitness and that ethics enshrine voluntary implementation mechanisms. This chapter recommends that the researchers and scholars should focus more on the jurisprudence approach while suggesting new ways to govern and resolve global environmental challenges.

References

Alsaad, A., Elrehail, H., and Saif-Alyousfi, A. Y. (2022). The interaction among religiosity, moral intensity and moral certainty in predicting ethical consumption: A study of Muslim consumers. *International Journal of Consumer Studies*, *46*(2), pp. 406–418.

An, N., Wang, E., Geng, X., Gao, Z., and Kiprop, E. (2021). Consumers' willingness to pay for ethical consumption initiatives on e-commerce platforms. *Journal of Integrative Agriculture*, 20(4), pp. 1012–1020.

Bain, M. (2021). Ethical Consumption Can't Be the Burden of Shoppers Alone. https://qz.com/2046991/ethical-consumption-cant-be-the-burden-of-shoppers-alone/ [Accessed 6 Aug. 2021].

Banović, D. (2021). About John Austin's analytical jurisprudence: The empirical-rationalist legal positivism. *ICLR*, 21(1), p. 242.

Berg, B. (2021). As the Pandemic Prompts Eco-Awareness, the Travel Industry Responds. https://www.washingtonpost.com/lifestyle/travel/green-travel-consumer-priorities-values/2021/03/04/d11b102c-76d7-11eb-9537-496158cc5fd9_story.html [Accessed 2 Aug. 2021].

Buckley, R. and Chauvenet, A. (2021). Six caveats to valuing ecosystem services. *Nature*, 592(7853), pp. 188–188.

Croci, E., Lucchitta, B., and Penati, T. (2021). Valuing ecosystem services at the urban level: A critical review. *Sustainability*, 13(3), p. 1129.

Delpy, F., Pedersen Zari, M., Jackson, B., Benavidez, R., and Westend, T. (2021). Ecosystem services assessment tools for regenerative urban design in Oceania. *Sustainability*, 13(5), p. 2825.

Dias, S.F. (2013). Ethics and consumerism: Legal promotion of ethical consumption?. In *The Ethics of Consumption*, pp. 141–146.

Dias, Á., Aldana, I., Pereira, L., da Costa, R.L., and António, N. (2021). A measure of tourist responsibility. *Sustainability*, 13(6), p. 3351. [online]. Available at: https://res.mdpi.com/sustainability/sustainability-13-03351/article_deploy/sustainability-13-03351-v3.pdf [Accessed 9 Jun. 2021].

Digital Marketing Institute. (2019). 16 Brands Doing Corporate Social Responsibility Successfully. Digital Marketing Institute. https://digitalmarketinginstitute.com/blog/corporate-16-brands-doing-corporate-social-responsibility-successfully [Accessed 2 Aug. 2021.].

Esposti, D.P., Mortara, A., and Roberti, G. (2021). Sharing and sustainable consumption in the era of COVID-19. *Sustainability*, 13(4), p. 1903.

Felipe-Lucia, M.R., Martín-López, B., Lavorel, S., Berraquero-Díaz, L., Escalera-Reyes, J., and Comín, F.A. (2015). Ecosystem services flows: Why stakeholders' power relationships matter. *PLOS ONE*, 10(7), p. e0132232.

Francis, D.C. (2012). An Evaluation of Ethical Consumer Behaviour in Relation to Ecotourism. A Dissertation, University of Leicester. https://www.academia.edu/8267734/An_Evaluation_of_Ethical_Consumer_Behaviour_in_Relation_to_Ecotourism [Accessed 7 Aug. 2021].

Gómez-Betancur, L., Vilardy Q. S. P., and Torres R. D. (2022). Ecosystem services as a promising paradigm to protect environmental rights of indigenous peoples in Latin America: the constitutional court landmark decision to protect Arroyo Bruno in Colombia. *Environmental Management*, 69(4), pp. 768–780.

Gram-Hanssen, K. (2021). Conceptualising ethical consumption within theories of practice. *Journal of Consumer Culture*, 21(3), pp. 432–449.

Himma, K. E. (2015). Conceptual Jurisprudence. An introduction to conceptual analysis and methodology in legal theory. *Revus. Journal for Constitutional Theory and Philosophy of Law/Revija za ustavno teorijo in filozofijo prava* 10(26), pp. 65–92.

Jax, K., Barton, D.N., Chan, K.M.A., de Groot, R., Doyle, U., Eser, U., Görg, C., Gómez-Baggethun, E., Griewald, Y., Haber, W., Haines-Young, R., Heink, U., Jahn, T., Joosten, H., Kerschbaumer, L., Korn, H., Luck, G.W., Matzdorf, B., Muraca, B., and Neßhöver, C. (2013). Ecosystem services and ethics. *Ecological Economics*, 93, pp. 260–268.

Jirojkul, S., Pongsakornrungsilp, S., Pianroj, N., Chaiyakot, P., Mia, S., Masst, T., and Techato, K. (2021). The Effect of Mindset on Tourist Behaviour and Mindful Consumption in a Community Enterprise in Krabi, Thailand. *Association for Information Communication Technology Education and Science*, 10(3), pp. 1082–1091.

Kaiser, J., Haase, D., and Krueger, T. (2021). Payments for ecosystem services: a review of definitions, the role of spatial scales, and critique, *Ecology and Society* 26(2), pp. 1–24. https://doi.org/10.5751/ES-12307-260212.

KTB (2016). *Sustainable Tourism Report*. Kenya Tourism Board, Government Printer: Nairobi, Kenya

Liu, Y., Zhang, R., and Yao, Y. (2021). How tourist power in social media affects tourism market regulation after unethical incidents: Evidence from China. *Annals of Tourism Research*, 91, p. 103296.

Mannarini, T., Procentese, F., Gatti, F., Rochira, A., Fedi, A., and Tartaglia, S. (2021). Basic human values and sense of community as resource and responsibility. *Journal of Community & Applied Social Psychology*, 31(2), pp. 123–141.

Market Business News. (2021) What is Jurisprudence? Definition and Examples. https://marketbusinessnews.com/financial-glossary/jurisprudence [Accessed 2 Aug. 2021].

Marselle, M.R., Hartig, T., Cox, D.T.C., de Bell, S., Knapp, S., Lindley, S., Triguero-Mas, M., Böhning-Gaese, K., Braubach, M., Cook, P.A., de Vries, S., Heintz-Buschart, A., Hofmann, M., Irvine, K.N., Kabisch, N., Kolek, F., Kraemer, R., Markevych, I., Martens, D. and Müller, R. (2021). Pathways linking biodiversity to human health: A conceptual framework. *Environment International*, 150, p. 106420.

Matzek, V. and Wilson, K.A. (2021). Public support for restoration: Does including ecosystem services as a goal engage a different set of values and attitudes than biodiversity protection alone? *PLOS ONE*, 16(1), p. e0245074.

Methorst, J., Arbieu, U., Bonn, A., Böhning, G.K., and Müller, T. (2020). Non-material contributions of wildlife to human well-being: A systematic review. *Environmental Research Letters*, 15(9), p. 093005.

Ministry of Tourism. (2021) National Strategy and Roadmap for Sustainable Tourism. Government of India. https://tourism.gov.in/sites/default/files/2021-06/Draft%20Strategy%20for%20Sustainable%20Tourism%20Ver%203%20June%202.pdf [Accessed 2 Aug. 2021].

Mokoena, L. (2019). Ethical tourism consumption: should businesses be concerned? *African Journal of Hospitality Tourism and Leisure*, 8, pp. 1–10.

Obiwulu, E.N.O., Oguh, C.E., Umezinwa, O.J., Ameh, S.E., Ugwu, C.V., and Sheshi, I.M. (2021). Ecosystem and ecological services; need for biodiversity conservation – A critical review. *Asian Journal of Biology*, 11, pp. 1–14.

Oh, R.R.Y., Fielding, K.S., Nghiem, L.T.P., Chang, C.C., Carrasco, L.R., and Fuller, R.A. (2021). Connection to nature is predicted by family values, social norms and personal experiences of nature. *Global Ecology and Conservation*, 28, p. e01632.

Paula, N. and Willetts, E. (2021). COVID-19 and Planetary Health: How a Pandemic Could Pave the Way for a Green Recovery. [online]. Available at: https://www.iisd.org/articles/covid-19-and-planetary-health-how-pandemic-could-pave-way-green-recovery [Accessed 2 Aug. 2021].

Rasool, S., Cerchione, R., and Salo, J. (2020). Assessing ethical consumer behavior for sustainable development: The mediating role of brand attachment. *Sustainable Development*, 28(6), pp. 1620–1631.

Ren, J., Su, K., Chang, Y., and Wen, Y. (2021). Formation of environmentally friendly tourist behaviors in ecotourism destinations in China. *Forests*, 12(4), p. 424.

Rulia, A., Sayema, S., Muhammad, M.M., Nusrat, J., and Al-Mamun, A. (2021). Consumers' environmental ethics, willingness, and green consumerism between lower and higher income groups. *Resources, Conservation and Recycling*, 168, p. 105274.

Saarinen, J. (2021). Is being responsible sustainable in tourism? Connections and Critical Differences. *Sustainability*, 13(12), p. 6599.

Sarathchandra, C., Abebe, Y.A., Wijerathne, I.L., Aluthwattha, S.T., Wickramasinghe, S., and Ouyang, Z. (2021). An overview of ecosystem service studies in a tropical biodiversity hotspot, Sri Lanka: Key perspectives for future research. *Forests*, 12(5), p. 540.

Schröter, M., Crouzat, E., Hölting, L., Massenberg, J., Rode, J., Hanisch, M., Kabisch, N., Palliwoda, J., Priess, J.A., Seppelt, R., and Beckmann, M. (2020). Assumptions in ecosystem service assessments: Increasing transparency for conservation. *Ambio*, 50(2), pp. 289–300.

Shiferaw, H., Alamirew, T., Kassawmar, T., and Zeleke, G. (2021). Evaluating ecosystems services values due to land use transformation in the Gojeb watershed, Southwest Ethiopia. *Environmental Systems Research*, 10(1), pp. 1–12.

Solomonsz, J., Melbourne-Thomas, J., Constable, A., Trebilco, R., Van Putten, I., and Goldsworthy, L. (2021). Stakeholder engagement in decision making and pathways of influence for Southern Ocean ecosystem services. *Frontiers in Marine Science*, 8(1), pp. 1–14.

Stålhammar, S. (2021). Assessing people's values of nature: Where is the link to sustainability transformations? *Frontiers in Ecology and Evolution*, 9, p. 145.

Stålhammar, S. and Thorén, H. (2019). Three perspectives on relational values of nature. *Sustainability Science*, 14(5), pp. 1201–1212.

Sudbury, R.L. and Kohlbacher, F. (2016). Ethically minded consumer behavior: Scale review, development, and validation. *Journal of Business Research*, 69(8), pp. 2697–2710.

Taekema, S. (2021) Methodologies of rule of law research: Why legal philosophy needs empirical and doctrinal scholarship. *Law and Philosophy*, 40, pp. 33–66.

Teng, Y., Ma, Z., and Jing, L. (2021). Explore the world responsibly: The antecedents of ethical tourism behaviors in China. *Sustainability*, 13(9), p. 4907.

The Investopedia Team. (2021). Natural L.w. https://www.investopedia.com/terms/n/natural-law.asp [Accessed 4 Aug. 2021].

The World Tourism Organization. (2020) Global Code of Ethics for Tourism. https://www.unwto.org/global-code-of-ethics-for-tourism [Accessed 10 Aug. 2021].

Tomşa, M.M., Romonţi, M.A.I., and Scridon, M.A. (2021). Is sustainable consumption translated into ethical consumer behavior? *Sustainability*, 13(6), p. 3466.

Toti, J.F. and Moulins, J.L. (2016). How to measure ethical consumption behaviors? *RIMHE*, 24(5), pp. 45–66.

Twining, W. (2005). Have concepts, will travel: Analytical jurisprudence in a global context. *International Journal of Law in Context*, 1(1), pp. 5–40. DOI: 10.1017/S174455230500102.

UNEP/FAO Factsheet. (2020). The UN Decade on Ecosystem Restoration 2021–2030. https://wedocs.unep.org/bitstream/handle/20.500.11822/30919/UNDecade.pdf [Accessed 4 May 2021].

University of Oulu. (2020). Ethical and Political Consumerism in Tourism. https://www.oulu.fi/geography/node/208382 [Accessed 2 Aug. 2021].

Urcuqui-Bustamante, A.M., Selfa, T.L., Hirsch, P., and Ashcraft, C.M. (2021). Uncovering stakeholder participation in payment for hydrological services (PHS) program decision making in Mexico and Colombia. *Sustainability*, 13(15), p. 8562.

Vallet, A., Locatelli, B., Levrel, H., Dendoncker, N., Barnaud, C., and Quispe Conde, Y. (2019). Linking equity, power, and stakeholders' roles in relation to ecosystem services. *Ecology and Society*, 24(2). [online]. Available at: https://www.ecologyandsociety.org/vol24/iss2/art14/ [Accessed 7 Jan. 2020].

Wang, B., Zhang, Q., and Cui, F. (2021). Scientific research on ecosystem services and human well-being: A bibliometric analysis. *Ecological Indicators*, 125, p. 107449.

Wardana, I.M., Sukaatmadja, I.P.G., Ekawati, N.W., Yasa, N.N.K., Astawa, I.P., and Setini, M. (2021). Policy models for improving ecotourism performance to build quality tourism experience and sustainable tourism. *Management Science Letters*. [online], pp. 595–608. Available at: http://growingscience.com/msl/Vol11/msl_2020_310.pdf [Accessed 11 Mar. 2021].

Weeden, C. and Boluk, K. (Eds.) (2017). *Managing Ethical Consumption in Tourism* (1st ed.) Routledge. London: UK

West, R. (2011). *Normative Jurisprudence: an introduction*. Cambridge University Press. New York: USA

Zagonari, F. (2020). Environmental sustainability is not worth pursuing unless it is achieved for ethical reasons. *Palgrave Communications*, 6(1), pp. 1–8.

Zollo, L., Yoon, S., Rialti, R., and Ciappei, C. (2018). Ethical consumption and consumers' decision making: The role of moral intuition. *Management Decision*, 56, pp. 692–710.

Part V

Management of Tourism Ecosystems Services Post-Pandemic

11 Impacts of COVID-19 Pandemic on Ecosystem Services at UNESCO World Heritage Site, Sundarbans

A Viewpoint on India and Bangladesh

Hiran Roy, Subhajit Das, Anisur R. Faroque, Vikas Gupta, and Mohammad Osman Gani

11.1 Introduction

The ecosystem services (ES) are defined as the services provided by nature and used by humankind (Margaryan et al. 2018). The MEA (Millennium Ecosystem Assessment) further defines ES as the benefits people obtained from ecosystems (Millennium Ecosystem Assessment 2005). Although the foundations of this concept have been laid out in the 1960s, its popularity has grown significantly since 1990 and was widely used in 2005 in the Millennium Ecosystem Assessment (Millennium Ecosystem Assessment 2005). However, the benefits of ecosystem services have been distinguished into three different categories: provisioning services (e.g., food, water, timber, and energy production), regulatory services (that affect the weather, floods, diseases, waste or water quality, erosion control, and carbon sequestration), and cultural services (e.g., tourism and recreation, well-being, spiritual or symbolic interaction with the natural environment) (CICES 2013; Grunewald & Bastian 2015; Millennium Ecosystem Assessment 2018). The last group of ecosystem services that included tourism and recreational services is included as subgroups of cultural services. Unlike the other two groups, this group of ecosystem services is embodied by intangible non-material values (Chan et al. 2011, 2012; Fish et al. 2016).

Likewise, other ecosystem services globally, Sundarbans mangrove ecosystem services, is an essential expression of the relationships between nature and societies. Sundarbans are highly productive mangrove wetland ecosystems and deliver a range of economic, social, and environmental benefits to the people, collectively called ecosystem goods and services (Millennium Ecosystem Assessment 2005). The economic potential of Sundarbans mangroves depends on three primary key sources: forest products, fisheries, and ecotourism (Paul et al. 2017). In addition, local and national economies and forest-dependent livelihoods are significantly influenced by Sundarbans mangrove forests. While for coastal protection and preservation of endangered species, Sundarbans mangroves play a critical role in the ecosystem services (Mukhtar & Hannan 2012; Sandilyan &

DOI: 10.4324/b23145-16

Kathiresan 2014). Further, Austin (2020) argued that Sundarbans mangrove forests act as highly effective carbon sinks, and it is found that mangrove forests can absorb 97.57 tons of carbon per hector, which is more than three times the absorptive capacity of the non-mangrove forest. In the tourism context, Sundarbans are heavily dependent on ecosystem services to develop their activities (Mieczkowski 1985). The tourist needs provisioning services (e.g., food, water, or energy), regulating services (e.g., ecosystem regulates the weather), and cultural ecosystem services (e.g., tourism and recreational services) (Pueyo-Ros 2018). Tourists are attracted by different cultural ecosystem services such as aesthetic appreciation (Urry 2002), recreational experiences (Ghermandi & Nunes 2013), or spiritual and religious experiences (Willson 2016); thus, tourism dependence on Sundarbans ecosystem services is critical. However, despite having importance of ecosystem services (ecological, social, and economic functions of mangroves), the Sundarbans ecosystem services are continuously threatened by anthropogenic activities and climatic vulnerability (Sannigrahi et al. 2019; Chaudhuri & Bhattacharyya 2021).

Since World War II, the deadly disease COVID-19 has emerged as one of the largest pandemics ever experienced by people globally (Chakraborty & Maity 2020). It was first reported in Wuhan, Hubei province, China, in late December 2019 and spread rapidly all over the globe and thereby declared an international public health emergency in a couple of weeks by the WHO (World Health Organization) (Chakraborty & Maity 2020). As of August 30, 2021, worldwide, more than 216 million cases were reported, and more than 4 million people died in all countries (WHO 2021). To curb the spread of this disease, drastic measures have been taken globally that included restrictions on international traveling and placing millions of people under lockdown situation (Hamzelou 2020; Zhang et al. 2021). This restriction of mobility has disrupted all types of businesses and has threatened human civilization by impacting all types of activities such as social, economic, industrial, and urbanization (Arora et al. 2020; Karlinsky & Kobak 2021). Although COVID-19 has been a significant threat to human existence, it has brought some direct (e.g., the reduction of air pollution, greenhouse gas emission, noise pollution, and clean beaches) and indirect (e.g., increases of both organic and inorganic waste due to the use of masks or gloves) changes in the environment and ecosystem in different countries because of halted transportation and industry activities (e.g., Bao & Zhang 2020; Wang & Su 2020; Zambrano-Monserrate et al. 2020).

The recent COVID-19 pandemic and its associated lockdown have also hard-hit Sundarbans mangrove ecosystem services in India and Bangladesh. The nationwide lockdown in India and Bangladesh has worsened the non-ecosystem generating activities of the ecosystem-dependent communities, leading to increased pressure on ecosystem services (hereby Sundarbans). Several studies from various countries have reported that lockdown has negatively impacted ecosystem services, mainly deforestation and forest degradation (FAO 2020; Troëng et al. 2020). For example, Amazon, the largest rainforest has experienced 30% higher deforestation during pandemics than last year (Spring 2020). Several African countries have also experienced the rise of bushmeat poaching and trafficking resulting from the induced lockdown, that initiated poverty due to the decreased tourism activities. A surge in poaching of wildlife and deforestation has been reported in Asian countries (e.g., India, Cambodia, Nepal, & China), where many people were forced to depend on forest resources for their livelihoods due to the pandemic (Ghosal & Casey 2020; Troëng et al. 2020). To this aspect, many countries (e.g., Nepal, Thailand, Pacific states) have assessed the impact of COVID-19 on deforestation, wildlife, biodiversity, forest, and forest-dependent people (Rondeau et al. 2020; Giri 2020). However, these studies did not argue for how to recover the loss from the pandemic due to the COVID-19 mobility restriction on the livelihood of forest-dependent people, where forests (hereby Sundarbans) provide major ecosystem services and are critical for human well-being and important for achieving sustainable development goals (Aryal et al. 2020).

The COVID-19 pandemic has brought substantial turmoil in Sundarbans ecosystem services, especially to the ecosystem-dependent people and surrounding communities. The result of the lockdown has impeded or altered the usual interaction of humans and ecosystem services. Therefore, it is necessary to explore this emergent evidence on how ecosystem services in the Sundarbans, the largest mangrove forest in the world and a UNESCO designated world heritage site, likely to be affected by COVID-19 pandemic over the short term is significant to the local community. This chapter also highlights a substantial impact on Sundarbans ecosystem services that might be a drawback for achieving sustainable development goals. Therefore, this chapter also highlights policy strategies that may be helpful for the policymakers and development planners of India, Bangladesh, and other countries to formulate appropriate policies to protect ecosystem services caused by the COVID-19 pandemic.

11.2 Methodology

11.2.1 Study Area

The study site Sundarbans is regarded as the world's largest delta of mangrove forest spread over India and Bangladesh. The Sundarbans are situated at the northern shoreline of the Bay of Bengal and have a total area of 10,000 km^2 (Iqbal 2020). It is estimated that the area of the Sundarbans in Bangladesh is 599,330 hectares which contains about 60% area, and the rest of the area (426,300 hectares, about 40%) is in West Bengal, India (Rahman et al. 1979; Sanyal 1983; Hossain et al. 2016). Sundarbans are also known for their most diverse and productive ecosystems in the world (Borrell et al. 2016). The mangrove forest is designated as a protected area and world heritage site in 1997 by the International Union for Conservation of Nature (IUCN) and the United Nations Educational, Scientific, and Cultural Organization (UNESCO). Hence,

Sundarbans gained popularity as a tourist destination for both domestic and international tourists (Amin 2018; Protect Planet 2018; UNESCO 2018). Sundarbans provide extensive protection of coastal communities from cyclones, tidal flooding, erosion, and other natural disasters (Payo et al. 2016). It is also a habitat of diverse floral and faunal species such as plants (334 species), mangroves (50 species), legumes (35 species), grasses (29 species), sedges (19 species), and euphorbias (18 species) (Chaffey et al. 1985). The Sundarbans hosts about 50 species of mammals, 320 species of birds, 53 species of reptiles, 11 species of amphibians, 177 species of fish, and 873 species of invertebrates (see also Islam & Bhuiyan 2018). The Sundarbans mangroves area supports the subcontinent's largest population of critically endangered species of Royal Bengal Tigers, which are well known for their reputation for being man-eaters and long-distance swimmers and for their extraordinary migrating behavior from island to island between India and Bangladesh (Mallick 2013). The ecosystem services of Sundarbans also created the backbones of coastal communities living and livelihoods due to its numerous contributions to agriculture, fishing, aquaculture, and ecotourism. Villagers of Sundarbans depend on the collection of natural resources, such as fish, shrimp, mollusks, crabs, wood, honey, building materials, and medicinal plants (Ekka & Pandit 2012; Rahman & Rahman 2013). Additionally, the Sundarbans ecosystem provides direct and indirect income and employment opportunities to 1.7 million people across the villages in Sundarbans (Inskip et al. 2013).

11.2.2 Data Collection and Analysis

This is a viewpoint article that captures the event on the novel coronavirus as they are unfolding now and examines the impact of COVID-19 on ecosystem services at Sundarbans. The viewpoint has valuable implications at the local as well as the regional/global level for enhancing livelihood and promoting tourism. The assessment draws from published academic research studies (e.g., scientific articles) and current emergent media sources (e.g., online media of newspapers, government, and non-government organizations) as the information from the academic literature is scarce. Thus, the online textual information from official online media has been used (Riff et al. 2014). Further several studies have noted that media information is an essential tool that can be used to obtain the problems and opinions of actors at the local level affecting the policy agenda and facilitating the discussion (Kleinschmit 2012; Rahman & Giessen 2014). For this study, academic research studies were collected from the database of Scopus, ScienceDirect, SpringerLink, PubMed, Taylor and Francis, Research Gate, and Google Scholar, however, not in a systematic manner. The data presented in the information are relevant to the objective of this study.

A content analysis is employed for this study to analyze the online textual information. Neumann (2003, p. 219) argued that "content analysis is a technique for gathering and analyzing the content of the text. The content refers to words, meanings, pictures, symbols, ideas, themes, or may message that can be communicated." Miles and Huberman (2002) stressed that there is no right and wrong standard way of conducting content analysis. Neumann (2003) further stated that both quantitative and qualitative methods could be applied in the content analysis; however, it depends on the purpose of the research study. In this study, the research findings are presented in the qualitative (descriptive) form. And this is believed to be appropriate, as the objective of this exploratory study is to provide some insights into an area in which there is a limited body of knowledge.

11.3 Results and Discussion

11.3.1 Environmental Impact of Ecosystem Services

The COVID-19 pandemic has brought a significant impact on the mangrove ecosystem of Sundarbans. Similar to the other parts of the world's ecosystem services during the COVID-19 lockdown, the Sundarbans mangrove ecosystem has also undergone socioenvironmental changes. The impact of the COVID-19 lockdown on air, water, and biodiversity of Sundarbans has been affected due to labor migration and tourism-related activities such as movements of motorboats, use of loudspeakers, and disposal of wastes. With restrictions on human interaction due to the pandemic, air quality has started improving, but other environmental aspects such as the quality of the rivers began showing adverse effects in the Sundarbans area (Chaudhuri & Bhattacharyya 2021). Because of the lockdown, there is a substantial reduction in the concentration of suspended particulate matter (PM), which is a major component of air pollution (Bera 2013). Mukherjee et al. (2020) pointed out that in April 2020, the CO_2 level in the air was recorded at a lesser concentration than the previous year. The river's water quality (e.g., River Hooghly and Ganges) became worse during the lockdown period of 2020 compared with the previous year's data of the same month (WBPCP 2020). This occurred due to the use of detergent, hand sanitizer, antibacterial chemicals, plastic masks, and gloves from hospital waste that are directly disposed of in the municipal waste and sewage system (Chaudhuri & Bhattacharyya 2021). Hence, different types of solid waste contaminated Sundarbans' mangrove habitats during the lockdown period (WBPCP 2020). Further, the COVID-19 restriction has also shown a significant improvement in water quality in the river of Ganges, following the eight weeks of nationwide lockdown. The pandemic lockdown has exposed the biodiversity of the aquatic systems positively due to the minimal input of wastes from several anthropogenic sources of industrial and domestic activities (Pal et al. 2020; Chaudhuri & Bhattacharyya 2021). This impact can be translated as many industrial units and commercial organizations were closed due to the pandemic, and water was not used by them, with a negligible discharge of industrial wastewater into the river. A similar trend has also been observed in the Sundarbans mangrove forests area.

11.3.2 Social Impact of Ecosystem Services

For many years, people have been living in proximity to mangrove ecosystem areas and many of these ecosystem services are the main sources of their income. It is estimated that there are 100 million people living within 10 km of significant mangrove ecosystem areas and many of them heavily rely on ecosystem services for their daily sustenance and well-being (Duke et al. 2014). The ecosystem of Sundarbans is rich in biodiversity and ecosystem services. It is the source of livelihood for local people and contributes directly to cultural, life support functions, and well-being (Uddin et al. 2013a, 2013b; Shameem et al. 2014). The life support functions for local people largely depend on Sundarbans ecosystem services such as fishing, honey, wax collection, and fuelwood/timber (Ekka & Pandit 2012). The majority of these people are poor and illiterate, and, therefore, their dependence on the Sundarbans ecosystem is greater (Hussain & Badola 2010; Iqbal 2020). Further, in the Sundarbans, 4.4 million people are classified as marginal workers (have guaranteed work for less than six months in a year), such as agricultural workers, cultivators, and household and daily wage workers (Ghosh et al. 2016), while the people in non-agricultural occupations in the region are involved in construction, driving, woodworking, and casual multipurpose labor. However, the impact of COVID-19 has significantly disrupted the lives and livelihoods of the vast majority of the Sundarbans people, especially economically weaker people (households mostly dependent on daily wages from the informal economy) and otherwise vulnerable groups (International Labour Organization 2020).

The COVID-19 pandemic created a substantial negative impact on the livelihoods of ecosystem-dependent people in the Sundarbans area. The national lockdown distressed the situation of these people by hindering their access to the natural resources and income generative activities from Sundarbans ecosystem services. Thus, the sharp rise in the economic vulnerability of these communities became prominent, in turn, increasing pressure on ecosystem products. In many cases, these vulnerable people suffer from food insecurity and income shortage. As a result, the Sundarbans forest is overexploited and its natural resources are degraded by the people due to the absence of economic opportunities (Irfanullah 2020; Ruszczyk et al. 2021). During the lockdown in Bangladesh, 222 hector forest were deforested in the first ten months of 2020 than in 2019 (8% more than 2019); however, deforestation in Sundarbans was relatively low (Rahman et al. 2021).

The COVID-19 outbreak has created a negative impact on Sundarbans ecosystem services. For example, the prolonged lockdown has increased threats and created substantial pressure on forests and biodiversity as a large number of people live beside the Sundarbans ecosystem areas and earn their livelihoods from the forests (Uddin Mahtab & Karim 1992; Salam & Noguchi 1998; WCS 2018). About 40% of rural poor people live near the forest area. As a result, Sundarbans forests have served as a safety cushion for these rural poor people and the forest-dependent communities (FAO 2020). With the imposed social and economic toll, these vulnerable communities near the forest have negatively impacted their livelihoods (Bhuiyan et al. 2021; Shammi et al. 2021), thus poverty and food insecurity have increased.

Like other South-Asian countries (Pakistan, Nepal, and Sri Lanka), illegal poaching, trafficking, trading of wild animals, and illegal logging have been observed predominantly during the lockdown of the COVID-19 pandemic (Badola 2020; Islam et al. 2020; Rodrigo 2020; Farand 2020). It is also evident that lockdown has provided new opportunities for the illegal poachers due to the restriction on movement and activities of law enforcement employees and agencies. The rate of wildlife poaching (28 per month) during the pandemic has been seven times larger than the five-year rate (4.15 per month) estimated by the Wildlife Conservation Society (2018).

A larger number of households are also dependent on aquaculture and fishing as a primary source of income in the Sundarbans region (Chand et al. 2012; Bhattacharya et al. 2018). The Sundarbans region is home to a significant number of crab species, white fish, small fish, shrimp, and Hilsa (*Tenualosa ilisha*). In the fiscal year, 1999–2000 to 2012–2013, the estimated average annual revenue for timbers and fisheries from Sundarbans was around US$ 744,000. About US$ 0.20 million of revenue was collected per year from fisheries over the same period (Uddin et al. 2013a), while the amount of crab harvested was about 444 tons per year during the period from 2001–2002 to 2009–2010 with average annual revenue of US$ 0.02 million. However, the majority of the fish farmers belong to low-income groups. As a result of the lockdown and the government's exemption of fishing, the demand and supply chain were hampered. Fishers, fish laborers, and other actors of the fisheries value chain encountered many problems due to the COVID-19 pandemic. Many fishers' livelihoods have relied entirely on fishing, which made them more vulnerable to the COVID-19 pandemic. A large number of fishing laborers from the Sundarbans area who are engaged in fish processing, harvesting, and marketing became unemployed, which created social uncertainty (Islam et al. 2017; Sunny 2017; Sunny et al. 2021). In addition, lack of access to the markets, labor shortages, and various logistical issues made these poor farmers more vulnerable to the risk associated with aquaculture and fishing losses and subsequently affected the trade and food security (Calvet et al. 2016).

11.3.3 Economic Impact of Ecosystem Services

The local and national economy and forest-dependent livelihoods are largely influenced by Sundarbans ecosystem services (Uddin et al. 2013a; Shameem et al. 2014). Sundarbans ecosystem services are recognized as a place of cultural importance (e.g., tourism, worship, educational research) by the local as well as national and international tourists. The Sundarbans ecosystem services (particularly in tourism) in Bangladesh account for 4.4% of the country's Gross Domestic Product (GDP) (Verma, Haque & Nishat 2018). The economic value of the cultural services of Sundarbans, regarding revenue collected from tourists, was estimated to be US$ 42,000 per year from 2001 to

2002 and 2009 to 2010 (Uddin et al. 2013a). Likewise, the total annual domestic tourist expenditure from the Bangladesh Sundarbans region recorded in 2015 was US$ 5 million (Khanom & Buckley 2015), while Indian Sundarbans region generated ecosystem services with an estimated value of US$ 0.6 million annually in 2014 (Verma et al. 2015) and US$ 704 million per year in 2012–2013 (Kavi Kumar et al. 2016). The impacts of the COVID-19 outbreak on the Sundarbans ecosystem services are significant and largely unexpected. To curb the spread of the COVID-19 disease, the nationwide lockdown has forced people to confine their lives and their livelihoods in the Sundarbans area. Sundarbans ecosystem services are significant in Bangladesh and India. More than 3.5 million people depend on the Sundarbans ecosystem services for their livelihood and income (Uddin et al. 2013a). Forest-based tourism, including Sundarbans mangrove forest, had been affected severely during the COVID-19 pandemic. According to the Bangladesh Tourism Board (BTB), there is an incurred loss of US$ 177 million in the first three months of the pandemic in the tourism industry in Bangladesh, including tourism and recreational services from Sundarbans (Hasan 2020).

Due to the COVID-19 lockdown, about 80% average household's livelihood has been affected by the lack of income generated from ecosystem services (Szabo et al. 2015). The ongoing lockdown also increases the poverty rate among these households to 40%, where 43.5% of the households' incomes are below the international poverty level (Godio 2020). Most families faced hunger and forced themselves toward alternative livelihoods, for example, forest-based products and seasonal honey collection from the Sundarbans mangrove forest. The local people earn income by selling it at local markets. Furthermore, people in Sundarbans live in vulnerable conditions, for example, cyclone-prone, monsoonal, and low-lying areas settled alongside the waterways and coastline (Chaudhuri & Bhattacharyya 2021). Many people from the Sundarbans region are working as migrant laborers outside of their division or states because of physically vulnerable circumstances. The people in this area are also unskilled laborers and migrate to the nearby city and other affluent places to earn money. It is estimated that about 30% of the families living in the Sundarbans region have at least one member working outside the division or the state, which means 250,000–300,000 migrant laborers had to return to their homes due to the lockdown (Basu 2020). As a result, many of them fall back to the forest for their livelihood collection. This has led to tiger attacks, and a significant number of people were killed for entering the forest illegally (Pramanik et al. 2021). However, affected families (e.g., tiger widows) were reluctant to inform the authorities to get any government compensation due to the engagement in illegal livelihood collection.

11.3.4 Government's Supports and Stimulus Packages

The Government of India has unveiled several plans for Sundarbans to help mitigate the unprecedented impacts of the pandemic. In addition, the

Government of India and the West Bengal state government have announced US$ 130 million and US$ 837,000 in aid to protect and rebuild the Sundarbans. The World Wildlife Fund (WWF) has joined with the West Bengal Forest Directorate to distribute food supplies around Sundarbans to tackle the aftermath of the COVID-19 pandemic (Borgen Magazine 2021). Likewise, the Bangladesh government has also unveiled a stimulus package of US$ 3.2 billion to assist the low-income group of people who are severely affected by the imposed lockdown (Dhaka Tribune 2021).

11.4 Conclusion

The assessment provides a brief overview of Sundarbans ecosystem services that are hard hit by the COVID-19 pandemic. It is argued that the pandemic has intensely and adversely impacted the Sundarbans ecosystem services. Despite the air quality improvement, which was due to the restriction of transportation and industrial activities, the analysis revealed that the pandemic caused an unprecedented situation to the socioenvironmental, socioeconomic, and tourism in general. The findings indicated that overexploitation, deforestation, and degradation of natural resources have increased because of lower surveillance, highlighting those inadequacies of the government's responses and preparedness level to cope with pandemic-induced risk and challenges (Phillips et al. 2020; Schwartz et al. 2020). The impact of the COVID-19 pandemic and its associated lockdown has made the economies more vulnerable and is likely to derail the United Nations Sustainable Development Goals (SDGs) pathways. For returning to normality from the pandemic outbreak and making the society resilient to similar kinds of shock, it urgently needs to develop a short- and long-term recovery strategy for Sundarbans ecosystems.

11.4.1 Post-COVID-19 Pandemic Recovery Strategies

Several post-COVID-19 pandemic recovery strategies have been proposed from the assessment. In the first recovery strategy, the government should embrace a sustainable mangrove ecosystem service that could provide many jobs and revenue annually (IUCN 2017) and help to achieve various sustainable goals (Aryal et al. 2020). Sundarbans' rich biodiversity of mangroves is highly attractive and can be used for recreational purposes and is a prime destination for ecotourism. So, emphasizing ecotourism could be the second recovery strategy for biodiversity conservation and rebuilding the national economy after the pandemic (Laudari et al. 2021). In addition, the promotion of domestic tourism would also be the best option to bounce back from the loss of economies tailored to the recreational needs of diverse people (Derks et al. 2020). The third recovery strategy is that the government needs to establish a post-COVID-19 recovery program that is focused on the livelihoods of mangrove-forest-dependent rural and indigenous communities and people involved in the ecotourism industry (Rahman et al. 2021).

The fourth recovery strategy is that the government should adopt a community-based approach for mangrove forests and biodiversity conservation (Laudari et al. 2021). The implementation of this strategy will not only help to tackle the problems of illegal logging and hunting (Oldekop et al. 2019) but will also make the conservation effort effective and efficient in the long term (Bajracharya et al. 2006). Further, community engagement in forest and biodiversity would also provide employment opportunities to the local vulnerable people even in the difficult time of pandemic that hinders permanent and temporary employment (Laudari et al. 2021). However, this strategy could be a challenge as the voluntary contribution of local people to mangrove forests and biodiversity conservation is becoming an obsolete business (Laudari et al. 2021). Therefore, the involvement of civil society organizations, private sectors, and other stakeholders is equally important. In addition, the forest department should use innovative ideas in using social media and modern technology to halt deforestation and wildlife poaching, and it should be a top priority in their policy agenda.

The fifth recovery strategy that can be undertaken by both the governments (India and Bangladesh) is to create a feasible joint economic opportunity for sustainable and integrated community-based tourism. For example, a collaborative effort of community-based river cruising services between the two countries can be developed to popularize the existing tourism routes and locations and identify the new routes for tourism for capturing the ecological, cultural, and historical uniqueness of Sundarbans.

References

Amin, M.R., 2018. Sustainable tourism development in Sundarbans, Bangladesh (a world heritage site): Issues and actions. *Journal of Business*, *39*(2), pp.31–52.

Arora, S., Bhaukhandi, K.D. and Mishra, P.K., 2020. Coronavirus lockdown helped the environment to bounce back. *Science of the Total Environment*, *742*, p. 140573.

Aryal, K., Laudari, H.K. and Ojha, H.R., 2020. To what extent is Nepal's community forestry contributing to the sustainable development goals? An institutional interaction perspective. *International Journal of Sustainable Development & World Ecology*, *27*(1), pp. 28–39.

Austin, D. E., Baro, M., Batterbury, S., Bouard, S., Carrasco, A., Gezon, L. L., ... and Walsh, C. (2020). *Terrestrial Transformations: A Political Ecology Approach to Society and Nature*. Lexington Books. London: UK

Badola, S., 2020. Indian wildlife amidst the COVID-19 crisis: an analysis of the status of poaching and illegal wildlife trade. Traffic, India office, New Delhi.

Bajracharya, S.B., Furley, P.A. and Newton, A.C., 2006. Impacts of community-based conservation on local communities in the Annapurna Conservation Area, Nepal. *Biodiversity & Conservation*, *15*(8), pp. 2765–2786.

Bao, R. and Zhang, A., 2020. Does lockdown reduce air pollution? Evidence from 44 cities in northern China. *Science of the Total Environment*, *731*, p. 139052.

Basu, J., 2020. People rush back to the Sundarbans. *The Third Pole*. https://www.thethirdpole.net/en/climate/people-rush-back-to-the-sundarbans-untested/

Bera, M.K., 2013. Environmental refugee: A study of involuntary migrants of Sundarban islands. In *Proceedings of the 7th International Conference on Asian and Pacific Coasts (APAC 2013)*, Bali, Indonesia, pp. 916–925.

Bhattacharya, M., Kar, A., Chini, D.S., Malick, R.C., Patra, B.C. and Das, B.K., 2018. Multi-cluster analysis of crabs and ichthyofaunal diversity in relation to habitat distribution at tropical mangrove ecosystem of the Indian Sundarbans. *Regional Studies in Marine Science, 24*, pp. 203–211.

Bhuiyan, A. K. M., Sakib, N., Pakpour, A. H., Griffiths, M. D. and Mamun, M. A. (2021). COVID-19-related suicides in Bangladesh due to lockdown and economic factors: case study evidence from media reports. *International journal of mental health and addiction*, *19*(6), 2110–2115.

Borgen Magazine, 2021. The Effects of COVID-19 and Cyclone Amphan in the Sundarbans. https://www.borgenmagazine.com/covid-19-and-cyclone-amphan/

Borrell, A., Tornero, V., Bhattacharjee, D. and Aguilar, A., 2016. Trace element accumulation and trophic relationships in aquatic organisms of the Sundarbans mangrove ecosystem (Bangladesh). *Science of the Total Environment*, *545*, pp. 414–423.

Calvet, G., Aguiar, R.S., Melo, A.S., Sampaio, S.A., De Filippis, I., Fabri, A., Araujo, E.S., de Sequeira, P.C., de Mendonça, M.C., de Oliveira, L. and Tschoeke, D.A., 2016. Detection and sequencing of Zika virus from amniotic fluid of fetuses with microcephaly in Brazil: A case study. *The Lancet Infectious Diseases*, *16*(6), pp. 653–660.

Chaffey, D. R., Miller, F. R. and Sandom, J. H., 1985. *A Forest Inventory of the Sundarbans, Bangladesh – Main Report*. Land Resources Development Centre. Tolworth Tower, Surbiton, Surrey, England.

Chakraborty, I. and Maity, P., 2020. COVID-19 outbreak: Migration, effects on society, global environment and prevention. *Science of the Total Environment*, *728*, p. 138882.

Chan, K.M.A., Goldstein, J., Satterfield, T., Hannahs, N., Kikiloi, K., Naidoo, R., Vadeboncoeur, N. and Woodside, U., 2011.Cultural services and non-use values. In *Natural Capital*. Oxford University Press, Oxford, UK, pp. 206–228.

Chan, K.M., Satterfield, T. and Goldstein, J., 2012. Rethinking ecosystem services to better address and navigate cultural values. *Ecological Economics*, *74*, pp. 8–18.

Chand, B.K., Trivedi, R.K. and Dubey, S.K., 2012. Climate change in Sundarban and adaptation strategy for resilient aquaculture. In *CIFRI Compendium on Sundarban, Retrospect and Prospects*. Central Inland Fisheries Research Institute, Kolkata, India, pp. 116–128.

Chaudhuri, P. and Bhattacharyya, S., 2021. Impact of COVID-19 lockdown on the socioenvironmental scenario of Indian Sundarban. In Ramanathan A.L., Sabarathinam, C., and Jonathan, M.P. (Eds.), *Environmental Resilience and Transformation in Times of COVID-19*. Elsevier, Amsterdam, pp. 25–36.

CICES, 2013. Towards a Common Classification of Ecosystem Services. http://cices.eu/

Derks, J., Giessen, L. and Winkel, G., 2020. COVID-19-induced visitor boom reveals the importance of forests as critical infrastructure. *Forest Policy and Economics*, *118*, p. 102253.

Dhaka Tribune, 2021. Govt announces 5 more stimulus packages worth Tk3,200cr for poor. https://www.dhakatribune.com/bangladesh/government-affairs/2021/07/13/pm-hasina-announces-five-more-stimulus-packages-worth-tk3-200cr-for-poor

Duke, N., Nagelkerken, I., Agardy, T., Wells, S. and Van Lavieren, H., 2014. *The importance of mangroves to people: A call to action*. United Nations Environment Programme World Conservation Monitoring Centre (UNEP-WCMC).

Ekka, A. and Pandit, A., 2012. Willingness to pay for restoration of natural ecosystem: A study of Sundarban mangroves by contingent valuation approach. *Indian Journal of Agricultural Economics*, *67*(3), pp. 1–11.

FAO, 2020. Lessons learned from COVID-19 crisis to the better management of forest and water resources. *FAO Rome*. http://www.fao.org/in-action/forest-and-waterprogramme/news/news-detail/en/c/1275837/

Farand, C., 2020. Forest destruction spiked in Indonesia during coronavirus lockdown. *Climate Home News*. https://www.climatechangenews.com/2020/08/18/forestdestruction-spiked-indonesia-coronavirus-lockdown/

Fish, R., Church, A. and Winter, M., 2016. Conceptualising cultural ecosystem services: A novel framework for research and critical engagement. *Ecosystem Services*, *21*, pp. 208–217.

Ghermandi, A. and Nunes, P.A., 2013. A global map of coastal recreation values: Results from a spatially explicit meta-analysis. *Ecological Economics*, *86*, pp. 1–15.

Ghosal, A. and Casey, M., 2020. Coronavirus lockdowns increase poaching in Asia, Africa. *ABC News*. https://abcnews.go.com/Technology/wireStory/coronavirus-lockdowns-increase-poaching-asia-africa-71377281

Ghosh, U., Bose, S., Bramhachari, R. and Mandal, S., 2016. Expressing collective voices on children's health: Photovoice exploration with mothers of young children from the Indian Sundarbans. *BMC Health Services Research*, *16*(7), pp. 119–130.

Giri, K. (2020). *Initial Assessment of the Impact of COVID-19 on Sustainable Forest Management Asia-Pacific States*. United Nations Forum on Forests Secretariat. Available at: https://www.un.org/esa/forests/wp-content/uploads/2021/01/Covid-19-SFM-impact-AsiaPacific.pdf

Godio, M.J., 2020. Bangladesh case study: Sundarbans mangrove forest communities further marginalized by COVID-19 measures as super cyclone devastates livelihoods. https://www.forestpeoples.org/en/covid19-impacts-case-study-bangladesh

Grunewald, K. and Bastian, O., 2015. Ecosystem services. More than just a vogue term? In Grunewald, K. and Bastian, O. (eds.), *Ecosystem Services – Concept, Methods and Case Studies*. Springer Nature, Berlin/Heidelberg, Germany, pp. 1–11.

Hamzelou, J., 2020. World in lockdown. *New Scientist*, 245(3275), p. 7.

Hasan, R., 2020. Tourism Sector: Operators incur loss of Tk 1,500cr. *The Daily Star*. https://www.thedailystar.net/frontpage/news/tourism-sector-operators-incur-loss-tk-1500cr-1890742

Hossain, M.S., Dearing, J.A., Rahman, M.M. and Salehin, M., 2016. Recent changes in ecosystem services and human well-being in the Bangladesh coastal zone. *Regional Environmental Change*, *16*(2), pp. 429–443.

Hussain, S.A. and Badola, R., 2010. Valuing mangrove benefits: Contribution of mangrove forests to local livelihoods in Bhitarkanika Conservation Area, East Coast of India. *Wetlands Ecology and Management*, *18*(3), pp. 321–331.

Inskip, C., Ridout, M., Fahad, Z., Tully, R., Barlow, A., Barlow, C.G., Islam, M.A., Roberts, T. and MacMillan, D., 2013. Human–tiger conflict in context: Risks to lives and livelihoods in the Bangladesh Sundarbans. *Human Ecology*, *41*(2), pp. 169–186.

International Labour Organization, 2020. Impact of lockdown measures on the informat economy: A summary. https://www.ilo.org/global/topics/employment-promotion/informal-economy/publications/WCMS_743534/lang--en/index.htm

Iqbal, M.H., 2020. Valuing ecosystem services of Sundarbans Mangrove forest: Approach of choice experiment. *Global Ecology and Conservation*, *24*, p. e01273.

Irfanullah, H.H., 2020. Will nature conservation remain a priority in post-corona Bangladesh? *The Daily Star*. https://www.thedailystar.net/opinion/news/will-nature-conservation-remain-priority-post-corona-bangladesh-1904146

Islam, S.D.U. and Bhuiyan, M.A.H., 2018. Sundarbans mangrove forest of Bangladesh: Causes of degradation and sustainable management options. *Environmental Sustainability*, *1*(2), pp. 113–131.

Islam, M.M., Shamsuzzaman, M.M., Sunny, A.R. and Islam, N., 2017. Understanding fishery conflicts in the Hilsa Sanctuaries of Bangladesh. In Song, A.M., Bower, S.D., Onyango, P., Cooke, S.J. and Chuenpagdee, R. (eds.), *Inter-Sectoral Governance of Inland Fisheries*. Blackwell Science, London: UK. pp. 18–31.

Islam, M.M., Sharmin, M. and Ahmed, F., 2020. Predicting air quality of Dhaka and Sylhet divisions in Bangladesh: A time series modeling approach. *Air Quality, Atmosphere & Health*, *13*(5), pp. 607–615.

IUCN, 2017. Can restoring mangroves help achieve the Sustainable Development Goals? https://www.iucn.org/news/forests/201703/can-restoring-mangroves-help-achieve-sustainable-development-goals

Karlinsky, A. and Kobak, D., 2021. Tracking excess mortality across countries during the COVID-19 pandemic with the World Mortality Dataset. *Elife*, *10*, p. e69336.

Kavi Kumar, K.S., Anneboina, L.R., Bhatta, R.C., Naren, P., Nath, M., Sharan, A., Mukhopadhyay, P., Ghosh, S. and Pednekar, V.D.C.A. S., 2016. *Valuation of Coastal and Marine Ecosystem Services in India. Macro Assessment*. Madras School of Economics, Chennai.

Khanom, S. and Buckley, R., 2015. Tiger tourism in the Bangladesh Sundarbans. *Annals of Tourism Research*, *55*(C), pp. 178–180.

Kleinschmit, D., 2012. Confronting the demands of a deliberative public sphere with media constraints. *Forest Policy and Economics*, *16*, pp. 71–80.

Laudari, H.K., Pariyar, S. and Maraseni, T., 2021. COVID-19 lockdown and the forestry sector: Insight from Gandaki province of Nepal. *Forest Policy and Economics*, *131*, p. 102556.

Mallick, J.K., 2013. Ecology, status and aberrant behavior of Bengal Tiger in the Indian Sundarban. *Animal Diversity, Natural History and Conservation*, *2*, pp. 381–454.

Margaryan, L., Prince, S., Ioannides, D. and Röslmaier, M., 2018. Dancing with cranes: A humanist perspective on cultural ecosystem services of wetlands. *Tourism Geographies 20*(5), pp. 1–22.

Mieczkowski, Z., 1985. The tourism climatic index: A method of evaluating world climates for tourism. *Canadian Geographer/Le Géographe Canadien*, *29*(3), pp. 220–233.

Miles, M. and Huberman, A., 2002. *The Qualitative Researcher's Companion*. Sage, Thousand Oaks, California: USA.

Millennium Ecosystem Assessment (MEA), 2005. *Ecosystems and Human Well-Being: Current State and Trends*. The Millennium Ecosystem Assessment Series XXI, Island Press, Washington. https://www.millenniumassessment.org/documents/document.766.aspx.pdf

Millennium Ecosystem Assessment (MEA). 2018. *Ecosystems and Human Well-being Synthesis*. Report of the Millennium Ecosystem Assessment, Island Press, Washington, DC, USA. https://www.millenniumassessment.org/documents/document.356.aspx.pdf

Mukherjee, P., Zaman, S. and Mitra, A., 2020. Covid-19 induced lockdown caused a reduction in atmospheric carbon dioxide level in the mangrove ecosystem of Indian Sundarbans: A spatio-temporal picture. In Mitra, A., Monruskin, M.C. and Chakrabarty, S.P. (eds.), *Natural Resources and Their Ecosystem Services – Webinar Proceeding on 'Ecosystem Services and United Nations Sustainable Development Goals' Celebrating the World Environment Day*, pp. 153–160.

Mukhtar, I. and Hannan, A., 2012. Constrains on mangrove forests and conservation projects in Pakistan. *Journal of Coastal Conservation*, *16*(1), pp. 51–62.

Neumann, W., 2003. *Social Research Methods: Qualitative and Quantitative Approaches*. Allyn and Bacon, New York: USA.

Oldekop, J.A., Sims, K.R., Karna, B.K., Whittingham, M.J. and Agrawal, A., 2019. Reductions in deforestation and poverty from decentralized forest management in Nepal. *Nature Sustainability*, *2*(5), pp. 421–428.

Pal, N., Barman, P., Das, S., Zaman, S. and Mitra, A., 2020. Status of brackish water phytoplankton during COVID-19 lockdown phase. *NUJS Journal of Regulatory Studies Special Issue 1*(1), pp. 83–86.

Paul, A.K., Ray, R., Kamila, A., Jana, S. (2017). Mangrove Degradation in the Sundarbans. In Finkl, C., Makowski, C. (eds.), *Coastal Wetlands: Alteration and Remediation. Coastal Research Library*, vol 21. Springer, Cham. Switzerland. https://doi.org/10.1007/978-3-319-56179-0_11

Payo, A., Mukhopadhyay, A., Hazra, S., Ghosh, T., Ghosh, S., Brown, S., Nicholls, R.J., Bricheno, L., Wolf, J., Kay, S. and Lázár, A.N., 2016. Projected changes in area of the Sundarban mangrove forest in Bangladesh due to SLR by 2100. *Climatic Change*, *139*(2), pp. 279–291.

Phillips, C.A., Caldas, A., Cleetus, R., Dahl, K.A., Declet-Barreto, J., Licker, R., Merner, L.D., Ortiz-Partida, J.P., Phelan, A.L., Spanger-Siegfried, E. and Talati, S., 2020. Compound climate risks in the COVID-19 pandemic. *Nature Climate Change*, *10*(7), pp. 586–588.

Pramanik, M., Szabo, S., Pal, I., Udmale, P., O'Connor, J., Sanyal, M., Roy, S. and Sebesvari, Z., 2021. Twin disasters: Tracking COVID-19 and cyclone Amphan's impacts on SDGs in the Indian Sundarbans. *Environment: Science and Policy for Sustainable Development*, *63*(4), pp. 20–30.

Protect Planet, 2018. Sundarbans West Wildlife Sanctuary. https://www.protect-planet.net/sundarbans-west-wildlife-sanctuary

Pueyo-Ros, J., 2018. The role of tourism in the ecosystem services framework. *Land*, *7*(3), p. 111.

Rahman, M.S. and Giessen, L., 2014. Mapping international forest-related issues and main actors' positions in Bangladesh. *International Forestry Review*, *16*(6), pp. 586–601.

Rahman, M.A. and Rahman, M.A., 2013. Effectiveness of coastal bio-shield for reduction of the energy of storm surges and cyclones. *Procedia Engineering*, *56*, pp. 676–685.

Rahman, N., Billah, M.M. and Chaudhury, M.U., 1979. Prepoaration of an up to date map of Sundarban forests and estimation of forest areas of the same by using Landsat imageries. In *Second Bangladesh National Seminar on Remote Sensing*, Dhaka.

Rahman, M.S., Alam, M.A., Salekin, S., Belal, M.A.H. and Rahman, M.S., 2021. The COVID-19 pandemic: A threat to forest and wildlife conservation in Bangladesh?. *Trees, Forests and People*, *5*, p. 100119.

Riff, D., Lacy, S. and Fico, F., 2014. *Analyzing Media Messages: Using Quantitative Content Analysis in Research*. Routledge, London, UK.

Rodrigo, M., 2020. In Sri Lanka, bushmeat poachers haven't let up during lockdown. Mongabay, Menlo Park, CA, US. https://news.mongabay.com/2020/05/in-srilanka-bushmeat-poachers-havent-let-up-during-lockdown/

Rondeau, D., Perry, B. and Grimard, F., 2020. The consequences of COVID-19 and other disasters for wildlife and biodiversity. *Environmental and Resource Economics*, *76*(4), pp. 945–961.

Ruszczyk, H.A., RaHMan, M.F., Bracken, L.J. and Sudha, S., 2021. Contextualizing the COVID-19 pandemic's impact on food security in two small cities in Bangladesh. *Environment and Urbanization*, *33*(1), pp. 239–254.

Salam, M.A. and Noguchi, T., 1998. Factors influencing the loss of forest cover in Bangladesh: An analysis from socioeconomic and demographic perspectives. *Journal of Forest Research*, *3*(3), pp. 145–150.

Sandilyan, S. and Kathiresan, K., 2014. Decline of mangroves – A threat of heavy metal poisoning in Asia. *Ocean & Coastal Management*, *102*, pp. 161–168.

Sannigrahi, S., Chakraborti, S., Joshi, P.K., Keesstra, S., Sen, S., Paul, S.K., Kreuter, U., Sutton, P.C., Jha, S. and Dang, K.B., 2019. Ecosystem service value assessment of a natural reserve region for strengthening protection and conservation. *Journal of Environmental Management*, *244*, pp. 208–227.

Sanyal, P., 1983. Mangrove tiger land, the Sundarbans of India. *Tigerpaper*, *10*(3), pp. 1–4.

Schwartz, M.W., Glikman, J.A. and Cook, C.N., 2020. The COVID-19 pandemic: A learnable moment for conservation. *Conservation Science and Practice*, *2*(8), pp. 1–8.

Shameem, M.I.M., Momtaz, S. and Rauscher, R., 2014. Vulnerability of rural livelihoods to multiple stressors: A case study from the southwest coastal region of Bangladesh. *Ocean & Coastal Management*, *102*, pp. 79–87.

Shammi, M., Bodrud-Doza, M., Islam, A.R.M.T. and Rahman, M.M., 2021. Strategic assessment of COVID-19 pandemic in Bangladesh: Comparative lockdown scenario analysis, public perception, and management for sustainability. *Environment, Development and Sustainability*, *23*(4), pp. 6148–6191.

Spring, J., 2020. Illegal loggers uncowed by coronavirus As deforestation rises in Brazil. Reuters. https://www.reuters.com/article/us-brazil-environment-idUSKCN21S1I1

Sunny, A.R., 2017. Impact of oil spill in the Bangladesh Sundarbans. *International Journal of Fisheries & Aquatic Studies*, *5*(5), pp. 365–368.

Sunny, A.R., Sazzad, S.A., Prodhan, S.H., Ashrafuzzaman, M., Datta, G.C., Sarker, A.K., Rahman, M. and Mithun, M.H., 2021. Assessing impacts of COVID-19 on aquatic food system and small-scale fisheries in Bangladesh. *Marine Policy*, *126*, p. 104422.

Szabo, S., Renaud, F.G., Hossain, M.S., Sebesvári, Z., Matthews, Z., Foufoula-Georgiou, E. and Nicholls, R.J., 2015. Sustainable development goals offer new opportunities for tropical delta regions. *Environment: Science and Policy for Sustainable Development*, *57*(4), pp. 16–23.

Tröeng, S., Barbier, E., Rodríguez, C.M., 2020. The COVID-19 pandemic is not a break for nature-let's make sure there is one after the crisis. *World Economic Forum*. https://www.weforum.org/agenda/2020/05/covid-19-coronavirus-pandemic-nature-environment-green-stimulus-biodiversity/

Uddin Mahtab, F. and Karim, Z., 1992. Population and agricultural land use: Towards a sustainable food production system in Bangladesh. *Ambio 12*(1), pp. 50–55.

Uddin, M.S., van Steveninck, E.D.R., Stuip, M. and Shah, M.A.R., 2013a. Economic valuation of provisioning and cultural services of a protected mangrove ecosystem: A case study on Sundarbans Reserve Forest, Bangladesh. *Ecosystem Services*, *5*, pp. 88–93.

Uddin, M.S., Shah, M.A.R., Khanom, S. and Nesha, M.K., 2013b. Climate change impacts on the Sundarbans mangrove ecosystem services and dependent livelihoods in Bangladesh. *Asian Journal of Conservation Biology*, *2*(2), pp. 152–156.

UNESCO, 2018. The Sundarbans. http://whc.unesco.org/en/list/798

Urry, J., 2002. *The Tourist Gaze*. Sage Publications, London, UK.
Verma, M., Negandhi, D., Khanna, C., Edgaonkar, A., David, A., Kadekodi, G., Costanza, R. and Singh, R., 2015. *Economic Valuation of Tiger Reserves in India: A Value+ Approach*. Indian Institute of Forest Management, Bhopal, 284.
Verma, M., Haque, A.K.E. and Nishat, B., 2018. Benefits of Cooperation: Focus on the Sundarban. Identification and Assessment. https://documents1.worldbank.org/curated/zh/384881587110084376/pdf/Benefits-of-Cooperation-Focus-on-the-Sundarban-Identification-and-Assessment-Lead.pdf
Wang, Q. and Su, M., 2020. A preliminary assessment of the impact of COVID-19 on environment – A case study of China. *Science of the Total Eenvironment*, *728*, p. 138915.
WBPCP, 2020. Water Quality Information System. http://emis.wbpcb.gov.in/waterquality/viewsampledatacitizen.do
WCS, 2018. Combating wildlife trade in bangladesh: Current understanding and next steps. Published by the Wildlife Conservation Society Bangladesh Program, Dhaka, Bangladesh, p. 50.
WHO, 2021. Coronavirus disease (COVID-19). https://www.who.int/emergencies/diseases/novel-coronavirus-2019?gclid=Cj0KCQjwg7KJBhDyARIsAHrAXaESCHzhj8Fke9B3sfjdXj3pFk30N1_CWL4BMu-sj-LhQFslaHsesUYaAqO6EALw_wcB
Willson, G.B., 2016. Conceptualizing spiritual tourism: Cultural considerations and a comparison with religious tourism. *Tourism Culture & Communication*, *16*(3), pp. 161–168.
Zambrano-Monserrate, M.A., Ruano, M.A. and Sanchez-Alcalde, L., 2020. Indirect effects of COVID-19 on the environment. *Science of the Total Environment*, *728*, p. 138813.
Zhang, J., Hayashi, Y. and Frank, L.D., 2021. COVID-19 and transport: Findings from a world-wide expert survey. *Transport Policy*, *103*, pp. 68–85.

12 Impacts of COVID-19 Pandemic on Tourism and Public Transport

Challenges and Prospects from Malaysia

Au Yong Hui Nee, Yip Chee Yin, and Abdelhak Senadjki

12.1 Introduction

Initially, COVID-19 was reported in China's Wuhan City in 2019, and since then, the disease has spread worldwide, and Malaysia was no exception. At the point of this writing, COVID-19 is still surging in many countries. COVID-19 pandemic affects almost every sector in most of the countries involved. Among these sectors, by far, transportation and tourism are the worst affected as these two sectors are intertwined. The aftermath of the decline of the transportation sector, especially public transport and tourism is fast spreading to other sectors and thereby causing recession. Advice by the authorities of the various countries on the use of public transport during the COVID-19 pandemic is different from one to another. This has resulted in different rates of recovery from the pandemic for different countries. Because of government measures to combat the spread of the COVID-19 pandemic through transportation, public preference has since then, shifted from using mass transit to private transport like cars and motorcycles. This change is mainly for minimizing infection risk exposure, notably keeping safe physical distance from one another, though there are many other related factors as well. Since transportation and tourism are interrelated and that they impact the livelihoods of many people adversely and thereby leading to economic recession and widespread unemployment, this chapter aims to provide more insightful information in the Malaysian context, for policy-makers to take appropriate measures to address more effectively the challenges brought about by the pandemic.

Air and land transportation are closely related to the progress or decline of economic activities, especially in the tourism industry. Traveling during the pandemic is more restrictive with the decline in service supply and worsened by the passenger's perception of public transportation as being unsafe compared with their own vehicles. Nevertheless, there are measures to mitigate the risks of these factors. The lockdown or MCO (movement control ordinance) which instated immediate closure of learning institutions and houses of worship was progressively implemented to control the movement of people who were not employed in essential services sectors.

DOI: 10.4324/b23145-17

COVID-19 has significant impacts on all economic sectors, particularly on the sectors of tourism, hospitality, leisure, and arts (Nientied & Shutina, 2020). The closure of tourist amenities has had a negative impact on the host economy. Though, these procedures can be short-term (Sohn et al., 2021). The UNWTO (World Tourism Organization, a United Nations specialized agency), in 2020, estimated that 100–120 million direct tourism jobs were lost in 2020, severely impacting the livelihood of millions of people, and disrupting economies. No country has been spared; the severity of the impact is directly related to the contribution of tourism to the national GDP. Given the impact of the COVID-19 pandemic, people around the world are facing relatively close spatial distances in affected countries and regions (Li et al., 2020).

However, there is one unique factor which is ignored unwittingly, i.e., traveling is closely related to ecosystems. Even though tourist arrivals bring economic prosperity, it is putting harm to the ecosystem due to the negligence in a societal and environmental sense. Uncontrolled tourism activities can disturb the balance of ecosystems. If the tourism ecosystem is exploited without due care for sustainable development, it will threaten the biodiversity that will impair its ecotourism value. Thus, a sustainable balance between the development of tourism activities and mitigating harm to the ecosystem must be always maintained since strong and healthy ecosystems mitigate ecological scarcity. Hence, it is paramount to target not only net economic benefits but also maintaining a balanced ecological and social harmony with ecosystem regeneration. Therefore, empowering local communities will strengthen social capital which in turn will support economic stability. For the post-COVID-19 pandemic, the tourism industry must be further developed in a more sustainable way. This can be done in a more constructive manner by looking into the linkage between the various ecosystem services.

12.2 Literature Review

The outbreak of the COVID-19 pandemic brings great impacts to the people and businesses, especially the travel and tourism industries. The dominant topics that came out of the qualitative study by Kaushal and Srivastava (2021) are "Human Resource Management", "Health and Hygiene", "(Business) Continuity", and "Concerns (toward revival of the industry and media roles)". Tourism is among the first sectors to be knocked by the COVID-19 pandemic and could be the last that will fully recover, since travel and leisure will be overwhelmed by the people's need for food, education, and security in a time of economic and social crisis (Nientied & Shutina, 2020). Transport has an impact on sustainable development due to travelers who access a tourist destination and move around the tourism destination zone (Wieckowski, 2021). Due to the COVID-19 pandemic, there was a change in demand from public to private transport. People placed higher concerns when selecting a transport mode during the pandemic (Abdullah et al., 2020). A decline in public transport usage was reported in Zhang & Hayashi (2020), which gave quite a vivid account of how the tourism industry is affected as a result.

Almost no transit systems (Liu, Miller & Scheff, 2020) stopped their decline in ridership prior to the local community spread of COVID-19. Thondoo et al. (2020) reported the reduction in car and motorcycle trips has resulted in a decline in premature deaths. Chen and Pan (2020) related the methods and strategies to encourage the use of private car travel or special vehicles, e.g., Didi, during the pandemic in China.

The absence of social interaction, isolation, and loss of salary is potentially leading to psychological consequences (Bhatt et al., 2020). The lockdown shows an absolute way that may bring back the environment and ecosystem at a very speedy rate (Mandal & Pal, 2020). In other words, the long-term changes in society expect changes toward improved sustainability (Zhang & Hayashi, 2020). Amidst the pandemic, community contactless logistics service shows the role of the Internet in logistics in China (Chen & Pan, 2020). There will be more shifts toward virtual spaces for the long-term changes in people's lifestyles (Zhang & Hayashi, 2020).

From the perspectives of public health, public knowledge, risk perception, and positive communication are important to disseminate the information to the community. Researchers such as Azlan et al. (2020), Bhatt et al. (2020), Lovrić et al. (2020), and Hanafiah and Wan (2020) report high levels of COVID-19 public knowledge, risk perception, and positive communication behavior. However, there are lesser practices of face mask usage (Azlan et al., 2020), while most of the respondents have concerns over misinformation (Bhatt et al., 2020; Hanafiah & Wan, 2020; Lovrić et al., 2020).

12.2.1 Linkage between Ecosystem Services

Pollution will affect environmental processes which are crucial for the ecosystem. For instance, noise pollution is one of the causes of environmental deterioration and affects the health of invertebrates (Solan et al., 2016), people, and the ecosystem (Zambrano-Monserrate & Ruano, 2019). Furthermore, tourism activities, such as the mass cruise tourism industry, disturb marine ecosystems and increase air pollution (Renaud, 2020). Consequently, there is a higher demand for tourism in protected natural areas, inspired by spreading the function of conserving the natural ecosystems (Cretu et al., 2020). The main consequential effects of the COVID-19 crisis on tourism and transportation on public welfare are the rising unemployment, falling income, and rising poverty and inequality. The economic effects of the COVID-19 movement control have raised worries of a rise in poaching, illegal fishing, and deforestation, with many jobs in the ecotourism and associated sectors at risk (Cherkaoui et al., 2020).

While these phenomena are encouraging signs of healing of the environment, they also underscore the increasing inequality between the rich and the poor, those who have been robbed of their livelihood by the pandemic, while the rich could enjoy and appreciate the respite of a better living environment – clean air, greatly reduced noise pollution and a more harmonious natural environment. These are among the many beneficial factors for human well-being, delivered by natural ecosystems which therefore need to be sustainably managed within their ecological limits when human society interacts with nature to satisfy the demand for economic development (Lin, 2012). The scenario points to the need to take into consideration of not only the economic gain of providing employment and livelihood but also the wider ecosystem service loss due to the impact on the environment and welfare of the stakeholders. Early environmental protection efforts attempted to exclude humans with the aim of maintaining uninterrupted ecosystems, a social-ecological systems view is increasingly used in conservation (Mace, 2014). Nevertheless, Stankov, Filimonau, and Vujičić (2020) suggest that the COVID-19 pandemic is a shift in the tourism ecosystem and offers an opportunity for travelers to think on their travel behaviors.

Therefore, tourism development in the future should review environmental aspects so that it will be an act of obtaining an equilibrium between the conservation of ecosystems and economic realities (Cherkaoui et al., 2020). The COVID-19 pandemic forces people to see how tourism can promote human health and well-being of their destination, the health of ecosystems, and a greater traveler experience. Recovery from the COVID-19 pandemic cannot be reached merely by one sector alone; collaboration is basic if we desire sustainable tourism and healthy ecosystems as thriving businesses are linked to the well-being of local people (Spenceley et al., 2021). It is vital to safeguard the equilibrium of ecosystems, stop the loss of biodiversity and enhance the capacity of people to manage and adjust to climate change (Cherkaoui, et al., 2020). Such an approach is pertinent in tourism development in view that the tourism industry involves the natural environment and people as well as public transport, a critical cause as well as a primary effect of tourism development. In destinations with exclusively precious natural environments, public transport tends to be desired over private transport (Wieckowski, 2021). At the same time, public transport is a vital component of economic progress that is sustainable and inclusive (Stjernborg & Mattisson, 2016).

12.2.2 Impacts of COVID-19 Pandemic on Tourism and Public Transport

The socioeconomic impacts of the pandemic on public transport expand to financial viability, social equity, and sustainable mobility. If public transport is deemed as weakly transformed with post-pandemic measures, public transport will be perceived as unhygienic (Tirachini & Cats, 2020). The results of Liu, Miller, and Scheff (2020) show substantial deviations from weekday hourly demand profiles of transit systems. The pandemic has caused a significant decline in mobility, changing traffic patterns and enhancing traffic safety. The public has opted for cars and active modes rather than public transport modes during the pandemic. This change in trends also reflected positive impacts on air emission and water discharge. Consequently, public transport especially air transport and tourism sectors are the worst affected (Muley et al., 2020). It has also been mentioned that the effect of the pandemic on public transport has a great chain effect on the tourism industry. Cochran (2020) found that the pandemic is aggravating many difficulties in public health, especially the mental aspects of it, which is a cumulative effect of lockdown and the serious downturn of the tourism industry and transportation and difficulties in obtaining assistance and performing daily living activities which people with disabilities regularly encounter. In contrast, various issues in developing countries are, e.g., problematic political governance, balance between travel demand management and public transport operation, promoting and safeguarding public transport via taxation and infrastructure improvements, standards for personal protective equipment for transit passengers and drivers, and enlightenments about physical distancing measures (Zhang & Hayashi, 2020).

12.2.3 Resilience to Ecosystem during the COVID-19 Pandemic

The seemingly optimistic reports of clearer skies in, otherwise, usually smog enveloped cities and smooth drives on streets devoid of the normally congested traffic during the period of lockdown or movement restrictions imposed by governments are stark reminders of the degradation of ecosystem services which the public has been experiencing. The return of certain fauna in popular tourism areas where it has long disappeared and reports of improved biodiversity have been recorded during this crisis. In other words, environmental issues such as pollution and shifts of ecosystem and biodiversity are displaying positive signs in this pandemic due to fewer traveler burdens on the environment (Bremer, Schneider & Glavovic, 2019; Coutts et al., 2010). Furthermore, growing the share of journeys accomplished via sustainable transport joins together the three origins of “proximity, slow tourism and green transport” (Wieckowski, 2021).

12.2.4 Research Problem and the Gaps

The COVID-19 pandemic is a crisis affecting countries worldwide, sparing no nation of its negative impacts. The nature of transmission of this deadly disease and the high rate of infectivity are closely related to the mobility aspect of humans and goods. Tourism and transportation are involved with the movement of people. Both are well-interrelated industries and are important economic sectors worldwide. Tourism is among the fastest-growing economic sectors and accounted for almost 7% of global trade in 2019. Transportation is one of the fundamental drivers of economic development and an essential factor for the growth of tourism. Transportation is both the cause and effect of tourism (Lohmann & Duval, 2011); to open tourist spots, transportation must first be able to provide the means of mobility and accessibility for the visitors. On the other hand, when tourism picks up, rising demand will spur the development of transportation, particularly air transportation. Public transit is a relatively more significant component compared to private transportation. It is a social utility providing transport means to the public, enabling shared mass transportation and operating on scheduled trips along fixed routes according to public demands. It helps to reduce traffic congestion on the road and often offers a more convenient and comfortable ride for commuters. And for those who could not afford or do not possess private transports, public transit fills the need for the people's mobility whether to work or to other social activities.

Physical distancing, stay-at-home, and no large public or private gatherings are the recommended practices to stop the spread of the COVID-19 virus. Authorities imposed massive restrictions on society including temporary suspension and restricting of business activities and manufacturing industries as pandemic control measures. Stay-at-home has been the basic order. In addition, the fact that COVID-19 is highly contagious and potentially fatal has struck the fear of taking public transit in most people. A significant decline in ridership for public transport has been reported worldwide, from air flights to buses for the general public (Bird, Kriticos & Tsivanidis, 2020). The closing of national borders and the strict entry requirements imposed by many countries to check the global transmission of the virus severely affected the tourism industry and the related chain industries, resulting in the drastic reduction of the number of international as well as domestic air travelers, crippling the airline industry.

On the other hand, public transport users are often those who have no alternative transport option, basically the poor who could not afford private vehicles. Many public transport plying routes would have to be stopped or frequency of trips reduced due to insufficient ridership, thus making it even more difficult for the poor, underprivileged groups to fill their mobility requirements. The result would be a growing number of the urban poor who will be marginalized in their accessibility to economic development and welfare, leading to greater inequality in society (Rodrigue, 2020).

However, the priority is to revive the tourism industry once the country is ready to reopen its national boundaries. But the question is how to bring back the tourists, when people are apprehensive about boarding the transport vehicle, be it the airplane, bus, or train, for fear of being infected by the COVID-19 virus. Steps have to be adopted to rebuild the commuters' confidence in using public transport. This is vital so that private transport would not become the preferred choice and lead to greater congestion of traffic and cause gridlocks in the cities. Additionally, an efficient public transport system needs the support of sufficient ridership and is attributed to good outcomes of traveling on public transit, satisfying the need of mobility and freedom, and thus contributing to the well-being of the general public.

In this chapter, essential issues related to the use of public transport during the pandemic were analyzed and understandings of the measures implemented are offered. The issues highlighted are related to the pandemic or movement control period where widespread instruments to minimize the spread of the COVID-19 were implemented by various levels of governments. One of which is the prevention of traveling by any means of transport. It has become crucial to deliberate the transformations in travel behavior, the resultant effect on social equity, and the policies executed throughout and after the pandemic (Yang et al., 2021).

This research is chosen based on the following reasons: one, the transportation sector is one of the most severely hit industries by this pandemic mainly because it is through air transportation that COVID-19 can spread to every corner of the world while land transportation is basically responsible for the pandemic spreading throughout a country. Secondly, transportation is an indispensable tool for tourism which brings in substantial revenue for many countries notably Italy, China, the United States of America, and most of the island nations including Mauritius, Maldives, and Sri Lanka. An investigation of the COVID-19 impacts on public transportation will provide insight into effective approaches to revive and improve the transport system to be more efficient and "pandemic resilient". A well-developed transport system that provides economic, social, and environmental gains is vital for the sustainability and well-being of society which will help to alleviate the severe impact on tourism. Thirdly, by this study, we will be able to measure how the ecosystem can be damaged and repaired by adopting resilient policies. Thus, this study can contribute and promote policies to control climate change.

12.3 Methodology

12.3.1 Study Design and Procedure

A phenomenology research design that focuses on the similarity of a group's lived experience is used in this study. The phenomenology research design is used in this study to understand the impact of the COVID-19 pandemic on public transportation from the people who have first-hand knowledge of public transportation. By applying the phenomenology research design, we have arrived with a deeper knowledge of the impact of the COVID-19 pandemic on public transportation by constructing a universal meaning of the event, situation, or experience via this method. In this study, interviews were perceived to be the most appropriate way to collect COVID-19 information in Malaysia. The call for participation was through personal contacts of researchers. All interview sessions were structured in a "Questions and Answers" category. The interviews aim to address the following two questions in-depth: "What are respondents' own thoughts about the impact of COVID-19 pandemic on public transportation?" and "What kinds of situations or settings have shaped respondents' perceptions of the impact of COVID-19 on public transportation?" The interviews were brief and one-on-one to ensure that each participant had enough time to adequately answer the questions. We have used recording means to make sure that all shared information is securely saved. Prior to recording, permissions to record the sessions were obtained from the respondents. All recorded sessions then were transcribed verbatim and systematically and then analyzed thematically. Due to the COVID-19 pandemic situation, all interviews' sessions followed the Standard Operating Procedures (SOPs) of the Malaysian government and strictly adhered to the safety protocol against COVID-19.

12.3.2 Data Collection Process

The purposive sampling method is applied to recruit the targeted respondents. We have taken into consideration respondents' working experience, working field, and the location of work. The targeted respondents/informants are those who work in public transportation. The targeted respondents are public transportation operators located in Klang Valley, Perak, and Penang states. We also have applied the snowball sampling to increase the recruitment and to guarantee an ample representation from all levels of positions (Naderifar et al., 2017). In October 2020, we had successfully interviewed five public transportation operators for very renowned public and private companies in Malaysia (see Table 12.1). The targeted respondents were informed and briefed on the purpose of the study. Before the commencement of the interviews, the researchers have detailed the interview procedure in order to avoid any miscommunications and/or understanding. Each respondent was allowed to take part in the interview with his/her own free will. For the data protection procedures, respondents were assured their information is treated with full confidentiality. To confirm that, letters were attached to each set of interview questions to explain what their data will be used for, and to get their consent.

The researchers took the responsibility of handling the interviews. All researchers involved in this study were present during the interviews to help and guide respondents in conducting the questions adequately and appropriately. To get accurate data and minimize biases, the interviews were conducted face-to-face. The interviews were conducted in the local dialects so that the respondents can understand very well all parts of the questions. Each interview took about 1–2 hours. Data collection ended when we reached the saturation level. After collecting data from the five respondents, no more new information was obtained and therefore we have concluded that the saturation level is reached and thus we have stopped the data collection.

The breakdown of the respondents is illustrated in Table 12.1.

Any social sciences research needs to consider ethical issues while collecting data (Punch, 1998). This study has taken all measures and ethical considerations prior to data collection. Researchers have assured respondents that their information will be treated with confidentiality and that anonymity will be guaranteed. The researchers have explained to the respondents the

Table 12.1 Breakdown of the Interviewees

No.	*Industry Type of Respondents*	*Gender Breakdown of the Respondents*	*Expertise*	*Number of the Respondents*
1	Airport Operator	Female	Engineering	1
2	Airport Operator	Male	Engineering	1
3	Bus Operator	Male	Operations	2
4	Tour Operator	Male	Management	1

purpose and nature of this research and ensured respondents that they have the right to withdraw at any time. Respondents were also informed and briefed about the expected duration of the study. The Universiti Tunku Abdul Rahman Ethics Committee approved the protocol, procedures, information sheet, and consent statement of the study (Ref. No. U/SERC/154/2020). Respondents who gave consent to voluntary participation would complete the interview.

12.3.3 Data Analysis Procedures

This study applied the content analysis method for qualitative data analysis. We have processed, categorized, classified, summarized, and tabulated the data. First, we printed out all the transcripts after transcribing the audios then we had gathered necessary materials, notes, and documents to assist in analyzing the data obtained from the interviews. Secondly, we have read the obtained data several times to get a deep understanding of the content of the data. We have kept notes and clarifications about emerged thoughts, ideas, and questions. In the third step, we created the initial codes. We applied the table method to create the codes. We have used highlighters, notes in the margins, sticky pads, and concept maps to create and generate the initial codes. Several codes were merged and reduced to be put together under one category. Again, we have read the data several times and highlighted keywords and phrases and developed and created more categories. In the fourth step, we reviewed the created codes and categories, and revised and combined the categories into wider themes. In the last step, three global themes were developed and created. These themes are 1) Financial Adversity, 2) Social Equity, and 3) Sustainable Mobility.

12.4 Results

On the effects of the pandemic, a new normal for the use of public transport involves (a) Hygiene, Sanitization, and Ventilation; (b) the Use of Face Mask; (c) and the Emergence of Physical Distancing. The staff of all sectors are briefed or trained to implement these COVID-19 SOPs. For instance, for the airport operator, there is an Airport Emergency Plan which consists of Business Continuation SOP, Management Team, and COVID-19 Outbreak Communication Team. The airport community also includes personnel from the Ministry of Health (MoH) from whom staff members who face mental stress may seek help. Other members of the Community are the Airlines, Fire Brigade, Immigration, Meteorology, and Police. Hence, it is important for the Airport to have good communication with staff amidst the COVID-19 pandemic. Under the new norm, the foremost importance is the entrance control to the Airport. At one time, the public is not allowed to pick up/send off passengers. Later on, frequent sanitization has been implemented, and a limited number of the public is allowed in designated areas with face masks and temperature checks. Referring to the thematic analysis, major positive and

Table 12.2 Summarized Inputs from the respondents

Themes	*Subthemes*	*Organization 1 (Airport Operator)*	*Organization 2 (Bus Transport Operator)*	*Organization 3 (Tour Operator)*
A. Positive				
1. Safety and Health	Hygiene Precautions	/	/	/
2. Regulation Compliance	Full Compliance to the SOP Regulations	/	/	/
	Educate the Workers about the Pandemic	/	/	/
3. Digital Transformation	New Projects Related to Digital Economy	/	/	/
	Digital Marketing	/	/	/
	Remote Office	/		
B. Negative				
1. Cost	Loss of Income	/	/	/
	Cost of Hygiene Kit	/	/	/
	Temperature Checking	/	/	/
	Client Demand for Refund			/
2. Human Resources	Worker Management/ Reassignment	/	/	/
	Retrain workforce to other positions	/	/	/
	Workers Layoff or Termination			/
3. Resource Available	Over Capacity	/	/	/

negative themes can be categorized. The inputs from the airport, bus, and tour operators are summarized in Table 12.2:

12.4.1 The Linkage Between Ecosystem Services and COVID-19 Pandemic in Tourism and Public Transport

According to Nientied and Shutina (2020), the impact of COVID-19 on tourism goes further than the considerations of the risk of health and a lost 2020 season. Due to movement control orders (MCOs) and quarantine measures, supply chain disruptions are affecting countries dependent on merchandise trade including Malaysia (United Nations, 2020).

The pandemic has brought negative impacts on the financial, social, and sustainability fronts in Malaysia. Since the outbreak of the pandemic, it has brought forth the greatest economic crisis experienced by the tourism and public transportation sectors. The acute reduction in demand for tourism and public transportation has been integrated with higher expenses in complying with the hygiene and cleaning standards in COVID-19 preventive measures. As a rule of thumb, hotels are expected to suffer losses below 50% occupancy rate. Under these circumstances, many public transportation providers are suffering financial losses, extending pressure to governments. An issue for public transportation providers looking for financial relief is that governments face more social needs asking for financial aids while experiencing a decline in tax revenues. Consequently, public transportation providers have to compete with several other social needs for government aids. The greatest issue to be confronted is due to the decline in demand and the financial stress is the risk of insolvency.

With the pandemic, many have stopped using public transportation, but the effect is uneven. The high-income earners have left public transportation in large numbers. In relation to the payment of tickets, new norms for its use with digital payment may incur capital investment. Some countries have sufficient funds to sustain public transportation, compared to other poorer nations. The financial circumstances of such public transportation providers and their employees rely very much on the recovery from the pandemic. As a result, the improvement of public transportation is also a social equity issue.

12.4.1.1 Impacts of COVID-19 Pandemic on Employment in Tourism and Public Transport

During the pandemic, services subsectors were severely hit, especially tourism and transportation which led to a high reduction in employment in this sector. The findings from the study show that services sector jobs are affected due to the pandemic in Malaysia. The reduction in employment was caused by a reduction in demand for services from subsectors because of measures such as the closure of airports and fewer international tourists' arrivals. The reduction of jobs in the services sector is also due to the laying off of Hotels and Restaurants employees.

12.4.1.2 Impacts of COVID-19 Pandemic on Public Demand in Tourism and Public Transport

The outbreak of the pandemic in Malaysia has led to people losing their jobs, causing an increase in the unemployment rate and also an increase in underemployment, which further reduced the demand for goods and services. The information from organizations sampled in this chapter affirms that transportation (Zhang & Hayashi, 2020) and tourism sectors were the worst affected by the pandemic, similar to Liu, Miller, and Scheff (2020) and Muley et al. (2020).

12.4.2 Ecosystem Services and Adoption of Resilient Policy

Humans and animals will flourish again post-COVID-19 pandemic with its sustainability-centered development. There is a need to reform the tourism and transportation sectors for sustainability. The COVID-19 pandemic is a formidable force to make human beings more capable to well settle international issues. The pandemic has impacted the economic development of the tourism and transportation sectors. It is the time to recon and to reshape the tourism and transportation sectors to make them more sustainable. Rebuilding the tourism and transportation sectors should focus not merely financially but also to reform the sector post-pandemic sustainably. The tourism and transportation sectors will recover post-pandemic by adapting the new normal. The plan of recovery for the tourism and transportation sectors post-COVID-19 has to concentrate on devising a sustainable business plan according to the global consciousness, similar to Zhang and Hayashi (2020).

Given the COVID-19 pandemic, mobility models have also been studied in relation to risk-taking attitudes (Przybylowski, Stelmak & Suchanek, 2021). In order to gain health benefits, sturdy actions are required to reduce movements via transportation. In terms of sustainable mobility, the purpose of reconstruction to present public transportation safe post-pandemic is to ensure a safe transportation that can assist in drawing more passengers mitigating pandemic concerns on hygiene reported (Abdullah et al., 2020; Tirachini & Cats, 2020). If buses are running with low occupancy rates during the pandemic, then the financial and environmental competence arguments for encouraging public transportation is harshly disputed. The public transport system is resilient in its ability to restore functionality while implementing COVID-19 measures. Businesses have to incorporate public health requirements into their planning, and effectively manage physical distancing to reduce public health risks in public transportation. Frequent sanitization is implemented to stop the virus spread in public transportation. Passengers are also reminded to wear face masks when entering public transport, similar to the findings of Zhang and Hayashi (2020).

Particularly, new normal tourism development has to be led by the dedication to provide travelers with spatial safety safeguards, sustained by more scattering of travelers via the development of “untact” destinations such as coastal areas, and better transport infrastructure, whereby travelers are urged to travel with private vehicles, instead of using public transportation to diminish the risk of COVID-19 (Sohn et al., 2021).

MySejahtera is a Malaysian government application to help in monitoring the COVID-19 outbreak by empowering users to evaluate their health risks against COVID-19. The operators provide the MySejahtera QR Code for use in contact tracing to minimize the virus spreading risk in tourism and public transport. Contact tracing using smartphones, Bluetooth, and mapping interfaces can speedily find the probable contacts of the infected people. Big data collected from these technologies can support scientists to improve their understanding of transmission patterns and take suitable measures (United Nations, 2020). Amidst the crisis, there is also a silver lining that brings new opportunities using the latest technologies.

12.4.3 Social–Economical Effects of the COVID-19 Outbreak on Public Transportation

12.4.3.1 Financial Adversity

The interviewees provide feedbacks on financial adversity consequences. For the Airport, only 2% of its normal capacity has been recorded during the lowest point during the pandemic. As of the Recovery Movement Control Order (RMCO) period, approximately more than 50% of domestic flights are back; however, international flights are still limited due to border closure. During the period, many of the flight arrivals are for transit purposes. Prior to the COVID-19 pandemic, the Airport received 100,000 passengers per day. During the pandemic, the Airport suffers losses of about RM110,000 per day. For the Stage Bus Operator, the worst impact was a 90% reduction in revenue during the MCO, and it gradually recovered to about 60% reduction in revenue during the RMCO. Some industry players of cross-border or KLIA-designated express buses are hit the hardest. For the Bus Operator, the drivers of express buses are reassigned to drive stage buses, as the drivers possess the same driver's license. Other industry players have discontinued the contract of service of drivers aged above 60. The express bus sector is represented by the Association of Express Bus. For the Tour Operator, the tours in forward in the next two months, i.e., February to April 2020 have been booked. For air freights, the airlines will refund direct customers (FIT), but only allow postponement for travel agencies (group booking). The issue is there are travel agencies' deposits withheld by the airlines. The hotel sector is the fastest hit by the pandemic. The impacts on the travel agent are about 50% of the manpower retrenched, including manager and tourist guides. For the roughly 40 local tour operators, some companies have closed. Some moved their offices to the first floor to save rental costs, while 60–70% of the agencies are still inactive. For the JB Branch, the business downturn is worse. The International Air Transport Association (IATA) predicts that the optimistic outlook will be normalized in 2021. There will be more change of F&B by the local and use of digital means (electronic commerce). The new opportunity is in F&B. The future of the marketing will be through social media such as Tik Tok.

12.4.3.2 Social Equity

One of the interviewees gives feedback on the social equity consequences. It has been highlighted that the severe long-term impact of the pandemic is poverty. In fact, the shocks of financial damages on the poor are expected to be extra severe and last longer (Yang et al., 2021).

12.4.3.3 Sustainable Mobility

The interviewees provide feedbacks on the consequences on mobility. The Airport Operator shares that the pre-COVID-19 passenger volume was 70 million, and it is estimated to recover fully in four years' time. From the Bus Operator, the recovery of the transportation sector will be the first to rebound after the introduction of the COVID-19 vaccines. Hence, the vaccines are the keys, and vulnerable groups, e.g., senior citizens will feel safe traveling via public transportation.

12.5 Discussion

The qualitative analysis conducted illustrates an impact on transport volume growth by the new COVID-19 cases in Malaysia. The study ascertained the extent to which the pandemic affected public demand in tourism and transportation in Malaysia. Aiming to block the chain of infection of COVID-19, Malaysia closed its international borders, directing to a lockdown on transportation.

The government of Malaysia has executed four phases of MCO to prevent the spread of the pandemic, with the initial phase, MCO 1.0, commencing 18 March 2020, MCO 2.0 from 13 Jan 2021, and MCO 3.0 (or Phase One of Movement Control under the National Recovery Plan, NRP), which is a full lockdown beginning from 1 June 2021.

Sectors such as public transportation have to shut down a majority of their operations due to the MCOs by the Malaysian government. There are trade-offs with respect to the pandemic. Activities essential for subsistence are more vulnerable to self-restriction social effects. Since shopping activities for essential items are required for subsistence, they must be carried out with regularity despite the perceived risk. Therefore, consumers would oblige to follow protective measures such as wearing masks and physical distancing. Inevitably, the transformations in public transport have a greater impact on the elderly than young people because the former group is more dependent on mass transit for mobility (Yang et al., 2021).

The tourism industry is one of the main contributors to GDP and economic growth globally. The impact of COVID-19 has restricted the "Visit Malaysia 2020" and "Malaysia-China Year of Culture and Tourism 2020" campaigns in Malaysia. The study established how tourism and transportation sectors have been affected by the pandemic in Malaysia. The chapter's outcomes expose the overall affected employment in the tourism and transportation sectors by the pandemic. The findings reveal that due to the decline in supply versus the low demand in the COVID-19 situation, these sectors have laid off their employees.

Ecosystem services are non-material benefits that people acquire from ecosystems directly and indirectly through recreation. Tourism experience may incur negative effects for the tourism business ecosystem. Excess exploitation of wildlife will threaten biodiversity, impair the natural equilibrium, and affect the ecotourism value of the natural landscape.

Post-pandemic, the development of eco-friendly tourism will contribute to the conservation of ecosystems and reduce the ecological footprint, especially mobility ecosystem which enables an integrated, efficient, and sustainable transportation services to access ecosystem, or community-based or rural tourism ecotourism which incline toward an equitable and sustainable society on ecosystem services provision. In short, ecosystems can regenerate while the public travel for recreational purposes.

To support the SMEs in the tourism sector, the Malaysian government provides tax incentives as well as PENJANA Tourism Financing (PTF). PTF was initiated by the Malaysian government, and aimed for the SMEs to assume the investments to stay viable post-pandemic. The purpose of PTF is to support working capital and/or capital expenditure.

12.6 Conclusions

Tourism was among the fastest-growing services segment before the COVID-19 pandemic. It has been discussed earlier that due to the pandemic outbreak, traveling sectors have suffered. Traveling is a possible channel for the rapid spreading of the virus if not conducted with proper COVID-19 SOPs. Once the COVID-19 crisis is brought under control, the suitability and cost-effectiveness of public transport for long-distance travel would be improved. This mode of traveling will be more appealing than driving, biking, and other means of travel (Yang et al., 2021). The easing of the movement restriction is implemented in Malaysia with effect from 11 October 2021. This chapter covers the knowledge, attitudes, practices, and impacts of COVID-19. The results reveal that the industry has sufficient knowledge of COVID-19. The industries have suffered tremendous losses due to the severe reduction in cash flows caused by the reduced capacity. The effect of the pandemic on the industries has been discussed. The pre-COVID-19 air travel volume is 70 million, and it is estimated to recover fully in four years' time. Poverty is the result of social costs suffered from the pandemic. And the new normal measures by the government are the key to aiding recovery from COVID-19. As a result of the pandemic, additional SOPs have been implemented to upgrade the facility's hygiene level. Some sectors may benefit from specific programs to recover from COVID-19. Certain manufacturing sectors such as information communication technology, and medical or health products such as gloves and face masks production will flourish. The onset of the pandemic has accelerated the transformation opportunity to e-business (domestic and cross border), digitalization of commerce, and digital marketing. Despite of the impact, *there is a limitation* in this chapter. The samples of the study were collected via a purposive sampling of the research team. Therefore, there is a possibility that some portions of the population may not have been able to participate in the study. A more diversified sampling is necessary to enhance the generalization of the findings.

12.6.1 Managerial Implications

It is important for the tourism and public transport sectors to pick up their activities from the danger of spill-over effects in related and adjacent sectors, e.g., KLIA Aeropolis. During the COVID-19 pandemic, the airport, bus, and tour operators have redesigned their services to cope with prevailing demand patterns and limitations of capacity by riding on the high growth of the digital economy.

For the airport operator, the manpower arrangement has been realigned. Some excess staff have been retrained for other functions that need manpower. Better services are to be provided to meet the opportunity of the increase of cargo for air transport with the rise in cross-border electronic commerce. The airport operator has also implemented creative packages to generate new income, such as hotel stay package with education and experiencing tour to the Fire Station at the airport.

For the bus operator, a new normal COVID-19 SOP has been adopted: face masks are required for passengers, and the enforcement of no standing rule on board buses, which is stricter than the government's SOP where social distancing does not apply in vehicles. The operator also further enhances the use of digital means which includes resuming the installation of the Touch-n-Go system and the introduction of e-wallet for cashless transactions. The next plan is a proprietary smart bus application with the aim to upgrade the service to the standards such as in Singapore and Hong Kong.

For the tour operator, a retrenchment measure has to be taken due to the total loss of revenue. However, new business opportunities have developed from the COVID-19 pandemic. The surge in e-commerce has brought forth a huge demand for parcel delivery and courier services. The operator has allowed flexibility to its driver to rent the company van to carry out deliveries for a courier operator (Poslaju). The company has also provided bus services to universities to send students who have been stranded on campuses due to the movement control restriction to return to their hometowns. Some other operators have ventured into agriculture and agro-food businesses such as prawn breeding.

12.6.2 Policy Recommendation

Due to the closure of borders, Malaysia experienced a sharp decline in the number of inbound travelers. In the first half of 2020, Malaysia recorded only 4.3 million tourist arrivals from overseas, a 68% drop compared to the same period last year; inbound tourist receipts was reduced by 71% to an amount of RM12.7 billion; most of which were recorded in January and February of 2020 (Prem Kumar, 2020). The federal and state governments have implemented COVID-19 relief package to businesses and the people.

However, from the industries, the government's sector-specific support is much needed toward post-COVID-19 recovery, including the study's sampled sectors, i.e., airport, bus, and tour operators. The operators have to provide additional measures in compliance with COVID-19 SOPs, and the operators wish that the government provides specific support to tourism and public transport, such as green travel bubbles, fast-track tax exemption approval to sell expiring duty-free products, incentives on station infrastructure upgrade or import duty and sales tax of buses, and specific unemployment allowance to the personnel of the travel and tourism industry.

Post-COVID-19, "Untact" tourism or "Non-contact tourism" is to be a new travel trend that prevents crowded places, prioritizing outdoor attractions with plenty of space. Such a new trend of tourism can be a solution for pandemics and environmental preservation, considering the overcrowded ecosystems since it enables the dispersal of travelers across ecosystems (Sohn et al., 2021). In other words, it may require the development of lower-impact and higher-quality tourism experiences for tourists, linked with a more holistic rural development (McGinlay et al., 2020). Nevertheless, Malaysia will have to minimize the widening digital divide between the urban and the rural in order to recover better (United Nations, 2020).

Acknowledgment

This work was supported by the Malaysian Ministry of Higher Education's Fundamental Research Grant Scheme (FRGS) for COVID-19.

References

Abdullah, M., Dias, C., Muley, D. & Shahin, M. "Exploring the impacts of COVID-19 on travel behavior and mode preferences", *Transportation Research Interdisciplinary Perspectives*, Vol. 8, p. 100255. 2020. Doi: 10.1016/j.trip.2020.100255

Azlan, A.A., Hamzah, M.R., Sern, T.J., Ayub, S.H. & Mohamad, E. "Public knowledge, attitudes and practices towards COVID-19: A cross-sectional study in Malaysia", *PLoS One*, Vol. 15, No. 5, p. e0233668. 2020. Doi: 10.1371/journal.pone.0233668

Bhatt, N., Bhatt, B., Gurung, S., et al. "Perceptions and experiences of the public regarding the COVID-19 pandemic in Nepal: A qualitative study using phenomenological analysis", *BMJ Open*, Vol. 10, No. 12, p. e043312. 2020. Doi: 10.1136/bmjopen-2020-043312

Bird, J., Kriticos, S. & Tsivanidis, N. "Impact of COVID-19 on public transport". International Growth Centre (IGC). 2020. https://www.theigc.org/blog/impact-of-covid-19-on-public-transport/.

Bremer, S., Schneider, P. & Glavovic, B. "Climate change and amplified representations of natural hazards in institutional cultures", *Oxford Research Encyclopedia of Natural Hazard Science*. 2019. Doi: 10.1093/acrefore/9780199389407.013.354

Chen, Q. & Pan, S. "Transport-related experiences in China in response to the Coronavirus (COVID-19)", *Transportation Research Interdisciplinary Perspectives*, Vol. 8, p. 100246. 2020. Doi: 10.1016/j.trip.2020.100246

Cherkaoui, S., Boukherouk, M., Lakhal, T., Aghzar, A. & Youssfi, L.E. "Conservation amid COVID-19 pandemic: Ecotourism collapse threatens communities and wildlife in Morocco", *I2CNP 2020. E3S Web of Conferences*, Vol. 183, p. 01003. 2020. Doi: 10.1051/E3sconf/202018301003

Cochran, A.L. "Impacts of COVID-19 on access to transportation for people with disabilities", *Transportation Research Interdisciplinary Perspectives*, Vol. 8, p. 100263. 2020. Doi: 10.1016/j.trip.2020.100263

Coutts, A., Beringer, J. & Tapper, N. "Changing urban climate and CO2 emissions: Implications for the development of policies for sustainable cities", *Urban Policy Research*, Vol. 28, No. 1, pp. 27–47. 2010. Doi: 10.1080/08111140903437716

Cretu, R.C., Hontus, A.C., Alecu, I.I., Smedescu, D. & ştefan, P. "Analysis of the ecotourist profile before the COVID-19 crisis and post-crisis forecasts", *Scientific Papers Series Management, Economic Engineering in Agriculture and Rural Development*, Vol. 20, No. 2, pp. 191–198. 2020.

Hanafiah, K.M. & Wan, C.D. "Public knowledge, perception and communication behavior surrounding COVID-19 in Malaysia", *Advance Social Sciences and Humanities.* 2020. https://covid19.researcher.life/article/public-knowledge-perception-and-communication-behavior-surrounding-covid-19-in-malaysia/219e8c38-e0c0-4d9b-8860-64f0fa26f301.

Kaushal, V. & Srivastava, S. "Hospitality and tourism industry amid COVID-19 pandemic: Perspectives on challenges and learnings from India", *International Journal of Hospitality Management*, Vol. 92, p. 102707. 2021. Doi: 10.1016/j.ijhm.2020.102707

Li, Z.Y., Zhang, S., Liu, X.Y., Kozak, M. & Wen, J. "Seeing the invisible hand: Underlying effects of COVID-19 on tourists' behavioral patterns", *Journal of Destination Marketing & Management*, Vol. 18, p. 100502. 2020. Doi: 10.1016/j.jdmm.2020.100502

Lin, Y.P. "Sustainability of ecosystem services in a changing world", *Journal of Ecosystem & Ecography*, Vol. 2, No. 2. 2012. Doi: 10.4172/2157-7625.1000e111

Liu, L., Miller, H.J. & Scheff, J. "The impacts of COVID-19 pandemic on public transit demand in the United States", *PLoS ONE*, Vol. 15, No. 11, p. e0242476, 1–22. 2020. Doi: 10.1371/journal.pone.0242476

Lohmann, G. & Duval, D.T. "Critical aspects of the tourism-transport relationship", *Contemporary Tourism Reviews*, pp. 1–37. 2011. http://hdl.handle.net/10072/54257.

Lovrić, R., Farčić, N., Mikšić, Š. & Včev, A. "Studying during the COVID-19 pandemic: A qualitative inductive content analysis of nursing students' perceptions and experiences education sciences", *Education Sciences*, Vol. 10, No. 7, p. 188. 2020. Doi:10.3390/educsci10070188.

Mace, G. M. 2014. Whose conservation? *Science* Vol. 345, No. 6204, pp. 1558–1560. https://doi.org/10.1126/science.1254704

Mandal, I. & Pal, S. "COVID-19 pandemic persuaded lockdown effects on environment over stone quarrying and crushing areas", *Science of the Total Environment*, Vol. 732, pp. 139281. 2020. Doi: 10.1016/j.scitotenv.2020.139281

McGinlay, J., Gkoumas, V., Holtvoeth, J., Fuertes, R.F.A., Bazhenova, E., et al. "The impact of COVID-19 on the management of European protected areas and policy implications", *Forests*, Vol. 11, 1214, pp. 1–15. 2020. Doi:10.3390/f11111214

Muley, D., Shahin, M., Dias, C. & Abdullah, M. "Role of transport during outbreak of infectious diseases: Evidence from the past", *Sustainability*, Vol. 12, No. 18, 7367, pp. 1–12. 2020. Doi:10.3390/su12187367

Naderifar, M., Goli, H. & Ghaljaie, F. "Snowball sampling: A purposeful method of sampling in qualitative research", *Strides in Development of Medical Education*, Vol. 14, No. 3, pp. 1–6. 2017.

Nientied, P. & Shutina, D. "Tourism in transition, the post COVID-19 aftermath in the Western Balkans", *Co-PLAN Resilience Series*. 14 May 2020. Doi: 10.32034/CP-PPRESI-P01-02

Prem Kumar, P. "Malaysia likely to remain closed to tourists into 2021: Minister", *Asia Nikkei*. 26 August 2020. https://asia.nikkei.com/Spotlight/Most-read-in-2020/Malaysia-likely-to-remain-closed-to-tourists-into-2021-minister.

Przybylowski, A., Stelmak, S. & Suchanek, M. "Mobility behaviour in view of the impact of the COVID-19 pandemic – Public transport users in Gdansk case study", *Sustainability*, Vol. 13, p. 364. 2021. Doi: 10.3390/su13010364

Punch, K.F. *Introduction to Social Research: Qualitative and Quantitative Approach*. London: Sage. 1998.

Renaud, L. "Reconsidering global mobility – Distancing from mass cruise tourism in the aftermath of COVID-19", *Tourism Geographies*. 2020. Doi: 10.1080/14616688.2020.1762116

Rodrigue, J.P. *The Geography of Transport Systems* (5th Ed.). New York: Routledge. 2020. Doi: 10.4324/9780429346323

Sohn, J.-I., Alakshendra, A., Kim, H.-J., Kim, K.-H. & Kim, H.-D. "Understanding the new characteristics and development strategies of coastal tourism for post-COVID-19: A case study in Korea", *Sustainability*, Vol. 13, p. 7408. 2021. Doi: 10.3390/su13137408

Solan, M., Hauton, C., Godbold, J.A., Wood, C.L., Leighton, T.G. & White, P. "Anthropogenic sources of underwater sound can modify how sediment-dwelling invertebrates mediate ecosystem properties", *Scientific Reports*, Vol. 6, No. 1, 20540. 2016.

Spenceley, A., McCool, S., Newsome, D., et al. "Tourism in protected and conserved areas amid the COVID-19 pandemic", *Parks*, Vol. 27, No. (Special Issue) March, pp. 103–118. 2021. Doi: 10.2305/IUCN.CH.2021.PARKS-27-SIAS.en

Stankov, U., Filimonau, V. & Vujičić, M.D. "A mindful shift: An opportunity for mindfulness-driven tourism in a post-pandemic world", *Tourism Geographies*. 2020. Doi: 10.1080/14616688.2020.1768432

Stjernborg, V. & Mattisson, O. "The role of public transport in society – A case study of general policy documents in Sweden", *Sustainability*, Vol. 8, p. 1120. 2016. Doi:10.3390/su8111120

Thondoo, M., Mueller, N., Rojas-Rueda, D., de Vries, D., Gupta, J. & Nieuwenhuijsen, M.J. "Participatory quantitative health impact assessment of urban transport planning: A case study from Eastern Africa", *Environment International*, Vol. 144, p. 106027. 2020. Doi: 10.1016/j.envint.2020.106027

Tirachini, A. & Cats, O. "COVID-19 and public transportation: Current assessment, prospects, and research needs", *Journal of Public Transportation*, Vol. 22 No. 1, pp. 1–21. 2020.

United Nations. The Impact of COVID-19 on South-East Asia. July 2020.

Wieckowski, M. "Will the consequences of COVID-19 trigger a redefining of the role of transport in the development of sustainable tourism?", *Sustainability*, Vol. 13, p. 1887. 2021 Doi: 10.3390/su13041887

Yang, Y.L., Cao, M.Q., Cheng, L., Zhai, K.Y., Zhao, X. & De Vos, J. "Exploring the relationship between the COVID-19 pandemic and changes in travel behavior:

A qualitative study", *Transportation Research Interdisciplinary Perspectives*, Vol. 11, 100450, pp. 1–13. 2021. Doi: 10.1016/j.trip.2021.100450

Zambrano-Monserrate, M.A. & Ruano, M.A. "Does environmental noise affect housing rental prices in developing countries? Evidence from Ecuador", *Land Use Policy*, Vol. 87, p. 104059. 2019. Doi: 10.1016/j.landusepol.2019.104059

Zhang, J. & Hayashi, Y. Impacts of COVID-19 on the transport sector and measures as well as recommendations of policies and future research: Analyses based on a world-wide expert survey (May 27, 2020). 2020. Available at SSRN: https://ssrn.com/abstract=3611806.

13 Post COVID-19 through Special Interest Tourism

Perspectives for South Africa and Zimbabwe

Unathi Sonwabile Henama, Apleni Lwazi, Bantu Mbeko Msengi, and Madiseng Messiah Phori

13.1 Introduction

Tourism is a major economic sector in the world and is a big business, a major employer and a contributor to gross domestic product (GDP). The perennial growth of tourism has led to many countries presenting themselves as tourism destinations. According to Jaafar et al. (2015) the World Travel and Tourism Council (WTTC), tourism is forecast to grow at the rate of 4% annually over the next 10 years and will account for 9.4% of the Gross World Product (GWP). Before the outbreak of the COVID-19 pandemic, tourism was a real economic success story, a major contributor of jobs and employment in many developed and developing countries. Tourism is a service located in the tertiary sector of the economy, and this had made tourism an economic sector that has diversified the economic prospects of many countries. Tourism is a known engine of job creation due to the labour-intensive nature of tourism employment. The low barriers to entry into many sectors of the tourism industry, make it an attractive avenue for entrepreneurship. The vast majority of businesses in the tourism industry are small businesses, which are associated with a growth in jobs and economic growth. These positive economic benefits associated with tourism reflect why tourism has been favoured by politicians and entrepreneurs. Tourism has been integrated into the economic development policies of South Africa and today tourism has been called the "new gold". The growth of tourism has been so impressive that it has shown growth rates similar to gold mining, which has dominated backward and forward linkages in the economy till today. The vast majority of companies listed on the Johannesburg Stock Exchange are linked to mining.

As mining fortunes have experienced a downward spiral, tourism growth had remained robust and resilient, achieving a year-on-year growth rate of 13% in 2015. This happens whilst the economy of South Africa has failed to grow at more than 1% per year for nearly a decade. In other words, tourism has been an economic messiah for South Africa, just like many countries on the African continent. Before the advent of COVID-19, tourism was the second largest employer after mining. Aref (2011) noted that tourism provides employment opportunities and tax revenues and supports economic diversity.

DOI: 10.4324/b23145-18

Tourism increases the tax coffers of the state, through the taxation of foreigners and additional taxes levied on tourists. One of the advantages is that the tourism product is a final product, produced and consumed at the destination area, which allows for the majority of the value-adding to occur at the destination area. This means that tourism can benefit a destination area profoundly if the local tourism industry is locally owned and designed to be pro-poor. This means that local agriculture, construction and retail can benefit for a tourism industry that is pro-poor and developed the locals and local economy. Tourism has a cross-cutting impact on the economy, with backward and forward linkages. The existence of tourists at a destination can lead to the development of public infrastructure such as roads, airports and communications at a faster pace to benefit and cater for tourists and locals.

Tourism is dependent on a responsive and capable state in supporting the private sector to lead in meeting the needs of tourists. This means that the tourism economy follows the neoliberal doctrine of a minimalistic state. African countries are disproportionally dependent on tourism as a means of diversifying their economies, which were dominated by primary industries and low presences in the services sector. Tourism is located in the tertiary sector of the services economy, the same services economy that is driving global growth and wealth. To African countries, tourism is an economic messiah, when you consider that the majority of tourists hardly visit Africa. Africa continues to attract less than 10% of global tourism receipts, which coupled with the sustained and perennial growth of tourism arrivals, increased the economic and political importance of tourism. Sustainable tourism is engulfed with ensuring that local businesses have a stake in the tourism industry. These pro-poor acts would ensure that tourism leakage is reduced, therefore disproportionally benefiting the local economy from the cash injection by tourists. A tourism industry that positively impacts the standard of living and Quality-of-Life of the residents and the local businesses would make the host community welcome tourists. Furthermore, this would motivate them to become better hosts because cordial relationships between tourists and the host community are paramount in the tourism experience. Tourism competitiveness has increasingly become man-made, and as part of post-COVID-19 tourism recovery, tourism recovery will be man-made by taking note of new tourism trends such as working from home (WFH) increasing leisure time, the growth of incentives to attract digital natives to zoom towns and vaccine tourism.

13.2 Theoretical Framework: Special Interest Tourism (SIT)

Special Interest Tourism (SIT or Niche Tourism) is a specialised form of tourism where tourists develop interest in a subject matter or activity and travel to participate in this common hobby. Niche tourism and SIT are used interchangeably. SIT is associated with specialised tourism activities, catering for the specific interests of individuals and groups. This can incorporate unusual hobbies, activities and themes, which will, in general, draw in speciality markets. SIT's growth can be associated with the need to customise tourism experiences, and the increased heterogeneous nature of tourism consumers, leading to the development of a plethora of tourism niches. Special interest tours have awesome possibilities of success and improvement in tourism sustainable development as they do not subscribe to mass tourism.

13.3 Theoretical Framework: Research Synthesis and Literature Review

A research synthesis and literature review is a meta-analysis research methodology followed in the crafting of the chapter. The metasynthesis is qualitative research that systematically reviews and integrates the findings from qualitative studies. According to Rrisko (2020), synthesizing multiple qualitative research studies on a topic is a valuable way to extend knowledge and theory. A synthesis analyses various pieces of literature to determine what is known and unknown about a particular topic. A literature review investigates and surveys books, journal articles and other relevant sources of literature on a particular issue, area of research or body of theory to create a summary and critical evaluation, in relation to the problem being investigated. The main purpose of the synthesis was to pull together the scattered opinion pieces, industry records and academic journals and books on the COVID-19 pandemic, with specific focus on how various tourism niches could assist tourism recovery in countries in Africa's Global South. According to Viljoen and Henama (2017), the literature review used was of an exploratory nature due to the limited body of knowledge with regard to the post-COVID tourism industry in the Global South. There is paucity of academic gaze on the topics researched, as the majority of the synthesis material had been from newspaper articles instead of journal articles. The synthesis adds to a gap in academic gaze on tourism markets associated with zoom towns in South Africa and vaccine tourism in Zimbabwe. This addition to the body of knowledge will be relied upon by other researchers as part of literature review in tourism studies about the impact of COVID-19 on the tourism industry and sector.

13.4 Zoom Towns: Opportunities for Visiting Friends & Relatives (VFR) Tourism

The changes brought by the COVID-19 pandemic included millions of people WFH due to stay at home orders, national lockdowns, closed borders and curfews. As millions of employees began experiencing the new normal of remote work, million others had to work as their services were critical from healthcare workers to grocery store employees and those in law enforcement. The office nowadays is a combination of physical spaces and virtual office resources for meetings, planning and collaborations between staff and stakeholders. Employers have embraced the changes that COVID-19 have done to the world of work and it's highly possible that some employees will exclusively work from home, with the physical office being used very rarely. There are several companies such as Deloitte that have indicated that their staff must never return to the office. This can be due to several reasons, such as an increase in productivity when employees work from home and the cost savings associated with fewer employees at the office. Stahl (2021) noted that there was a 400% increase in employees that work from home at least once a week. The changes to the world of work as a result of remote work meant that employees were not bound by geographic locations. This meant that commuting time to and from work diminished almost completely – a cost saving for millions of people. Those with children also had to experience this new normal of schooling from home almost completely and/or reduced contact classes for learners on several days of the week, instead of the whole week. Remote work meant that people could work from almost anywhere and this opened up new tourism consumption opportunities linked to remote workers. The almost nonexistent requirement that employees be at work meant their flexibility to work from anywhere was increased, providing an excellent opportunity for tourism consumption, due to increased flexibility. The immediate result was that many workers chose to live a little further away from the congested central business districts and the office. The biggest beneficiary had been outlying areas on the urban fringe and amenity-rich areas which were once regarded as places of play, instead of permanent stay and residence.

Cities seeking to recover their distressed tourism value chain have presented themselves as zoom towns and cities, attracting those WFH. This has been in the form of two trends:

- Cities attracting WFH professionals for a long-term stay that would not exceed one year; and
- Cities attracting WFH professionals to permanently relocate to these new cities and making these cities their primary residence.

The trend of moving to zoom towns is already reflected in the Global North. According to Inquirer (2020), a third of U.S. office employees work entirely remotely and are moving to stay in zoom towns. The zoom towns are located in outer-lying smaller towns and cities, which are not crowded. In the Global South, South Africa is an excellent example of how zoom towns have developed on the existing geography of second homes destinations, characterised by having high amenity value. Rogerson and Hoogendoorn (2014) noted that second homes stimulate VFR travel to the second homes. The emergence of zoom towns has stimulated the housing markets in those towns and cities located on the urban fringe, avoiding the crowded urban centres and cities. Migrants to zoom towns seek less crowded destinations, with less traffic congestion, which offer much more safer environments and a better Quality-of-Life and a standard of living. In the case of the Global North, several countries have offered incentives for remote workers to stay a year in foreign lands, as a way of attracting tourists. This was all part of tourism recovery, as tourism was one of the most battered industries due to national lockdowns and travel bans. In Tusla, which is Oklahoma's second-largest city, grants are made available for digital nomads to move to the city. According to Expart Network (2021), Tusla Remote attracts digital nomads by offering free desk space, networking events and $10,000 either as a cash lump sum towards a new home or a monthly stipend. Stahl (2021) noted that Tusla Remote's stipend was to cover housing and living expenses. These financial incentives offer an additional motivation to consider moving to a zoom town, and the towns would over time recoup the money in terms of rates and taxes, attracting skilled residents that may set up businesses and more expenditure through the arrival of "new money" that would become residential in nature. Jumping on to the zoom town bandwagon has meant that these cities and towns are attracting people instead of businesses.

The financial contribution they would make to these zoom cities and towns through their residence would be a financial boost from home improvements, rates and taxes and monthly purchases made within the local economy. In the case of zoom towns in the Global South, an established tourism destination such as Cape Town is dominating the zoom migration, especially among foreign travellers. It is not surprising considering that Cape Town is an established second home destination amongst many South Africans and foreigners. Cape Town is the leading city in terms of tourism arrivals and remains firmly a bucket list destination for many people. According to Smith (2021), Cape Town has made the "Best cities for remote working" list on the travel website Big 7 Travel's "50 Best Places for Remote Working in 2021". CapeTownEtc (2020) noted that the City of Cape Town launched a "Digital Nomad" initiative, seeking to attract remote workers from South Africa and abroad. The City of Cape Town has lobbied the government to introduce a "Remote Working Visa". This would make it much easier to attract foreigners that are digital natives and this would help in activating the recovery of the tourism industry in South Africa. Outside Cape Town, the town of Hermanus an established second home destination has been embraced as a zoom destination, through inward migration of new residents due to the flexibility offered by WFH. Muller (2020) noted that coastal enclaves in the Eastern Cape, typically an hour's commute from East London or Port Elizabeth, including St Francis Bay, Chinsta and Kenton-on-Sea, are becoming increasingly sought after relocation destinations. The experience of WFH motivated migration between the Global North and Global South shares some similarities and some unique differences. Firstly, the move to smaller towns and cities for zoom relocation is the same across regions. Secondly, what is divergent is that the relocation to zoom towns in the Global South, in the case of South Africa has primarily been to established second home destinations which are amenity-rich in nature. The relocation due to flexibility offered by the flexibility of WFH, has led to the development of zoom towns and cities in existing second homes destination. Relocating to zoom towns and cities in the case of South Africa further strengthens the tourism potential of these localities which serve additionally as second home destinations. This represents an interesting avenue for the growth and development of domestic and Visiting

Friends & Relatives (VFR) and a unique experience on zoom town migration in the Global South in South Africa.

The new world of work and the new normal have meant that those that migrate to zoom towns would become permanent residents instead of being temporary visitors. This has meant that major urban cities are losing not just residents but also ratepayers. Additionally, they are losing skills to cities and towns on the urban fringe and losing many entrepreneurs who have also migrated. The vast majority of digital nomads that migrate to zoom towns are financially well off than the host community and this may lead to competing for scarce local resources. This is felt immediately as housing stock prices increase, above the affordability of many locals who may not be affluent. The boom of zoom towns has seen reverse migration as much more smaller towns and cities are experiencing a property boom due to digital nomads. The change of focus towards zoom towns can be motivated by the realisation that the post-COVID-19 tourism industry would be completely different from what we have always experienced. The focus on domestic tourism and international tourism through zoom tourism is a strategy to mitigate the decline of business travel, which was a major component of tourism transactions in many countries. This has had a detrimental impact on the robustness and sustainability of many tourism destinations. Business travel is picking up again, but it would never go back to pre-COVID-19 times and this means destinations must adapt to other markets and niches, such as promoting zoom tourism. Internet connectivity and a greater national fibre infrastructure have facilitated the increased connectivity in secondary towns, making them the biggest beneficiary in the development of zoom towns. Targeting digital nomads has been considered a post-COVID-19 recovery strategy by a plethora of destinations. In the case of South African destinations, zoom towns are located in already-established second home tourism destinations. The second homes tourism towns and cities transitioned into zoom towns further entrenching VFR in those cities and towns. The development of second home towns stimulated VFR tourism and zoom towns are expected to exhibit similar characteristics. It's interesting that the development of zoom towns in the case of South Africa occurred almost naturally without financial and non-financial incentives as exhibited by countries in the Global North. This is what makes the experience of zoom tourism completely divergent in the Global South when compared to countries in the Global North. This can be associated with the fact that countries in the Global South are developing countries with a backlog on developmental indicators such as poverty eradication, development of infrastructure, provision of education and healthcare. Extending financial incentives to promote the development of zoom tourism could face still political opposition in the case of South Africa, considering that these finances would have to be extended by the states and spheres of the state. Client demand to relocate to zoom towns that have primarily served as second homes has been the South African experience with zoom towns and cities.

13.5 Vaccine Tourism: The Case of Zimbabwe

The tourism industry is important to African countries, used as a means to diversify their fragile economy. It's a major attractor of foreign exchange and it's one sector where many African countries have a trade surplus, as they receive more inbound tourists than outbound tourists. The tourism industry accounts for 7% of Africa's GDP and Africa's travel and tourism sector employed more than 24 million people in 2019 as noted by Monnier (2021). The growth of tourism in the past 20 years has been remarkable, from 26 million tourists in 2014 to over 60 million tourists before the pandemic. African countries are disproportionally dependent on tourism, even though Africa continues to receive less than 10% of global tourism receipts. Tourism has been incorporated into the economic development policies of many African countries, as almost all destinations have jumped on the tourism bandwagon. Tourism is one of the industry's leading changes in the African continent, by providing a plethora of opportunities for job creation, small business development, economic growth and improvements in public infrastructure used by locals and tourists. Tourism's growth before the pandemic has always experienced an upward spiral and Africa's growth rates have been impressive. African countries have jumped on the tourism bandwagon, and tourists bring much needed foreign exchange. In many African countries without sizeable extraction industries and agriculture, tourism has emerged as the major foreign exchange earner. The major tourism destinations on the African continent include South Africa, Egypt, Algeria, Uganda, Morocco, Tunisia, Kenya, Mauritius, Botswana, Tanzania and Zimbabwe. African destinations sell themselves for the nature experience, with the abundance of bush and animals. News24 (2021) elucidates that the travel and tourism sector is among the hardest-hit sectors globally as countries implemented various levels of lockdowns and travel restrictions to limit the spread of the COVID-19 pandemic. Zimbabwe is a tourist destination that relies more on international travellers, a market which has been severely affected by the COVID-19 pandemic (Woyo, 2021). Al Jazeera (2021) underscores that Zimbabwe's tourism fraternity is estimated to have lost $1 billion in potential revenue in 2020 due to the COVID-19 pandemic. Zimbabwean Tourism Authority highlighted that tourism contributed 7.2% and 6.5% of the country's GDP in 2018 and 2019 respectively. But with business slowing down in 2020, Zimbabwe's tourism sector is estimated to have lost at least $1 billion in potential revenue (Al Jazeera, 2021).

The emergence of the COVID-19 pandemic has changed the world forever. The infectious nature of the COVID-19 pandemic resulted in lockdown regulations and closure of national borders, and almost all international travel ceased to exist. Advances in medical technology led to the possibly the fastest development of the COVID-19 vaccine, which the vast majority of developing countries have pre-procured. This has led to accusations about vaccine apartheid, where the vast majority of Western countries had been able to pre-buy the vaccine, leading to mass vaccination programmes, resulting in the return to normal in many Western countries. Countries that have advanced vaccination programmes had been able to offer themselves as vaccine tourism destinations, offering visitors a vaccine on arrival. Such countries had been able to recover tourism arrivals using vaccines as an enabler of tourism recovery. Du Preez (2021) noted that it has however become a lucrative business proposition to a (seemingly still) price-insensitive niche market. The huge demand for vaccine tourism is due to the huge demand for vaccines and the desire to preserve living after being vaccinated.

Vaccinated citizens that have medical proof of being inoculated are welcomed with open arms by many countries.

> One way to restart the travel industry is thru country-wide, fast-paced vaccination drives. However, supply cannot keep up with the demand in some countries and the long wait fuels vaccine tourism. Some countries with excess supply see this as an opportunity to attract tourists. Many travel agencies began offering lucrative travel packages to places where travellers can be vaccinated
>
> Mariana (2021: 1)

Kamal (2021) contends that there is no official definition of vaccine tourism. Vaccine tourism favours those who have the financial means to undertake international travel to vaccine tourism destinations. Mariana (2021) noted that the following countries had presented themselves as a vaccine tourism destination, namely Maldives, Russia, the United Arab Emirates and the United States. It was to be expected that countries in the Global North would dominate vaccine tourism, as they already dominate international tourism arrivals, and also benefit disproportionally from tourism consumption. Gillespie (2021) noted that Florida, in the United States, was one of the first states to make the vaccine available to everyone aged 65 and over and many people flew into the Sunshine State for their shot from other states, and even from other countries, like Argentina and Canada. Newton (2021) noted some U.S. cities and states are going beyond tacitly allowing travellers to get vaccinated and are actively encouraging the vaccine-tourism trend, and Alaska announced by June 2021 that it would provide free COVID-19 vaccines upon arrival to any visitor landing at one of the state's four major airports.

In the case of the Global South, Zimbabwe has pioneered vaccine tourism in the African continent. On the African continent, Zimbabwe has presented itself as a vaccine tourism destination and it has attracted many vaccine tourists from the African continent. The attraction of Zimbabwe as a vaccine destination had been the slow vaccination process on the African continent, which created the means to become a vaccine tourism destination. The slow vaccination programmes in Africa became a blessing for developing a vaccine tourism industry in Zimbabwe. For a fee of US$100, a vaccine tourist would get both shots. According to the Sowetan (2021), Zimbabwe is seeing a boost in vaccine tourism as foreign nationals have started arriving in the country to get vaccinated against COVID-19, for a fee. Zimbabwe has so far authorised the use of four COVID-19 vaccines: China's Sinopharm and Sinovac vaccines, Russia's Sputnik V and Covaxin from India. This initiative was informed by the fact that other African countries are slow when it comes to COVID-19 vaccination. These countries include but are not limited to South Africa, Angola and Mozambique to mention just a few. The vaccine tourism initiative has not augured well with everyone in Zimbabwe. For instance, Zimbabwe Coalition on Debt and Development (ZIMCODD) has raised concerns. ZIMCODD view this as something that may lead to the prioritisation of the rich who pay for the service at the expense of citizens who receive vaccine for free (Harris, 2021). Since the country its vaccine tourism initiative, it has been a boost as foreign nationals come to be vaccinated against COVID-19. However, ZIMCODD contends that an ethical framework should be developed to guide vaccine distribution and prioritisation of the vulnerable should remain key to make sure that the available COVID-19 vaccines benefit both the rich and poor (Harris, 2021). Interestingly though, Zinyuke (2021) notes that Zimbabwe has surged ahead of all mainland Southern African Development Community (SADC) countries in terms of procurement and administration of COVID-19 vaccines with the top-rated national vaccination in the region.

Zimbabwe's efforts to curb the spread of the COVID-19 pandemic have been commended by the World Health Organisation including other international organisations (Zinyuke, 2021). Another interesting aspect about Zimbabwe is how it has strategically opted to use Victoria Falls to vaccinate people. Victoria Falls is often referred to as one of the major tourist hubs of Zimbabwe and it shares this waterfall with neighbouring Zambia. Zimbabwe focused on vaccinating all the residents of Victoria Falls including employees of tourism establishments to achieve herd immunity so that this vaccination success story could be used in marketing outlays to attract tourists and vaccine tourists to Zimbabwe's Victoria Falls. Development Reimagined (2021) puts forward that the tourism industry has taken a hit during the pandemic, with a reported job loss rate of 50%. Vaccine tourism can be regarded as the "means to an end", to bring back the former glory of Zimbabwean tourism, which suffered due to international sanctions and the negative perception associated with the land reform process and its associated violence under former President Robert Mugabe. The government of Zimbabwe and the tourism authorities must be commended for pioneering vaccine tourism in the African continent. The vast majority of tourists to Zimbabwe originate from neighbouring South Africa and vaccine tourism will attract a plethora of South Africa. This can be attributed to the slow vaccination program in South Africa, which will stimulate vaccine tourism from South Africa to benefit the tourism value chain in Zimbabwe.

13.6 Conclusions and Policy Recommendations

COVID-19 has changed the trajectory of the tourism industry by leading to widespread lockdowns and a ban on travel that impacted the world's largest industry. Tourism is big business and the stay at home orders had a more disastrous impact on the tourism industry.

The tourism industry was the hardest hit by the COVID-19 pandemic and considering the contribution of tourism to the world economy, a decline resulted in economic ruin for millions of employees and entrepreneurs. The industry is labour-intensive and for tourism consumption to occur physical contact between guests and service providers is a prerequisite due to the unique characteristics of the tourism industry. History is amassed with examples of the resilience of the tourism industry and tourism was expected to be the first industry to recover after the COVID-19 pandemic. Tourism is a major contributor to GPD, jobs, attracting foreign exchange and has deep linkages with other economic sectors such as agriculture, manufacturing, healthcare, construction and transportation. Restarting tourism was a top priority for many countries and destinations, beginning with domestic tourism when cross border international tourism was still banned in many countries. The absence of business had a major dent in tourism transactions. Inward gazing for solutions for tourism recovery has been centred on domestic tourism primarily. Countries with advanced vaccination programs have been able to reduce lockdown restrictions, open up their economic activities and allow international travel from countries with advanced vaccination programs.

In the case of Zimbabwe in the Global South, the advances made in public vaccination provided an avenue to provide a vaccine tourism programme in the major tourism region of Zimbabwe, Victoria Falls. The slow vaccination programme in many African countries in the Global South had been a major challenge for recovering their tourism industries. Vaccine tourism is pioneered by Zimbabwe on the African continent and Zimbabwe shall benefit from a plethora of tourists that would seek to get the jab, by paying the costs associated with the jab. Unfortunately, vaccine tourism would benefit only the financially well-off residents in many African countries, instead of the majority of the population. This raises challenges of class and inequality associated with tourism consumption, during a global pandemic excluding those without the financial means to acquire the vaccine in a foreign land. Needless to say, the tourism industry in Zimbabwe is a major contributor to the economy and the return of using vaccine tourism can reignite the tourism industry in Zimbabwe. The creativity of tourism officials in Zimbabwe by geographically concentrating the vaccine tourism project in Victoria Falls was a strategy to once again revive tourism recovery on Zimbabwe's most important tourism value chain.

The concept of WFH and the creation of digital natives have meant that people could work from almost anywhere in the world. Working from any geographical location has promoted the development of zoom towns across South Africa, which have traditionally been second home destinations. Zoom towns and cities in South Africa provide an opportunity to tap into this lucrative market, and countries in the Global South must ensure they have reliable electricity and internet access to make it easy to connect in the new world of WFH. Many African destinations should jump on the WFH bandwagon to become zoom destinations by presenting themselves as zoom destinations. The residential nature of digital nomads at zoom towns makes them an attractive market for tourism that could be regarded as sustainable. Tourism recovery is being led by countries in the Global North which continue to dominate tourism consumption as the vast majority of tourism occurs between countries in the Global North. Developing countries that seek to increase their share of tourism arrivals need to adapt to new tourism niches that have developed in the post-COVID-19 tourism industry. Tourism competitiveness is about agility and adaptation in meeting the needs of consumers and consumer markets.

References

Al Jazeera. (2021). *Could Vaccinating an Entire Resort Town Revive Zimbabwe Tourism?* Retrieved from: https://www.aljazeera.com/news/2021/3/29/could-vaccinating-a-resort-town-revive-tourism-in-zimbabwe. [Accessed 22 June 2021].

Aref, F. (2011). The Effects of Tourism on the Quality of Life: A Case Study of Shiraz, Iran. *Life Sciences Journal*, 8 (2): 26–30.

CapeTownEtc. (2020). *Cape Town's "Exotic" Lifestyle Attracts Remote Workers around the World.* Retrieved from: https://www.capetownetc.com/cape-town/cape-towns-exotic-lifestyle-attracts-remote-workers-around-the-world. [Accessed 16 June 2021].

Development Reimagined. (2021). *How Can China Support Vaccine Deployment in Zimbabwe?* Available from: https://developmentreimagined.com/2021/04/30/how-can-china-support-vaccine-deployment-in-zimbabwe/. [Accessed on 22 June 2021].

Du Preez, E. (2021). *Expert Opinion: Are We Ready to Jab the Economy with Vaccination Tourism?* Retrieved from: https://www.up.ac.za/news/post_2980587-expert-opinion-are-we-ready-to-jab-the-economy-with-vaccination-tourism. [Accessed on 22 June 2021].

Expart Network. (2021). *Zoom Towns – The Countries and States That Will Pay You to Move There.* Retrieved from: https://www.expatnetwork.com/zoom-towns-the-countries-and-states-that-will-pay-you-to-move-there/. [Accessed on 22 June 2021].

Gillespie, C. (2021). *What Is Vaccine Tourism, and Is It Legal? Here's What You Need to Know.*Retrieved from: https://www.health.com/condition/infectious-diseases/coronavirus/what-is-vaccine-tourism. [Accessed 22 June 2021].

Harris, L.B. (2021). *Vaccine Tourism Immoral: ZIMCODD.* Retrieve from: https://cite.org.zw/vaccine-tourism-immoral-zimcodd/. [Accessed 22 June 2021].

Inquirer. (2020). *'Zoom Towns': US Remote Workers Are Headed to Smaller Cities. Science, Health and Research.* Retrieved from: https://technology.inquirer.net/zoom-towns-us-remoe-workers-are-headed-to-smaller-cities. [Accessed 16 June 2021].

Jaafar, M., Ismail, S., & Rasoolimanesh, S.M. (2015). Perceived Social Effect of Tourism Development: A Case Study of Kinabalu National Park. *Theoretical and Empirical Researches in Urban Management*, 10 (2): 5–20.

Kamal, H.M. (2021). *Vaccine Tourism: Is It Legal and Ethical to Fly Out of COVID-19 Shots? All You Need to Know.* Retrieved from: https://www.firstpost.com/india/vaccine-tourism-is-it-legal-and-ethical-to-fly-out-for-covid-19-shots-all-you-need-to-know-9636521.html. [Accessed 22 June 2021].

Mariana, K. (2021). *These Countries Try to Revive Travel with Vaccine Tourism.* Retrieved from: https://www.traveldailymedia.com/these-countries-try-to-revive-travel-with-vaccine-tourism/. [Accessed on 22 June 2021].

Monnier, O. (2021). A ticket to recovery: Reinventing Africa's tourism industry. International Finance Corporation, World Bank Group. Available at: https://www.ifc.org/wps/wcm/connect/news_ext_content/ifc_external_corporate_site/news+and+events/news/reinventing-africa-tourism

Muller, J. (2020). *'Zoom Boom' Park City Exodus. Features*. Retrieved from: https://www.businesslive.co.za/fm/features/zoom-boom-sparks-city-exodus. [Accessed 16 June 2021].

News24 (2021) Jobs vs Lives: The COVID Dilemma, City Press. Available at: https://www.news24.com/citypress/voices/editorial-jobs-vs-lives-a-covid-19-dilemma-20210711

Newton, J. (2021). *A Global Search for Vaccines Fuels a Different Kind of Tourism*. Retrieved from: https://www.afar.com/magazine/vaccine-tourism-is-on-the-rise. [Accessed 22 June 2021].

Rogerson, C.M. & Hoogendoorn, G. (2014). VFR Travel and Second Home Tourism: The Missing Link? The Case of South Africa. *Tourism Review International*, 18: 167–178.

Rrisko, J.W. (2020). Qualitative Research Synthesis: An Appreciative and Critical Introduction. *Qualitative Social Work*, 19 (4): 736–753.

Smith, C. (2021). *Cape Town Ranked One of the World's Best Cities for Remote Working*. News24. Retrieved from: https://www.news24.com/fin24/companies/travelandleisure/cape-town-ranked-one-of-the-worlds-best-cities-for-remote-working. [Accessed 15 June 2021].

Sowetan. (2021). *Vaccine Tourism: South Africans Cross Border to Zimbabwe for COVID-19 Jabes*. Retrieved from: https://www.sowetanlive.co.za/news/south-africa/2021-05-08-vaccine-tourism-south-africans-cross-border-to-zimbabwe-for-covid-19-jab/. [Accessed 22 June 2021].

Stahl, A. (2021). *The 7 Best Cities for Remote Workers in 2021*. Retrieved from: https://www.forbes.com/sites/ashley/2021/03/19. [Accessed 15 June 2021].

Viljoen, J. & Henama, U.S. (2017). Growing heritage tourism and social cohesion in South Africa. *African Journal of Hospitality, Tourism and Leisure*, 6 (4): 1–15.

Woyo, E. (2021). The sustainability of using domestic tourism as a post COVID-19 recovery strategy in a distressed destination. In Wörndl, W., Koo, C., & Stienmetz, J.L. (Eds). *Information and Communication Technologies in Tourism 2021*. Springer, Cham. Retrieve from: https://doi.org/10.1007/978-3-030-65785-7_46. [Accessed 22 June 2021].

Zinyuke, R. (2021). *Zim Tops SADC in COVID-19 Vaccination*. Retrieved from: https://www.herald.co.zw/zim-tops-sadc-in-covid-19-vaccination/. [Accessed 22 June 2021].

14 The Impact of COVID-19 on Cultural Ecosystem Services Nexus

A Vulnerability Assessment of Simien Mountains National Park, Ethiopia

Prakash Chandra Rout, Alelign Takele, Afera Gebremedihn, Kindieneh Awoke, and Kidanu Melese

14.1 Introduction

Cultural Ecosystem Services (CES) are the immaterial pleasure human receive through the natural world. These include aesthetic enjoyment, physical and mental health benefits, spiritual experiences, and recreation (MEA, 2005). The role of CES is vital for overall human health and well-being. Taking CES theory into account, it offers safeguarding the ecosystem from deterioration, by combining the ecosystem and landscape approach in spatial planning and nature conservation, from the perspective of sustainable development (COELC, 2000; MEA, 2005). Recreation and tourism have long been recognized as important tools for countries and communities to harness CES in order to achieve economic development goals (Emerton et al., 2006; Kettunen & ten Brink, 2013). CES play an important role in tourism, particularly, in destinations closer to protected areas and wildlife. They create demand for tourists to visit protected areas blessed with larger CES-based values and activities. Further, they support employment and generate income for destination communities, indigenous people, operators, and governments. Often these stakeholders safeguard and maintain CES-based values in regions that have been designated as protected and conserved (Ceballos-Lascurain, 1996).

The COVID-19 pandemic has had a substantial impact on regions that are protected or conserved. Protected and conserved areas have suffered both immediate and long-term repercussions as a result of the pandemic. The global pandemic has forced the closure of parks and protected areas for tourism and recreation in several countries, which leads to a myriad of impacts on the associated nexus. Many protected area agencies had cut their staff duties and were even being laid off. This may have a severe effect on the conservation of key habitats and species. Further, the pandemic has forced to cut protected area operational budget due to a reduction in revenue from tourism (IUCN, 2020). Indigenous peoples and local groups who rely heavily on these areas have seen their economies shattered and their livelihoods vulnerable (Hockings et al., 2020). Prior to the COVID-19 pandemic, 8 billion people

DOI: 10.4324/b23145-19

visited the world's terrestrial protected areas each year. These visits generated over US$ 600 billion in direct in-country spending and US$ 250 billion in consumer surplus per year (Balmford et al., 2015).

COVID-19 has had a significant impact on CES and their associated components in African protected areas. Loss of revenue has a significant impact on protected area management across many countries of Africa. The global pandemic has affected collaboration between PAs and their strategic partners, according to the International Union for Conservation of Nature (IUCN). Local communities, governmental and non-governmental organizations, researchers, and landowners all fall into this category. Unexpected closures of tourist-dependent community companies resulted in job losses, hindered engagement with investors, and jeopardized several development initiatives that relied on tourism revenue. Unexpected closures of the global tourism industry have resulted in job losses in the tourist-dependent community. Further, it has hampered engagement with investors and risked several development initiatives that relied on tourism revenue (Waithaka, 2020). This has opened up a debate, that how to address these issues closely associated with the CES nexus of protected areas. In line with this, the global initiative through the *United Nations Decades on Ecosystem Restoration 2021–2030* promises a healthier ecosystem for the planet to support poverty alleviation, combat climate change, and avert a mass extinction (UN, 2019).

In this study, three aspects are studied through the lens of COVID-19 and its impact on CES nexus framework by taking Simien Mountains National Park (SMNP) as a case: (1) the extent COVID-19 has impacted the CES around SMNP, (2) the vulnerability of the associated components and, (3) the possible way forward to restore the CES nexus in the post-pandemic world.

14.2 The Framework of Cultural Ecosystem Services Nexus

The pandemic had a connected impact on the CES nexus, in relation to tourism. Whereas, a nexus approach necessitates a system thinking in order to appreciate the multiple linkages and response mechanisms in the social-ecological system which is core to building integrated solutions that address critical areas of sustainable development (Liu et al., 2015). The nexus approach between CES and tourism has only been studied in a few cases. A number of major nexus studies have been done to examine how ecosystem services can help with landscape planning, management, and biodiversity protection (Tengberg et al., 2012; Furst et al., 2017; Rozas-Vasquez et al., 2017). To comprehend CES and the dynamic nature of human-environment interactions, as well as potential synergies and choices between cultural, supporting, provisioning, and regulating ecosystem services, interdisciplinary mechanisms are required (Tengberg et al., 2012). The nexus approach of CES in tourism refers to the set of connected activities that take place at the destination scale among tourists, nature, and locals. Take the Millennium Ecosystem Assessment (2005) into account, which divides CES into aesthetic, spiritual, educational, and recreational categories. Aesthetic values refer to

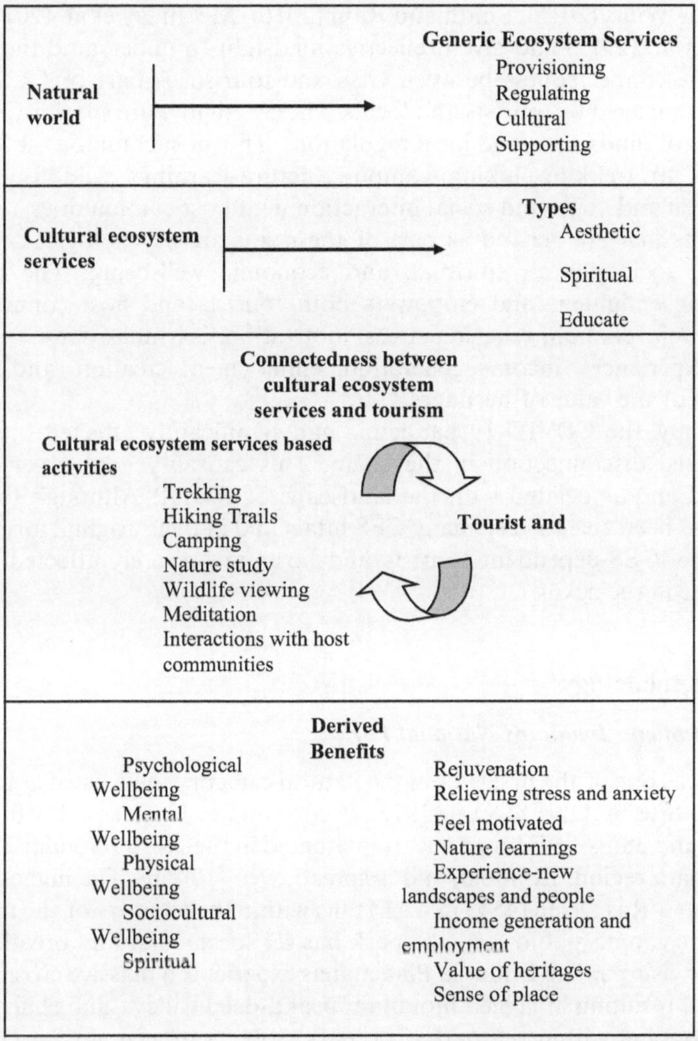

Figure 14.1 Cultural ecosystem services nexus.

Source: Authors.

the beauty or aesthetic enjoyment derived from the ecosystem. Spiritual values indicate the benefits obtained from the ecosystem attributes such as sacred landscape, sacred forests, and sacred species (Figure 14.1).

Educational values are part of environmental education for both formal and informal societies. Recreational values are drawn from ecotourism-related activities that take place in the landscape and often generate economic values for the host communities. The nexus approach in the study is understood through the connectedness between CES and tourism activities. Earlier studies by Willis (2015), Smith and Ram (2016), McGinlay et al. (2018), and Daniel et al. (2012) had given reflective hindsight to understand the nexus, where the connectedness between CES and tourism is part of CES-based activities engaged by tourists and locals. These activities are specific to different types of landscapes and local regulations. The most common CES-based activities are trekking, hiking, camping, nature learnings, wildlife viewing, meditation and yoga, and social interaction with host communities.

Benefits that are derived as part of the nexus are psychological, mental, physical, sociocultural, spiritual, and economic well-being. The derived well-being enlightens and empowers both tourists and host communities through rejuvenation, relieving stress, motivation, learnings, sense of place, newer experiences, income generation, employment creation, and understanding of the value of heritages.

Certainly, the COVID-19 pandemic had significantly affected the nexus and caused disconnection in the chain. This causality is also very much observed and associated with the landscape of SMNP. Although the pandemic has been able to keep many CES intact and in their original form in the landscape, CES-dependent tourists and hosts are severely affected by the pandemic in the nexus.

14.3 Methodology

14.3.1 Simien Mountains National Park

SMNP was one of the first sites in the natural category to be listed as a World Heritage Site by UNESCO in 1978. It coordinates between 13° 10′ 59″ N latitude and 38° 4′ 0″ E longitude. It is situated in the North Gondar Zone of the Amhara region, Ethiopia, and is spread over 412 km^2. The highest point of Ethiopia-Ras Dejen (4543 m.a.s.l.) falls within the territory of the national park. The geomorphology of the park has set spectacular and breathtaking views for visitors. The Ethiopia Plateau has experienced massive erosion over the years, resulting in jagged mountain peaks, deep valleys, and sharp precipices dropping within the park. The park's observation points at Gidirgot and Metgogo, in particular, provide breathtaking views of the high plateau and lowland areas (Figure 14.2).

The Simien Mountains have two seasons, namely dry season and wet season. The dry season is observed from the month of October to April and the wet season is from May to September. More than half of the precipitation falls between June and September. There is a great temperature variance between day and night. The annual daytime temperature ranges between 11°C and 18°C and the night temperature ranges between −2°C and 6°C. The park is also home to globally endemic and endangered plant and animal species with paramount biophysical features. The study by Hurni (1986) suggests that the park has approximately 253 species of plants which belong to 176 genera and 100 families. Of which, nearly 20 species of plants are endemic to the country, and four to five plant species are nearly endemic to Simien

Figure 14.2 (a) Walia ibex; (b) Park landscape; (c) Camping within the park; (d) Gelada baboon.

Photo source: Author (Afera Gebremedihn).

Mountains. SMNP is home to 21 mammal species, of which ten are endemics. The Walia ibex (*Capra walie*) is an endangered wild mountain goat found nowhere else except the SMNP. The Ethiopian wolf (*Canis simensis*), one of the rarest and most endangered canines in the world, is found in the park. More than 182 species of birds are recorded in the park. Of these, six are endemic to Ethiopia. These include the Abyssinian catbird (*Parophasma galinieri*), Abyssinian longclaw (*Macronyx flavicollis*), Ankover sirin (*Serinus ankoberensis*), black-headed siskin (*Serinus nigriceps*), spot-breasted plover (*Vanellus melanocephalus*), and Abyssinian woodpecker (*Dendropicos abyssinicus*) (African Wildlife Foundation, 2015).

14.3.2 Cultural Ecosystem Services of the Park

To take CES into account, the world heritage site offers the following value-based services:

- ***Aesthetic***: To visitors, SMNP provides great aesthetic values through its varied landscape and wilderness. The scenic beauty of the landscape spread through vast open space, unique landforms, vegetation, and water features. Particularly, the observation points at Gidir Got and lmet Gogo of the park provide a strong aesthetic beauty. Apart from it, the landscape is the center of many unique flora and fauna.
- ***Educational:*** SMNP offers strong educational values to tourists and researchers. A guided tour within the park explains details about the trail, geomorphological setups and changes, weather pattern, history, park bird, animal, and plant species (including endemic and endangered), history and practices of wildlife trade, local culture and lifestyle, and traditional medicines.
- ***Cultural:*** SMNP is interlinked with human culture and history because in the history of Ethiopia people used to live in the highlands of the country before the lowland – where ancient civilization had existed. SMNP represents a unique cultural setting that is linked with the stories of great saints and churches. As part of the local culture, tourists can experience local foods and buy local crafts and souvenirs.
- ***Recreational:*** A wide variety of recreational and ecotourism activities are carried out within SMNP. These include trekking, hiking, camping, spotting landscape flora and fauna, horse riding, interacting with locals, experiencing local culture with highlanders and lowlanders, swimming in the Ensiya River, campfire, local dance, song, and farm-based activities. Recreational services generate economic opportunities for the locals who are engaged in tourism.
- ***Social:*** Travel and visit to SMNP allow tourists/visitors to be social. While trekking, camping, and cooking, interactions with other visitors and local settlers give enough time and space to become social.
- ***Spiritual and Religious:*** SMNP and its peaceful and serene environmental settings provide a place to relax and rejuvenate oneself. The place refreshes the spirit of travelers. As the landscape is far away from modern conveniences, it provides inner solace to be spiritual.

14.3.3 Method

This research is based on an explorative study method. In this, the study tries to explore various components of CES in and around SMNP and, further, to understand how the COVID-19 global pandemic has impacted the cultural ecosystem service and its associate nexus at SMNP. Due to the global pandemic and geopolitical constraints, it was difficult to access the study site. Data and information were collected over telephonic interviews. The interviews were conducted in Amharic language and understanding of the information was decoded into English. A total of 126 (n = 126) interviews were conducted, in which 111 informants were male (Male –111) and 15 informants were female (Female –15). The sample informants are park-dependent stakeholders, who were directly or indirectly associated with SMNP prior to the pandemic. Their major association with the park were in the form of tour guides, scouts, cooks, souvenir sellers, and members of the ecotourism association. Questions of the telephonic interviews were based on to underscore the following:

- Major cultural ecotourism services of SMNP.
- Most impacted components of CES nexus matrix.
- Suggestions on revival and strengthening of CES in a post-pandemic situation.

14.4 Results

Local stakeholders were asked to prioritize different CES offered by the park, based on their experiences, types of service demanded, and earlier interaction with tourists and park visitors. The priority level was determined over a seven-point scale, where, 1 – Not a priority, 2 – Low priority, 3 – Somewhat priority, 4 – Neutral, 5 – Moderate priority, 6 – High priority, and 7 – Essential priority. Analysis of frequency was run in SPSS (Statistical Package for the Social Sciences).

The output of the result indicates, as in Table 14.1, that recreation as an ecosystem service has a greater demand among tourists and travelers. At the priority level, it receives 68.8% (F-88) responses as an essential priority, 15.1% (F-19) responses as a high priority, 8.7% (F-11) responses as a moderate priority, 3.2% (F-4) responses as a neutral, 1.6% (F-2) responses as somewhat priority and low priority, respectively. The second sought-after ecosystem services in the park at the priority level is Aesthetic with 57.9%

Table 14.1 Cultural ecosystem service priority level at SMNP

Priority Level	*Aesthetic*		*Educational*		*Cultural*		*Recreational*		*Social*		*Spiritual and Religious*	
	F	*%*	*F*	*%*	*F*	*%*	*F*	*%*	*F*	*%*	*F*	*%*
Essential priority	73	57.9	52	41.3	33	26.2	88	68.8	22	17.5	6	4.8
High priority	19	15.1	18	14.3	8	6.3	19	15.1	17	13.5	3	2.4
Moderate priority	9	7.1	13	10.3	31	24.6	11	8.7	8	6.3	5	4.0
Neutral	7	5.6	8	6.3	17	13.5	4	3.2	30	23.8	7	5.6
Somewhat priority	9	7.1	10	7.9	28	22.2	2	1.6	14	11.1	8	6.3
Low priority	8	6.3	10	7.9	9	7.1	2	1.6	25	19.8	36	28.6
Not a priority	1	0.8	15	11.9	–	–	–	–	10	7.9	61	48.4

n = 126, F – Frequency, and % – Percentage.

(F-73) – essential priority, 15.1% (F-19) – high priority, 7.1% (F-9) – moderate, 5.6% (F-7) – neutral, 7.1% (F-9) – somewhat priority, 6.3% (F-8) – low priority, and 0.8% (F-1) as not a priority response from the respondents. The third important CES within the park are educational, with an essential priority level of 41.3% (F-52), high priority level of 14.3% (F-18), moderate priority level of 10.3% (F-13), neutral priority level of 6.3% (F-8), somewhat priority level, and low priority level of 7.9% (F-10) each, respectively, and 11.9% (F-15) as not a priority level response. Culture – as a cultural ecosystem service – has received the fourth key priority among the park visitors with essential priority level 26.2% (F-33) highest to 7.1% (F-9) lowest priority. The fifth prioritized CES among the visitor were social. It has received the level of 17.5% (F-22) as essential priority, 13.5% (F-17) as high priority, 6.3% (F-8) as moderate priority, 23.8% (F-30) as neutral, 11.1% (F-14) as somewhat priority, 19.8% (F-25) as low priority, and 7.9% (F-10) as not a priority. Further, in this category, spirituality and religion had received the lowest and least priority as CES with the priority level responses of 48.4% (F-68) as not a priority and 4.8% (F-6) as an essential priority.

In order to understand the impact of the pandemic on the CES nexus, the study tries to find out the major vulnerable components of the nexus. Based on the literary findings and interviews, it was found that tourism-dependent stakeholders and the natural ecosystem of the park are more vulnerable components impacted due to COVID-19. Following are the associated impacts:

- ***Loss of Job and Occupation***
  International travel restrictions to Africa in early 2020 had severely impacted the jobs and occupations of tourism operators, tour guides, porters, lodge staff, cooks, and souvenir sellers. These park-dependent stakeholders almost completely lost their jobs and occupations due to the pandemic.
- ***Loss of Income***
  Loss of job and occupation, and salary cut off had made the park-dependent stakeholders more susceptible to facing economic hardship. This has forced many of them to search for another livelihood option in nearby cities like Gondar and others. Whereas, others joined their family for agricultural activities.
- ***Impact on Conservation Initiatives***
  The shutdown of tourism has suddenly cut off the flow of benefits to the park and dependent stakeholders. Sehmi (2020) from African Wildlife Foundation reported that many of the ongoing conservation-based initiatives and activities are severely affected due to the pandemic.
- ***Threat to Natural Resources***
  Further, the Sehmi (2020) from African Wildlife Foundation reported that the shutdown has lessened the patrolling within the park and increased the illegal removal of natural resources from the park. Further, grazing by domesticated livestock within the restricted zone was also observed.

The findings indicate that the pandemic has disconnected the cultural ecosystem nexus of the park and threatened the overall livelihoods of park-dependent stakeholders. Recreation, aesthetics, education, and culture are priority ecosystem services having higher demand among the park visitors. This creates an interconnectedness and interdependencies between the park and locals. Another finding of the study indicates the participation of females in the tourism and allied business are less in the area and they are over-dependent on their male counterparts and family members. This reveals that due to the pandemic, males are more vulnerable in terms of job and income loss. This has a cascading impact on the male-dependent family members. Interviews reveal that the minimal participation of female in tourism businesses in the park are due to the rough geophysical settings of the park, which requires strong physical endurance. This opens up the scope for female participation in park-based tourism in the future. As a whole, the overall findings of the study translate that the pandemic has a multitude of impacts on local livelihood and economy due to loss of job, occupation, and income. The pause in conservation initiatives and less patrolling within the park have negative impacts on the overall ecosystem of the park.

14.5 Conclusion

The restrictions on travel due to COVID 19 are still in play in major parts of the world at the time of writing this article. Thus, it is important to reflect on what has really happened in the national park's context to reflect and provide a way forward for the long-term management of such a crisis, which is associated with local vulnerability and conservation management. Based on the findings of the study, the following recommendation can be considered at the policy and practice level: (i) develop a mechanism for visitor management upon opening of the park for tourists and visitors, (ii) careful planning at the local level must be carried to address issues related to social and economic equity, (iii) incentives or other livelihood-generated opportunities must be in place to avert such crisis in the future, (iv) stronger conservation mechanisms must be carried out to avoid illegal trades of natural resources, and (v) upon the resumption of international travel and reopening of the park, stronger management mechanisms shall imply to restore the CES and strengthen the disconnected nexus.

This study has explored the impacts that the COVID 19 pandemic has presented to the CES of SMNP and associated local communities with the park. The global health crisis has impacted the CES and made local livelihoods and national park management more complex. This has opened up an opportunity to reengage with new audiences, rebuilt the connectedness, and manage the landscape for the public through proper planning and implementation.

References

African Wildlife Foundation 2015, *Simien Mountains National Park Grazing Pressure Reduction Strategy*, viewed 30 July 2021 <https://whc.unesco.org/document/140897>.

Balmford, A, Green, JMH, Anderson, M, Beresford, J, Huang, C, Naidoo, R, Walpole, M & Manica, A 2015, 'Walk on the wild side: Estimating the global magnitude of visits to protected areas', *PLoS Biology*, vol. 13 (2), pp. 1–6.

Ceballos-Lascurain, H 1996, *Tourism, Ecotourism and Protected Areas*, IUCN, Gland, Switzerland, p. 301.

Council of Europe Landscape Convention 2000, Florence, 20 October 2000, European Treaty Series, No. 176, p. 28.

Daniel, TC, Muhar, A, Arnberger, A, Aznar, O, Boyd, JW, Chan, KMA, Costanza, R, Elmqvist, T, Flint, CG, Gobster, PH, Gret-Regamey, A, Lave, R, Muhar, S, Penker, M, Ribe, RG, Schauppenlehner, T, Sikor, T, Soloviy, I, Spierenburg, M, Taczanowska, K, Tam, J & von der Dunk, A 2012, 'Contributions of cultural services to the ecosystem services agenda', *Proceedings of the National Academy of Sciences*, vol. 109 (23), pp. 1–8.

Emerton, L, Bishop, J, & Thomas, L 2006, *Sustainable Financing of Protected Areas: A Global Review of Challenges and Options*, Best practice protected area guidelines series #13.IUCN, Gland, Switzerland, viewed 30 July 2021, <https://portals.iucn.org/library/efiles/documents/PAG-013.pdf>.

Fürst, C, Luque, S & Geneletti, D 2017, 'Nexus thinking – How ecosystem services can contribute to enhancing the cross-scale and cross-sectoral coherence between land use, spatial planning and policy-making', *International Journal of Biodiversity Science, Ecosystem Services & Management*, vol. 13 (2), pp. 412–421.

Hockings, M, Dudley, N, Elliott, W, Ferreira, MN, MacKinnon, K, Pasha, MKS, Phillips, A et al. 2020, 'COVID-19 and protected and conserved areas', *PARKS*, vol. 26 (1), pp. 7–24.

Hurni, H 1986, *Management Plan Simien Mountains National Park and Surrounding Rural Area*, Ministry of Agriculture, Natural Resources Conservation and Development Main Department, Wildlife Conservation Organization, UNESCO, Ethiopia.

International Union for Conservation of Nature (IUCN) 2020, *Conserving Nature in a Time of Crisis: Protected Areas and COVID-19*, viewed 30 July 2021, <https://www.iucn.org/news/world-commission-protected-areas/202005/conserving-nature-a-time-crisis-protected-areas-and-covid-19>.

Kettunen, M & ten Brink, P (Eds.) 2013, *Social and Economic Benefits of Protected Areas: An assessment guide*, Routledge, Adbingdon, UK.

Liu, J, Mooney, H, Hull, V, Davis, SJ, Gaskell, J, Hertel, T, Lubchenco, J, Seto, KC, Gleick, P, Cremen, C, & Li, S 2015, 'Systems integration for global sustainability', *Science*, vol. 347 (6225), pp. 1–9.

McGinlay, J, Parsons, DJ, Morris, J, Graves, A, Hubatova, M, Bradbury, RB, & Bullock, JM 2018, 'Leisure activities and social factors influence the generation of cultural ecosystem service benefits', *Ecosystem Services*, vol. 31 (2018), pp. 468–480.

Millennium Ecosystem Assessment 2005, *Ecosystems and Human Well-Being: Synthesis*, Island Press, Washington, DC, viewed 30 July, 2021, <https://www.millenniumassessment.org/documents/document.356.aspx.pdf>.

Rozas-Vasquez, D, Fürst, C, Geneletti, D, & Munoz, F 2017, 'Multi-actor involvement for integrating ecosystem services in strategic environmental assessment of spatial plans', *Environ Impact Assess Review*, vol. 62, pp. 135–146.

Sehmi, H 2020, *Nurturing resilience among wildlife tourism-dependent communities*, viewed 1 August, 2021, <https://www.awf.org/news/nurturing-resilience-among-wildlife-tourism-dependent-communities>.

Smith, M & Ram, Y 2016, 'Tourism, landscapes and cultural ecosystem services: A new research tool', *Tourism Recreation Research*, vol. 42, pp. 1–7.

Tengberg, A, Fredholm, S, Eliasson, I, Knez, I, Saltzman, K, & Wetterberg, O 2012, 'Cultural ecosystem services provided by landscapes: Assessment of heritage values and identity', *Ecosystem Services*, vol. 2 (2012), pp. 14–26.

United Nations 2019, *Decade on Ecosystem Restoration: There Has Never Been a More Urgent Need to Restore Damaged Ecosystems Than Now*, viewed 10 October, 2021, <https://www.decadeonrestoration.org/>.

Waithaka, J 2020, *The Impact of Covid-19 Pandemic on Africa's Protected Areas Operations and Programmes*, viewed 30 July, 2021, <https://www.iucn.org/sites/dev/files/content/documents/2020/report_on_the_impact_of_covid_19_doc_july_10.pdf>.

Willis, C 2015, 'The contribution of cultural ecosystem services to understanding the tourism–nature–wellbeing nexus', *Journal of Outdoor Recreation and Tourism*, vol. 10, pp. 38–43.

15 Restoration of Ecosystems Services in Post-COVID-19 Kenya

A Cost–Benefit Analysis of Green Financing

Isaac Kimunio and Francis Gitagia

15.1 Introduction

Tourism is one of the global major economic sectors (Lew, 2011). In 2019, it accounted for 7% of global export trade making it the third-largest in the export category (Bhaduri & Pandey, 2019). However, COVID-19 has taken a heavy toll on the tourism sector largely due to travel advisories issued to contain the pandemic (UNCTAD, 2021). Accordingly, global tourism recorded the worst performance in 2020 as the international arrivals dropped by 74% (UNCTAD, 2021). This parallels the 4% decline recorded in the global financial crisis of 2009. Globally, the destinations received 1 billion fewer international arrivals in 2020 compared to 2019 (UNCTAD, 2021). The estimated loss resulting from the ban on international travels amounted to USD 1.3 trillion with a direct tourism job loss of between 100 and 120 Million (UNWTO, 2021). The international travel bans put in place to curb the spread of the COVID-19 pandemic affected the regional economies including the tourism sector in Sub-Sahara Africa (Rogerson & Baum, 2020). Similarly, the Ministry of Tourism and Wildlife report of 2020 indicates that the Tourism sector in Kenya was adversely affected with the Meetings, Incentives, Conferences and Exhibitions (MICE) sector losing roughly KSh163.56 billion in the first half of the year 2020 (GoK, 2020b). IUCN (2021) posits that there is a nexus between tourism activities and ecosystems by virtue of recreational values offered by the ecosystems.

According to the Kenya National Climate Change Action Plan for 2013–2017 (GoK, 2010), ecosystems in Kenya are highly degraded leading to extreme climate change which could cost the economy as much as $500 million a year. With this, the role of tourism management in the restoration of ecosystems cannot be overemphasized. Consequently, the United States Agency for International Development (2011) indicates that Sustainable tourism management has the potential to not only alleviate the possibly harmful consequences of visitation to a natural area, but it can also act as a potent means of supporting conservation and restoration of the ecosystems upon which it relies on. In this regard, World Economic Forum (2021) asserts that green financing investments play an important role in delivering several Sustainable Development Goals (SDGs) including environmental sustainability. This type of investment

DOI: 10.4324/b23145-20

is an important element that could be utilized as a tool for sustainable tourism management to restore the Kenyan ecosystem.

Lack of attention to social dimension of ecological interventions poses a serious shortcoming to the restoration of ecosystems. An unequal playing field often excludes women from participation in relevant decision-making processes around restoration projects and from receiving direct benefits (Bioversity International, 2019). Indeed, ignorance of gender in the restoration of ecosystem may create exclusions for vulnerable populations who depend on such an ecosystem. This chapter introduces the concept of green financing investments as elements of sustainable tourism management. Green financing takes into consideration the environmental sustainability which in turn supports the tourism industry.

15.2 Nature of Ecosystem Services and Their Relationships to Tourism

An ecosystem is a geographic area containing biotic or living parts, as well as abiotic, nonliving parts, and maybe a big area like an ocean or a small area like an oasis in a desert (Walter & Breckle, 2013). Ecosystems can broadly be categorized into terrestrial and aquatic ecosystems.

The critical function of every ecosystem is the regulation of essential ecological processes that support life systems and thereby rendering stability of those ecosystems through the various food chains, ecological pyramids, and food webs that are active and live within the area. The various food chains include, for example, plants – grasshopper – mice – cobra – eagle. Ecosystems are not just significant in supporting various forms of life in an area but they harbor enormous economic potential that may transform the lives and welfare of people if proper policies are put in place about them. For instance, the biotic and abiotic parts of ecosystems have immense potential for tourism with consequent positive economic impacts. Ecosystem tourism in the world generates a lot of revenue and taxes for businesses and governments including creation of various types of employment.

In view of the recognized significance of ecosystems in supporting various forms of life systems, the United Nations, in 2001, launched the global Millennium Ecosystem Assessment (MA). The MA observed that human activity is creating a world that is likely to be degraded substantially for future generations. Humans' alteration of the natural environment has resulted in major societal advantages. These advantages, however, have been matched by rapidly rising costs as a result of environmental degradation. Because it affects a chain of diverse sorts of biotic parts, ecosystem degradation could have devastating and long-term consequences.

15.3 Scope and Characteristics of Ecosystem Tourism in Kenya

Kenya has abundant ecotourism sites covering both terrestrial and aquatic ecosystems as well as cultural tourism. The country has designated about

7.5% of its land mass as wildlife conservation areas. Thus, there are 23 National Parks and 28 National Reserves, six World Heritage Sites plus six marine reserves. The conservation areas have a wide range of wildlife including lions, elephants, zebras, and over 1,070 bird species. The 536 kilometers of Kenya's coastline are largely beautiful sandy beaches with coral reefs. Most of the coastal reefs offer important ecosystem services and tourism features, especially sports and diving. The country is also bestowed with savannah grasslands; beautiful geographic landscapes; and forests and freshwater lakes that are an important anchor to tourism.

Tourism is one of the major industries in Kenya and the country now has over 100 eco-rated tourism facilities in line with its Kenya Vision 2030 economic development blueprint that was launched in 2008. The Kenya Tourist Board which is vested with the responsibility of overseeing tourism development in the country works on four principles toward that front. The four principles are environmental conservation; education and empowerment; social responsibility; and culture and heritage preservation. Due to resolute efforts by the Kenya Tourist Board, the sector has recorded positive progress over the years and received 2.05 million international tourists in 2019 compared to 2017 when only 1.47 international tourists visited the country.

However, the outset of the COVID-19 pandemic drastically cut down on those gains due to the global lockdown that was meant to control the spread of the disease. The country stopped all leisure activities; hotel rooms lost revenue estimated at USD 511 million; 36,800 airline jobs were lost with associated revenue losses estimated at USD 125 million; and, the MICE sector that was impacted by the cancellation of gatherings was estimated to have lost Ksh. 163.56 billion (GoK, 2020a). The COVID-19 pandemic has affected not only the tourism industry in Kenya but that of the whole world.

15.3.1 Kenya's Ecosystem Services

Kenya's ecological zones and habitats include lowland and mountain forests, forested and open grasslands, semi-arid scrubland, dry woodlands, inland aquatic, and coastal and marine environments. Agriculture, livestock production, energy production (through hydroelectric developments), fisheries, and local and international tourists all benefit from Kenya's wetlands. Although Kenya's biodiversity is well protected, several unprotected places are causing the country's status to rapidly deteriorate owing to pressures that have resulted in numerous conservation issues.

Forests underpin Kenya's tourist and agriculture sectors, making them the backbone of the country's economy. They also help people make a living by providing food, medicine, wood for construction and fuel, as well as services like water catchment areas and tourism attractions. Kenya's freshwater and saline habitats cover around 8% of the country's land area. Biodiversity, food production, hydrological stability, tourism, mineral cycling, and socioeconomic development are all essential factors. Freshwater and saline lakes and wetlands of this type are important migratory routes for wild animals and thousands of birds, attracting tourists. The Kenyan coast's marine waters and mangrove areas are noted for their abundant biodiversity, much of which is still untouched, except in encroached areas. These are vital resources for the tourism business in the country.

15.3.2 Main Threats to Kenya's Ecosystem

Urbanization, climate change, poaching, high population pressure, escalating poverty, community conflicts and political instability, poor land-use practices, insufficient laws, policies, institutional framework, poor education, and inadequate community involvement are the major threats to biological diversity in Kenya (Otianga-Owiti et al., 2021). Invasive species prevalent in Kenyan lakes, such as the Nile perch and water hyacinth, as well as rising pollution, soil degradation, and poor land-use practices, are all dangers. Furthermore, with the construction of new hotels and access infrastructure, the tourism sector is encroaching on delicate marine habitats and coastal areas.

15.4 How Ecosystem Services Support Tourism Production and Consumption in Kenya

Kenya is a popular destination in sub-Saharan Africa for eco-tourism. Eco-tourism is defined as responsible travel to natural destinations that conserve the natural environment and improves the well-being of the surrounding communities. In Kenya, ecotourism is supported by the public and private initiatives buoyed by its rich wildlife and biodiversity.

15.4.1 Integrating Ecosystem into Tourism Industry Growth and Building Eco-Tourism Conservation Skills

Ecosystem and tourism performance are closely related. In fact, the success of the tourism industry directly depends on the performance of biodiversity. In addition, several tourism businesses and activities are based on ecosystem services. Genuine eco-tourism, in the end, refers to tourism that has no harmful impact on eco-systems and contributes positively to the destination's social and environmental well-being (Fennell & Cooper, 2020). For instance, the recreational value of the ecosystem is one of many cultural services that the ecosystem provides to tourists, along with spiritual and educational values. If tourism activities are carried out in a sustainable and environmentally friendly manner, the economy will gain much through:

- Increased economic output and economic development opportunities.
- Increased fund generation for conservation and environmental protection through donations and visits to parks.
- Increased biodiversity conservation awareness and support of social activities for the surrounding communities.
- Provision of a solution to reduced international tourism earnings as a result of COVID-19.

15.4.2 Integrating Ecosystem into Tourism Growth

Integrating biodiversity issues in planning and operational decisions for tourism areas like hotels is important for increased conservation activities. It helps to achieve financial sustainable growth from tourist attraction features. The hospitality industry's success depends on a healthy ecosystem because those biodiversity – and the wildlife, habitats, landscapes, and natural attractions that comprise them – are often the very thing that attracts tourists to the destination in the first place.

15.4.3 Building the Ecotourism Business Skills of Conservation Organizations

The ecotourism approach is important in encouraging general conservation. As a result of creating value for an ecosystem, animals, or landscape, the ecotourism approach can generate funds to protect and conserve these important natural resources. Through ecotourism, countries can reduce the poverty rate, generate employment and income for surrounding communities, and offer alternative means of livelihood. Overall, this helps reduce natural resource overuse and dependability. If well-conceived and managed, ecotourism can accelerate sustainable development around the globe.

15.5 Economic Valuation of Ecosystem Goods and Services

Ecosystem services are the benefits that people obtain from ecosystems whereas Human well-being refers to people's ability to live a life they value (McMichael et al., 2005). The flow of ecosystem services and the level of human well-being are liked in both directions. Thus, it is important to put some monetary value on the environmental commodities in our ecosystems. If monetary value is not attached to the environmental commodities, they will remain implicitly zero which shall not aid in policy formulation on such commodities. Values of our ecosystem commodities may broadly be classified into two, that is, use values and non-use values. Use values are those which people derive from direct use of the ecosystem goods such as timber from trees. Non-use values are those values that people assign to the ecosystem goods (including public goods) even if they never have and never will use them and include option values, bequest values, existence values, and altruistic values. Environmental experts have developed how these different values may be estimated in monetary terms and such methods include the following.

15.5.1 Hedonic Pricing Method

Hedonic pricing is a mechanism for estimating economic values for ecosystem or environmental services that have a direct impact on market prices. Environmental quality, such as air pollution, water pollution, or noise; and environmental amenities, such as scenic vistas or closeness to recreational sites, can all be estimated using the hedonic pricing method. The hedonic pricing method's main idea is that a marketed good's price is related to its features or the services offered.

15.5.2 Travel Cost Method

The travel cost approach is used to estimate the economic usage values associated with ecosystems or recreational locations. The method can be used to calculate the benefits or costs associated with changes in recreational site access costs, the deletion of an existing recreational site, or the inclusion of a new recreational site. The travel cost method's main idea is that the time and travel costs that people incur to visit a place represent the price of access to that site. As a result, the number of trips people make at various transport costs can be used to determine their willingness to pay to visit the location. This is similar to gauging people's willingness to pay for a sold commodity by looking at the quantity demanded at various pricing.

15.5.3 Contingent Valuation Method

The Contingent Valuation Method (CVM), which has its roots in Consumer Theory, is the most extensively utilized valuation technique. The major theoretical notions underlying non-market valuation of private and public goods are Willingness to Pay (WTP) and Willingness to Accept Compensation (WTC). In theory, the CVM approach directly estimates the true Hicksian welfare measures of these changes by eliciting explicit statements of how much income consumers are WTP to ensure a welfare gain (or avoid a welfare loss) or how much income they are WTC to ensure a welfare loss (or forego a welfare gain). In principle, this is equal to explicitly assessing the real Hicksian welfare measurements of these changes, which evaluates welfare change as the money income adjustment required to maintain a constant level of utility before and after a change in the provision of environmental goods/services.

15.6 Green Financing

Green financing is a broad term that encompasses investment in green environment that promotes emissions of less greenhouse gases as well as channeling of funds to sustainable development (PWC, 2013). However, United Nations Environmental Programs (UNEP) defines green financing as the use of all funds – public, private, and micro-lending – to support investment in sustainable development that is environmentally friendly to fight climate change (UNEP, 2021).

Green financing was established by 194 countries to encourage investments that reduce and or limit greenhouse gas emissions in most developing countries as well as assist the vulnerable ones to be resilient to the effect of climate change and respond to climate change through the fund platform. The system also plays a vital role in delivering a number of SDGs by assisting countries to align regulatory frameworks so that countries comply with green lending requirements.

The United Nations, through its environmental agency, has been working with countries, donors, and financial partners to align funds to the Sustainable Development Agenda 2030 through the financial markets due to the globalization of the economy to help shape individual behavior, particularly in production and consumption patterns both today and tomorrow (UNEP, 2021). The fund mainly focuses on enhancing the capacity of the public sector to creating a conducive environment for green investment, encouraging public–private partnerships (PPP) on financing modalities by incorporating both green bonds and green bills, and lastly, building the capacity of the community on micro-lending (Bjerborn Murai & Kirima, 2015). The report further states that in order for the platform to expand, countries need to change their frameworks, harmonize financial needs, escalate green financing to different sectors within the economy, and also align financing decision priorities with SDG implementations. Further, countries can engage in clean and green technology and green economies and also promote climate-smart blue economy through the use of green bonds.

In Kenya, the program was launched in 2017 by developing domestically green bond markets to promote financial markets as well as sectors with an issuance of about 4.3 billion shillings to avail over 5,000 families with affordable houses within Nairobi which are environmentally friendly. International Finance Corporation is focusing exclusively on the private sector in the economy to develop and leverage the private sector to come up with viable ways that ensure environmentally and socially sustainable development (Bjerborn Murai & Kirima, 2015). The program forms a basic part of the environment, social, and governance system that supports the growth of the economy in a clean, resilient and sustainable way geared toward reducing pollution, adaptation, and mitigation of the damages caused to the environment.

Kenya's economy largely depends on agriculture and tourism. Therefore, failure to incorporate environmentally friendly investment championed by the green financing program may lead to a financial loss of about USD 500 million per year of Gross Domestic Product. The loss, that is due to intense climate change that the green financing is focusing to address, is attributed to declining in economic activities in fishery, energy, manufacturing, agriculture, and tourism.

Even though the platform has achieved several milestones in the realization of sustainable development that is environmentally friendly, the capacity of the system to mobilize adequate funds to facilitate realization of the goals has been faced with microeconomic constraints such as bearing the burden of the externalities caused by environmental polluters, particularly developed nations that fail to internalize the external externalities (OECD, 2017). Failure by members to share information perfectly, incapacity to carry out analytical evaluation of the greenhouse gas effect, and failure to clarify what exactly the term green means are some of the challenges green financing implementation faces in most countries.

15.6.1 Green Financing Prospects

Most of the time, investors and business people prioritize economic growth at the expense of the environment, ecosystem, and social justice. However, the climate crisis has exposed most tourism businesses to greater risk, while an expected boost from the government is not forthcoming. So, what role can sustainable finance play in mobilizing funds for restoration, climate, and other green investments?

Sustainable or green finance involves making investment decisions that consider not only the financial returns of tourism but also social and corporate, environmental, and governance aspects as expressed in Table 15.1. These consist of the environmental considerations approach, that is, climate adaptation and mitigation measures, the conservation of ecosystem, and the circular economy (Gücklhorn, 2021).

Climate finance is a subset of sustainable finance and aims at providing mitigation and adaptation actions that will address climate change. It can be financed by the central government or private partners.

15.6.2 Methodology

The study adopted qualitative descriptive phenomenology research design where description analysis of the patterns and relationship between humans and ecosystem on cost–benefit economic valuation of green financing were considered. A desktop review of previous researches and published papers was done during the period of study in order to obtain data that were used to evaluate the economic ecosystem that exists as a result of green financing.

15.6.3 Cost–Benefit Analysis of Green Financing

Green financing is concerned with investment in sustainable development which is environmentally friendly and results in the betterment of the well-being of the households by encouraging social equity. Sectors focused on by green financing are the agriculture, fishery, housing, forestry, transport, and coastal resources.

Table 15.1 Environmental, Social, and Governance Aspects of Sustainable Finance

Environmental	*Social*	*Governance*
• Climate • Biodiversity • Greenhouse gas emissions • Water consumption • Pollution • Waste management • Use of renewable energy	• Inequality • Human rights • Labor relations • Community impacts • Supplier relationships • Working conditions • Health and safety • Diversity	• Executive compensation • Transparency • Board independence • Shareholder rights • Taxation • Anti-corruption • Political contributions

Source: Gücklhorn (2021).

Green financing ensures that investments in fishery and coastal sectors of the economy are of benefits and improves the well-being of the households. Fishermen carry out fishing without adhering to regulations put in place by the government to maintain the ecosystem, leading to overfishing and hence depleting the ocean, seas, and lakes which are very important resources. Failure to follow these rules calls for government intervention by imposing the rules and regulations as well as ensuring they are followed by the fishermen, and this involves a cost incurred by the government agencies. The cost will both be in short run and long run, in the short run, the government will enforce the regulations on the fishermen to ensure compliance while, at the same time, fishing activities continue to generate income for the fishers, and this constitutes benefits. However, in the long run, fishermen will comply with the rules and regulations and therefore, the benefits realized outweighs the costs accrued by the government in ensuring sustainable development within the coastal and fishery sectors of the economy maintaining the ecosystem within the sectors leading to continuous growth and existence of the sea and lake resources.

The agricultural sector plays a key role in the economic development of the developing countries like Kenya. Agriculture provides food to the rural dwellers; the society also receives income as well as employment opportunities from the sector. Even though the sector ensures the country is food secure and assists in poverty reduction, improvements in the sector faces a number of challenges to ensuring green growth and sustainable development. The sector is affected by two key agents – farmers and pastoralists. Farmers would want to increase the size of their farmland by carrying out deforestation; the process leads to a reduction in forest cover and hence interferes with the forest ecosystem that the green financing focuses on maintaining. Similarly, pastoralists would encroach the gazette forest areas for greener pasture for their livestock. The two activities benefit both the farmers and the pastoralists as the farmers increase the quantity of output which they sell to generate income (benefit), pastoralists, on the other hand, ensure that the livestock feed on greener pasture inside the forest and becomes healthy, which they later sell to receive benefits in the form of income. The government incurs a cost of protecting the forest by fencing around to keep both the farmers and the pastoralists away as well as employing forest guards for security purposes; this persists to the future and hence, the costs of maintaining the forest ecosystem outweigh the benefits derived by both the farmers and the pastoralists. Green financing subsidizes the cost incurred to ensure green growth as well as sustainable development in the use of forest resources that is socially beneficial to all by reducing pollution, minimizing emissions of greenhouse gases, and, at the same time, efficiently using the forest resources that maintain biodiversity.

Public transport is the most preferred means and an eco-friendly way of moving people from one place to another. This is because it can entice people to get out of their automobiles and offer a cost-effective means of combating climate change. The most used modes of transportation are the buses, trains, and motorcycles which normally utilize fossil fuels that undergo incomplete combustion, emitting a large amount of greenhouse gases that pollute the environment and hence interfering with the ecosystem. Green financing funds means of transport which are environmentally friendly such as shifting to the use of solar drive engines and also electric power drive engines that do not produce pollutant gases; this ensures biodiversity in the transport sector and green transportation which is environmentally friendly thereby maintaining the ecosystem as it has low impacts on the environment. In the short run, the costs are more than the benefits of green transportation but in the long run, the benefits received through the reduction of greenhouse gases that pollute the environment exceed the costs and hence it is more beneficial to adopt green transportation means rather than brown transportation in both the private and public means.

15.6.4 Importance of Green Financing

Considering forest cover and agriculture make up over 30% of solutions to climate change, it is ironic that they receive less allocation from climate financing funds. The climate fund will need to spend USD 600 to 800 billion per year to slow and reverse the biodiversity crisis by 2030. Raising such an amount will require a coordinated effort from both private and public entities to fill the climate finance gap. The cases below illustrate the importance placed by governments around the world on green financing. They allocate part of their budget to fund green ecosystems.

In Kenya, the government has shown great importance in green financing. This is demonstrated by its allocation of green financing in the budget as expressed in Table 15.2.

Table 15.2 shows that Green financing in Kenya is implemented by the Ministry of Environment and Forestry under a program-based budget where mitigation and management of the ecosystem allocated USD 2,507,258 for goods and services in the financial year 2020/2021. This was increased to USD 2,996,696 in 2021/2022 and projected to increase to USD 3,129,779 and USD 3,269,703 in 2022/2023 and 2023/2024, respectively. This is to ensure

Table 15.2 Green Financing Expenditures in Kenya

	Green Financing-Kenya			
Categories	*Baseline Estimates (USD)*	*Estimates (USD)*	*Projected Estimates (USD)*	
	2020/2021	2021/2022	2022/2023	2023/2024
Goods and Services	2,507,258	2,996,696	3,129,779	3,269,703
Social Benefits	147,934	64,348	62,544	64,348
Water Tower rehab & Cons	5,666,606	5,628,721	5,628,721	5,628,721

Source: GoK (2021).

that goods and services provided are of great benefit to human society within the ecosystem they provide. It is estimated that if we fail to check on the ecosystem through green financing, then about 3.2 billion people are expected to lose their well-being and we could have a 10% global Gross Domestic Product (GDP) loss every year (Hourcade et al., 2021).

The ministry also allocated USD 147,934 in the 2020/2021 financial year to social benefits. This is estimated to reduce to USD 64,348 in the 2021/2022 financial year, then further projected to decline to USD 62,544 in 2022/2023 (GoK, 2021). This may be due to non-commitment by the government to support the livelihood of the public as well as the efforts made to protect the ecosystem to ensure sustainable development fostered by green financing. On the same note, the ministry did allocate USD 5,666,606 in 2020/2021 and USD 5,628,721 in 2021/2022 for water tower rehabilitation and conservation. The allocation is projected to remain constant in the 2023/2024 financial year (GoK, 2021). In Kenya, this constant budgetary allocation is due to insufficient funds caused by the poor performance of the economy necessitated by the COVID-19 pandemic which resulted in the closure of key economic sectors such as agriculture, tourism, and hospitality as well as social service sectors putting constraints on the budget. The emergence of the COVID-19 pandemic interrupted most environment-friendly investments that results in low carbon emissions that are climate-resilient; therefore, calling the developing countries like Kenya to scale-up financial decisions during the budgeting process to align their financial obligations with sustainable development is the goal of green financing.

The green funds are mainly sourced from issuance of green bond targeting the provision of environmentally friendly houses which are affordable within Nairobi and its environment and also focuses on involving both the local and international potential investors to invest in sustainable green investment in the country.

Table 15.3 shows the Green Climate Fund (GCF) initiated project in Lao to support the vulnerable community suffering from perennial floods by strengthening the capacity of the ecosystem around by controlling the flow of water as well as reducing the exposure of the public to climate effect. In order to achieve this, the GCF offered a grant amounting to USD 10 million and co-financing to a tune of USD 1.5 million to help curb economic losses as a result of floods majorly borne by the public residing in these areas. The GCF through the funding program aims at changing the public mentality from gray infrastructure like dams to green infrastructure that conserves the environment as well as improves the well-being of the households. The funds also redirect the run-offs from the source rather than constructing drainage systems; this allows the development of the urban centers in Laos city geared toward sustainable development which is a climate-resilient flood management system and, at the same time, does not interfere with the local ecosystem. Similar projects were also done in Benin and Namibia with grants amounting to USD 10 million and USD 9.1 million, respectively, to maintain ecosystems in these countries.

Table 15.3 Some Sources of Green Finance

Objectives	*Grant (USD Million)*	*Co-financing (USD Million)*	*Total (USD Million)*
Building resilience of urban population with ecosystem-based solutions in Lao PDR	10.0	1.5	11.5
Green Climate Fund ecosystem-based adaption program in the Western Indian Ocean	35.8	29.8	65.6
Climate proofing food production investments in Imbo and Moso basins in the Republic of Burundi	10.0	21.7	31.7
Ecosystem and ecosystem services	540.3	–	540.3
Toward ending drought emergencies: Ecosystem-based adaptation in Kenya's arid and semi-arid rangelands	23.2	11.4	34.6

Source: Hourcade et al. (2021).

In the Western Indian Ocean, green financing funded a project amounting to USD 65.6 million of which USD 35.8 million was grant while USD 29.8 million was co-funded. The project is in the four Indian Ocean Islands (Mauritius, Madagascar, Comoros, and Seychelles) to protect them against extreme climatic changes and maintain ecosystems in the area. The islands fully depend on the ecosystem and the resources around the region that requires maintenance and development in accordance with green financing goals. The funding allows the islands to adapt to changes in climate which is ecosystem-based by developing resilient as well as supporting the rehabilitation and conservation of the ecosystem. Climatic changes in the Indian Ocean result in increase in water volume that submerges the Islands, leading to low economic activities as a result of diminishing land cover in these countries and hence the need for maintenance and sustainable development that is environmentally friendly for the benefits of the households in the Islands. Some of the challenges the project seeks to address are changing rainfall patterns, frequent rises in temperature, increase in sea level, and intensified storms which affect the socio-economics, environment, and ecosystem of the households. Since the Islands vary in terms of geographical, cultural, and climatic conditions, they share similar economic statuses such as that of agriculture, fishing, and tourism and also face similar vulnerabilities (Canales, Atteridge & Sturesson, 2017). This calls for GCF to intervene for the benefit of the entire Island's sustainable development and improvement in the ecosystem.

In the Republic of Burundi, a project branded "climate-proofing food production investment" funded by GCF amounts to USD 31.7 million to address food security in the country occasioned by adverse climate change affecting the majority of the rural population. The country is annually faced with droughts and floods in the plains parts of Imbo and Moso hindering agricultural activities and resulting in reduced production of about 5–25%, impacting the economic growth of the vulnerable country to about 2.4% (GCF Report, 2021). Funding of the project ensures that the vulnerable farmers in these two regions become resilient to climate change, and, at the same time, helps them improve agricultural productivity and ensures that the two regions are food secure and the entire country by adopting agroecosystem practices to ensure soil and water conservation is carried out in the poor landlocked country. The green financing will further ensure that three key activities in line with the green financing mandate are implemented; the first is that both water and soil management improves by adopting the best agroecosystem land-use practices. Also, GCF team trains the farmers on agroecosystem practices that ensure water and soil conservation measures are practiced in the region. Lastly, the households develop a green environment for the conservation of the significant resources in the livelihood of the households.

In Kenya, a project aimed at increasing livestock resilience against climate change and reinstating land use in Arid and Semi-Arid Lands (ASALs) via proper control of the ecosystem received funds amounting to USD 34.6 million from GCF. Grants constituted USD 23.2 million while USD 11.4 million was co-funded. The fund focused on 11 counties in ASALs to improve their abilities to adapt to climatic changes.

15.6.5 The Main Providers of Green Finance

Corporations are the largest source of climate-related funds; they do this through direct contributions to climate funding and investments and especially in renewable energy, transportation, and infrastructure development. Banks also account for a large percentage of the financial resources available for green investments and mitigation. Many countries, through policy and regulatory tools, central banks, and regulatory authorities, can also influence the finance sector's progress toward green investments. Similarly, international financial institutions can also help scale up green investments by using instruments like green bonds and influencing global financial regulation to support the restoration of ecosystem services.

Green investment institutions and some development banks, for instance, provide funds for projects related to economic sustainability and development, respectively. Furthermore, international institutions such as the United Nations can only contribute limited funds but may set the international agenda on environmental sustainability issues and assist coordinate climate funding sources.

Climate funds, such as the Climate Investment Fund, the Adaptation Fund, and the GCF, are multilateral funders that support climate adaptation and mitigation through individual state contributions. Green investment finance and institutional assistance are also determined by national and regional governments. They can also help build and construct dedicated domestic investment sources like national climate and environmental providers and funds. Lastly, private-sector financiers of the green economy are institutional investors, such as pension funds, sovereign wealth funds, and insurance companies. Furthermore, stock markets might specialize in environmentally friendly and sustainable investments. This is in addition to the tourism sector's and stakeholders' green investment funds and investment may help the restoration of ecosystem services.

15.6.6 How Can Sustainable Finance Help Support Ecosystem Restoration?

Ecosystems actively contribute to economic output as an input to the economic activity process, as well as indirectly through their effect on the productivity of other inputs of production. Industries that produce a clean and healthy natural environment also promote growth. Ecosystems supply the raw elements for economic commodities and service creation. Ecosystems have a wide range of effects on both labor quantity and quality (Everett et al., 2010). There's also evidence that having more green areas makes it more likely for people to engage in and maintain physical activity, which is a vital component of both physical and mental health. Finally, a clean and healthy atmosphere may be an effective strategy for attracting and retaining capital.

Government intervention is needed to come up with policies to manage the provision and use of ecosystem resources in a way that supports improvements in prosperity and well-being, for current and future generations. A range of policies are available including:

- Encouragement of eco-tourism.
- Market-based instruments, such as the use of taxation to improve the welfare of local communities and payments for environmental stewardship.
- Public spending and technology programs such as electric fences in parks, development of flood infrastructure, and public procurement of sustainable products.
- Providing public information and other measures to remove barriers to behavior change.
- Measures to improve the efficiency of the use of available resources by companies.
- Guidelines for the correct pricing of ecosystem resources.
- Investments that help the economy adapt to climate impacts, such as afforestation.
- Establish appropriate financing mechanisms: These may include the establishment of a national forest or environmental funds or public incentive systems.

References

Bhaduri, K. and Pandey, S., 2019. Sustainable smart specialization of small-island tourism countries. *Journal of Tourism Futures* 6(2), 121–133.

Bioversity International, 2019. https://www.bioversityinternational.org/e-library/publications/detail/bioversity-international-financial-statements-2019/

Bjerborn Murai, C., and Kirima, W. 2015. Aligning Kenya's Financial System with Inclusive Green Investment: Current Practice and Future Potential to Mobilize Investment in a Sustainable Economy. International Finance Corporation, Washington, D.C. © World Bank. https://openknowledge.worldbank.org/handle/10986/26321 License: CC BY-NC-ND 3.0 IGO.

Canales, N., Atteridge, A. and Sturesson, A., 2017. *Climate finance for the Indian Ocean and African small island developing states*. Working Paper 2017-11, Stockholm Environment Institute.

Everett, T., Ishwaran, M., Ansaloni, G.P. and Rubin, A., 2010. Economic growth and the environment. MPRA Paper No. 23585, posted 01 Jul. 2010 00:37 UTC Online at https://mpra.ub.uni-muenchen.de/23585/

Fennell, D.A. and Cooper, C., 2020. Sustainable tourism. In *Sustainable Tourism*. Channel View Publications.

GCF Report. (2021). Climate proofing food production investments in Imbo and Moso basins in the Republic of Burundi. Available online at https://www.greenclimate.fund/project/sap017

Government of Kenya, 2010. National Climate Change Action Plan. 34. Available online at www.environment.go.ke

Government of Kenya, 2020a. Impact of COVID-19 on Tourism in Kenya, the Measures Taken and Recovery Pathways. Report. Available online at https://www.tourism.go.ke/wp-content/uploads/2020/07/COVID-19-and-Travel-and-Tourism-Final-1.pdf\

Government of Kenya, 2020b. *COVID-19 and Tourism in Kenya: Impact, Measures Taken and Recovery Pathways*. Ministry of Tourism and Wildlife National Tourism Risk and Crisis Management Committee.

Government of Kenya, 2021. 2021 Budget Review and Outlook Paper. Available on the website at: www.treasury.go.ke

Gücklhorn, A., 2021. How to mobilize funds for the planet. https://news.globallandscapesforum.org/40996/40996-what-is-sustainable-finance/

Hourcade, J.C., Glemarec, Y., de Coninck, H., Bayat-Renoux, F., Ramakrishna, K. and Revi, A., 2021. *Scaling Up Climate Finance in the Context of COVID-19: A Science-based Call for Financial Decision-makers*. South Korea: Green Climate Fund.

Lew, A.A., 2011. Tourism's role in the global economy. *Tourism Geographies* 13(1), 148–151.

McMichael, A., Scholes, R., Hefny, M., Pereira, E., Palm, C., and Foale, S., 2005. Linking Ecosystem Services and Human Well-being. In Capistrano, Doris, Samper K., Cristián, Lee, Marcus J., and Raudsepp-Hearne, Ciara, (eds.) *Ecosystems and Human Well-being: multi-scale assessments. MIllenium Ecosystem Assessment Series*, Vol. 4. Island Press, Washington DC, USA, pp. 43–60.

OECD, 2017. Green financing: Challenges and opportunities in the transition to a clean and climate resilient economy. *Financial Market Trends* 2016(2), pp. 63–78.

Otianga-Owiti, G.E., Okori, J.J.L., Nyamasyo, S. and Amwata, D.A., 2021. Governance and Challenges of Wildlife Conservation and Management in Kenya. *Wildlife Biodiversity Conservation*, pp. 67–99

Pricewaterhouse Coopers Consultants, 2013. Exploring green finance incentives in China. *Final Report*. https://www.pwccn.com/en/migration/pdf/green-finance-incentives-oct2013-eng.pdf

Rogerson, C.M. and Baum, T., 2020. COVID-19 and African tourism research agendas. *Development Southern Africa* 37(5), 727–741.

UNCTAD, 2021. COVID-19 and Tourism - an Update. United NationsConference on Trade and Development, Geneva, Switzerland, June30 2021.

UNEP, 2021. Green financing. https://www.unep.org/regions/asia-and-pacific/regional-initiatives/supporting-resource-efficiency/green-financing

United States Agency for International Development. 2011. *USAID Evaluation Policy*, United States Agency for International Development, Washington DC.

UNWTO, 2021. UNWTO World Tourism Barometer and Statistical Annex. *UNWTO World Tour. Barom., 19*, pp. 1–42.

Walter, H. and Breckle, S.W., 2013. *Ecological Systems of the Geobiosphere: 2 Tropical and Subtropical Zonobiomes* (Vol. 2). Springer Science & Business Media, Verlag Berlin Heidelberg.

World Economic Forum Annual Meeting, 2021. https://sdg.iisd.org/events/world-economic-forum-annual-meeting-2021/#:~:text=Davos%2C%20Graubunden%2C%20Switzerland-,World%20Economic%20Forum%20Annual%20Meeting%202021,fair%2C%20sustainable%20and%20resilient%20future

16 When the Music's Over

Destination Ecosystem Services – Post-Pandemic Solutions and Challenges

Crispin Dale, Ade Oriade, and Neil Robinson

16.1 Introduction

The impact of the Covid-19 pandemic on tourism destinations across the globe (Hall et al., 2020; Haywood, 2020) has caused stakeholders to seek methods to ensure tourist safety and security (Dale et al., 2021). Furthermore, the global pandemic has resulted in tourism destinations having to appraise their ecosystem services and consider the adoption of new models that ensure their sustainability. Building upon the discussions raised by Gowreesunkar et al. (2021) in the management of destinations in a post-pandemic context and more specifically by Dale et al. (2021), in the context of safety and security, the chapter will explore the application of ecosystem services in ensuring the protection of residents (hosts) and tourists (guests).

The chapter will initially outline tourism ecosystem services in a pre-pandemic context before going on to discuss the impact of the Covid-19 pandemic on the tourism industry. In the light of the need for post-pandemic safety and security measures, the chapter will discuss the potential for tourism destinations to become self-serving and self-sustaining ecosystem services that thrive with limited external interaction. Thus, with a growing focus on localisation and systems that support this, the chapter will discuss a tourism ecosystem bubble that can potentially secure tourist health and safety in a post-pandemic environment. The chapter will draw upon examples that develop services that act as tourism ghettos or bubbles that minimise external interaction with the aim of protecting the host and the guest. In this context, the notion of the "leisure-casual" consumer will also be presented. The chapter will acknowledge how the introduction of revised ecosystem services models can result in additional charges occurring at a time when consumer expenditure is impacted upon. In pursuit of their survival, businesses within overly dependent tourism destinations then have the potential to fall into modes of prohibited activity such as crime and extortion. The chapter thus reviews ecosystem services theory in its application of measures that could enable the sustainability of tourism destinations. The chapter adopts the common international classification of ecosystem services (CICES) for understanding ecosystem services and their application to an all-inclusive context (Czucz et al., 2018).

DOI: 10.4324/b23145-21

16.2 Ecosystem Services in Tourism

According to Gretzel et al. (2015, 558) ecosystems are acknowledged "as communities of interacting organisms and their environments and are typically described as complex networks formed because of resource interdependencies". Pueyo-Ros (2018) reviews the role of tourism in the ecosystems framework and notes the need to acknowledge the interrelationship between cultural, economic and environmental perspectives. The impact tourism has on each of these perspectives, both positively and negatively, cannot be ignored. The Covid-19 pandemic has crystallised the challenges destinations encounter when they become over-reliant on tourism and the role it plays in supporting communities in the ecosystem's framework, culturally and economically. However, the tensions in conceptualizing tourism from a cultural, economic or environmental ecosystems perspective are confused in the literature (Pueyo-Ros, 2018). Thus, such confusion can generate parochialism amongst those who discuss ecosystem services and the interrelationship between these perspectives. Furthermore, the notion of "eco" can be conflated with ecological connotations that denote environmental issues and concerns. This is further compounded when the term is prefixed with tourism and the eco-labelling of products in the tourism system (Atieno and Njoroge, 2018).

Previous research has noted the challenges with implementing an ecosystems policy in nations including definitional and cross-sector issues (Matzdorf and Meyer, 2014). Nevetheless, ecosystem services are integral to maintaining sustainability and enhancing human welfare (Jaung, Carrasco and Bae, 2019). However, in destinations where there are competing uses for land, particularly in coastal regions, these resource pressures can be exacerbated (Lange, 2015). This is important when tourism activities such as snorkelling, diving and fishing rely on a healthy coastal ecosystem (ibid). Though in the context of Zanzibar, Lange (2015) further notes the prevalence of all-inclusive resorts which undermine the coastal ecosystem when tourist behaviour is geared towards activities that lack active marine use such as remaining on the beach or boat rides. Prime tourism locations are often located in coastal regions thus generating significant pressure on the existing ecosystems of the areas where they are situated (Drius et al., 2019). From this perspective, there is some caution with commodifying ecosystem services and assigning them an exchange value for the purposes of safeguarding nature (Outeiro et al., 2019). Nevertheless, in terms of employment and in a natural landscape context, it has been noted that ecosystem services contribute to the multiplier effect within economies (Laterra et al., 2019).

From a cultural perspective, ecosystems are less explored and potentially more challenging to address (Drius et al., 2019). From a cultural perspective, ecosystems are less explored and potentially more challenging to address, though the importance of these ecosystems services should be acknowledged (Outeiro et al., 2019). Indeed, "cultural values can be connected with public values since many ecosystem services not only support cultural activities but also offer benefits for the public" (Jaung, Carrasco and Bae, 2019).

This generates challenges when cultural ecosystems may not reflect the culture of the host population and are constructed for the purposes of tourist consumption. Hence, in a cultural context, it is argued that tourism should be recognised as an outcome rather than a service (Drius et al., 2019).

16.3 Impact of Covid-19 Pandemic on Tourism

The impact of the Covid-19 pandemic on the global economy and specifically service industries has been significant (Suneson, 2020). Tourism as a service industry, and the related businesses that it comprises, has been one of those that has suffered extensively from the fallout of the Covid-19 pandemic (ibid). According to the World Travel and Tourism Council (WTTC) (2021), there has been a loss of US$4.7 trillion in 2020 with the sector contribution to the global GDP declining from 10.4% to 5.5% in a single calendar year. The subsequent impact on employment has been just as large with the loss in 2020 of 62 million jobs across the sector (WTTC, 2021). In some respects, further job losses have been mitigated by job retention schemes in nation-states that have supported continued employment. However, as these schemes come to an end, continued job losses are likely to occur. In terms of visitor expenditure, there has been a decline in domestic and international spending by 45% and 69% respectively (WTTC, 2021). These figures are substantial when it is acknowledged that prior to the pandemic tourism was argued to account for one in four new jobs and 10.6% of all global employment (WTTC, 2021). Similarly, the United Nations World Tourism Organisation (UNWTO) (2021) does not foresee a return to the 2019 levels of international tourism until 2024 or later. Studies have also argued that the tourism growth may potentially be hampered for a further 15 years (Fotiadis, Polyzos, and Huan, 2021).

The impact on local economies has been profound, particularly on those destinations that have had a high dependency upon tourism. Examples have included destinations in India, Thailand, Turkey and Costa Rica (Arora, 2021; Girma, 2021). In times of crisis and for the means of survival, tourism ecosystem services can turn to areas of activity that support corruption and extortion (Dale et al., 2021; Klien, 2021). Furthermore, the pandemic has impacted destination ecosystems where tourism is integral to their conservation and preservation. This has included the Galapagos Islands which require a consistent flow of visitors for their continued sustainability (Díaz-Sánchez and Obaco, 2020). It also cannot be ignored that in destinations, such as Venice, the pandemic has facilitated a renewal of the natural ecosystems that tourism depends upon (Dale et al., 2021; Sengel, 2021).

The pandemic has provided an opportunity to rethink the world we live in and how we interact with it (Lashua, Johnson and Parry, 2021). It has already generated several disruptive effects across the sector. One of these effects has been the need to ensure the safety and security of tourists (Dale et al., 2021) with health and hygiene being viewed as a key future trend in tourism (WTTC, 2021). These safety and security measures will potentially vary depending upon the type of visitor/tourist travelling, their demographic, the destination location being travelled and so on. From a demographic perspective, there is likely to be an increased desire for safety. Therefore, the assurance of protected ecosystem services will continue to become paramount.

The pandemic influences decision making and risk aversion behaviour (Kim et al., 2021). Travel behaviour towards local and domestic destinations has occurred in addition to the changing means of transaction through cashless purchasing. Indeed, the growing role of technology and digitisation in tourism particularly when facilitating a seamless and personalised experience has become foremost. Technology has facilitated the opportunity for a different mode of transaction that can potentially be perceived as risk-reducing. For example, when using robots for the means of service interaction this can increase visit intention due to a perceived lowering of virus risk (Wan et al., 2021). During the pandemic, virtual tours have also played a significant role in the tourism ecosystem (El-Said and Aziz, 2021). The introduction of an augmented pre-trip experience and the scope for visitors to preview the ecosystem services in the destination. Considering the post-pandemic shock and from a culinary ecosystem's perspective in Jamaica, Milwood and Crick (2021) note the consistent use of online responsiveness to ensure resilience.

16.4 Methodology

A multi-method approach was utilised in gathering resources to develop this chapter. The chapter method of data collection comprises observation, conversation with managers, case study analysis and review of extant literature. The observation was carried out by the first author during a short holiday getaway break. Also, this observation forms part of the case study analysis which was used in shaping the all-inclusive tourism ecosystem bubble section. The case analysis coupled with the review of extant literature formed the basis of developing recommendations for sustainable options and moving towards sustainability.

16.5 The All-Inclusive Tourism Ecosystem Bubble

It has become apparent that tourism businesses and destinations need to develop resilience to be sustainable in future scenarios when a crisis may occur (Huang and Farboudi Jahromi, 2021). Developing economic resilience is important when destinations, as previously mentioned, have become overdependent on tourism and thus impacted their sustainable economic growth (Lee et al., 2021).

Huang and Farboudi Jahromi (2021) develop a conceptual model that notes the resilience-building strategies and resources that can equip service firms for survival. The elements of the model include market orientation, supply chain optimisation, strategic corporate reorganisation, innovation and business model transformation. These elements are underpinned by resilience-building resources which include financial, human, social and technological capital (ibid). Tourism destinations have the opportunity to follow a similar path to ensuring resilience in times of crisis.

The all-inclusive tourism resort context has been much criticised for being exploitative of local tourism and hospitality entrepreneurs who become limited in their ability to attract visitors to the use of their facilities such as restaurants, taxis and attractions. Due to competition between operators, margins can also be suppressed, resulting in the knock-on effect of low returns for businesses throughout the supply chain and the low wages for employees who are often local residents in the destination (McVeigh, 2014). Resentment by residents in the destination can occur as the benefits of tourism do not filter out to the local community (Ulrich, 2015). Though these concerns are acknowledged in a post-pandemic environment it could be argued that the all-inclusive resort context offers a resilience-building approach to developing ecosystem services.

This notion and development of "tourism bubbles" or "enclaves" has grown in the literature. From a spatial perspective and in an urban context, Ioannides et al. (2019) explore how AirBnBs have contributed to the development of tourism bubbles. Though these bubbles may have a disconnect to the location their situated as Ioannides et al. (2019) continue to observe they are not gated off in the same way an all-inclusive resort may be. Therefore, the challenge of isolating the tourist from external threats will always predominate as they venture into the wider community to visit attractions, shop or dine in local restaurants, cafes and hotels. In his book *The Holiday Makers*, Krippendorf (1987) discussed the notion of "tourism ghettos" which replicate an environment where tourists are catered for their every need. Though it is acknowledged that the notion of ghettos and ghettoization has negative associations (Schwartz, 2019), in this context, tourists are isolated from the external environment resulting in an entirely manicured experience. This includes the sourcing of the food through to the production of the culinary experience where lateral flow and polymerase chain reaction (PCR) testing is normalised throughout. The adoption of an all-inclusive ecosystem service model in a post-pandemic world can potentially protect tourists from further virus outbreaks. In a self-contained tourism context such as this, vaccinating tourism workers should be viewed as a first priority in highly dependent tourism nations, which was the case in Greece and the Dominican Republic (Girma, 2021). This is particularly important when there are disparities in communities and vaccine equity can be challenged (Girma, 2021).

This approach to a tourism bubble was adopted by the tour company Sunweb when they facilitated an organised eight-day excursion for 178 tourists to an all-inclusive resort in the Greek island of Rhodes. The tourists were tested pre- and post-trip, prohibited from leaving the resort and were required to quarantine on their return (BBC News, 2021).

16.6 The "Leisure-Casual" Consumer

Though the tourist ghetto and the all-inclusive model have been critiqued (Tourism Concern, 2014) as being a multi-faceted and an over-commercialised entity that advocates excess, mass consumption and possibly loose moral values/hedonism, we must at least pay homage to its conception. The tourist enclave has some merits to its inception and should its evolution and growth be monitored we have the opportunity to reengineer safe and ethically compliant ecosystems that offer some hope for travellers longing for a safe and virus-free leisure environment. It seems quite ironic that the urban tourist jungle, once the go-to location for the leisure casuals, now might offer some hope and stability for future holidaying generations. Indeed, the gentrification of the urban tourist jungle gives a glimmer of hope if correctly designed, managed and controlled with visitor management capacity constraints at the forefront of such strategies to ensure success.

It could be argued that the age-old elephant in the room here (the tourist destination/ghetto) has been historically mismanaged and abused by its capitalist zoo keepers, hell-bent on wealth creation via unregulated entrepreneurship and at any costs to the social and economic fabric of the surrounding ecosystem, not to mention the physical health of its human inhabitants, fauna and wildlife. Undeniably parallels here can be drawn with the goose that laid the golden egg, with its greedy human keeper not fully appreciating the benefits of prudence, sustainability and slow passage in exchange for cutthroat expansionism and ultimately the death of the tourism goose. If it had not been for the global Covid-19 pandemic, society would have possibly continued in its time-honoured and disrespectful consumption patterns with little thought of one's social footprint or appreciation of the magnitude of our impacts on a global setting.

Whilst society never wished for the Covid-19 pandemic we as global inhabitants have to accept that much of the pandemic and its spread could have been limited had better contingency planning taken place. Take, for example, the naive and cavalier manner by which the travel and tourism infra and superstructure continued to operate turning a possible blind eye to practices that exacerbated its spread. Certainly, the inability of airline providers and transport hubs to limit, control, monitor and profile their travelling clients is in itself verging on the criminally negligent. Such practices often saw overseas travellers allowed to travel unobserved between differing regional and international travel hubs unchecked and once safely ensconced at the final destination, allowed to simply melt into the flotsam and jetsam of everyday life, unmonitored and possibly exacerbating the spread of Covid-19. In addition, practices at holidaying destinations did little to limit or curtail the spread of the pandemic between holiday makers and hosts, enabling its terrible mutation to be exported globally by the trappings of poorly prepared travel, tourism, hospitality and recreation multinationals.

The importance of contingency planning for every eventuality appears to have been lost on the travel executives, who should have been better prepared for dealing with such a spread. Whilst hindsight is a wonderful thing, mother nature to her credit did give us some prior warning on how things might develop and this can be seen in the Severe Acute Respiratory Syndrome (SARS) outbreak and subsequent spread through the Asian subcontinent in

2003 and the many resulting deaths (Robinson and Dale, 2003). On this occasion, the world was partially spared the terrible death count as SARS was predominately confined to South East Asia and its spread was not as viral or on the same scale as its near relative the Covid-19 virus.

The annals of tourism history are littered with examples of mismanaged destination ventures that have merely paid lip service to sustainable development. Take, for example, the proposed idyllic destination of Cancun in Mexico, a World Bank-financed project of the 1960s and blueprint for future safe holidaying and sustainable ventures (de Kadt, 1984). Such buoyant and some might say farfetched optimism was never really realised and the development gave birth to a leviathan-like creature which was unregulated and uncontrolled, resulting in mass-scale development and ultimately the ruining of a venture that should have been sustainable and blueprint for all future tourism destinations and developments alike.

The potential for future global pandemics on the scale of Covid-19 appearing in the future should not be underestimated. As the liberalisation of international airways and maritime waters continues and destinations become increasingly more accessible due to enhanced social affluence and social mobility, it can possibly be expected to see future pandemics as the possible new norm. This should not be greeted with absolute pessimism but a possible heads up on how we as a new world order (with sentiments of peace, hospitality and social inclusivity) must collectively as developed nations not only help and assist our lesser developed geographical neighbours, but also put contingency plans in place for the inevitable occurrence. The Covid-19 pandemic has presented society with many challenges but also opportunities to recalibrate our emergency response provision for such future events, as the old adage goes, forewarned is forearmed. Future emergency planning at the global/macro level needs to be suitably funded and coordinated based on a war-type footing scenario, key contenders who might oversee such a coordinated response need to be big political entities with suitably funded resources. This includes the United Nations (UN), the North Atlantic Treaty Organization (NATO), the European Union (EU), the Warsaw Treaty Organization (The Warsaw Pact) and related African, Asian, South Pacific and Australasian counterparts must ensure that any effort to combat such pandemics is a global effort irrespective of the geographical impact on the country of origin. If we are to look back into the annals of history, we must better appreciate that the human species has been roaming this planet for approximately 200,000/300,000 years, compared to dinosaurs who roamed the earth for over 160 million years; this then surely shows the fragility of people kind who have possibly taken long life and prosperity for granted. In the same way that a cataclysmic event brought the end to those previous species that roamed the earth, we as a global entity must do more to ensure that the human species is not made extinct or that the difference between the haves and the have-nots does not bring about anarchy and insurrection, ultimately resulting in major climatic changes and a form of geographical/social apartheid.

16.7 A Sustainable Option/Towards Sustainability

So, what should the future blueprint look like for this new eco-friendly and virus-free tourism environment? Well, probably not too dissimilar from its distant relative the all-inclusive resort of yesteryear. Agreed, the previous model (1.0) did little for supporting the local values of cooperative employment, locally sourced materials/ingredients and denying opportunities for full tourist spending outside the confines of the hotel's airconditioned interior and its annexed golden facing exterior that only allows access to the privileged paying guest. That said certain components of this model had some benefits that designers could take forward and develop into a version (2.0). Take, for example, the International Organisation for Standards (ISO) that is often a key part of corporate international hotel chains, that dictates key performance standards, continuity in staff training, health and safety provision, internationally recognised hygiene standards and scientifically managed operational standards of food production and key performance indicators (KPIs) for all staff to follow.

Ecosystem services as a concept has the capacity to delineate externalities and enable policy formulation to internalise the value of given externalities in service delivery and decision making (de Groot, Wilson and Boumans, 2002). The discussion of ecosystem services in tourism has sustainability at its core given that ecosystem services are critical in emphasising the role of nature in human subsistence, well-being and long-term economic sustainability. This is amplified with the Covid-19 pandemic and the subsequent need for destination managers to identify and adopt a sustainable option to managing organisations post Covid-19. The concept of sustainability continues to gain ground in the tourism industry and allied sectors. Even though the definition of the concept is highly contested, as a multi-faced concept, it continues to flourish among practitioners and academics (Oriade et al., 2021) and persists to be a phenomenon idealised even if not immediately achieved. Sustainability is prevalently conceptualised as having three main dimensions, namely: social, economic and environmental. Economic viability, social development and environment conservation are mutually dependent and are key building blocks of sustainable development; these dimensions should be harnessed to promote responsible business operations over time. The three perspectives as depicted in Figure 16.1 have implications for the discussion in this section. The mutual dependency of the three key building blocks warrants the discussion to take place *pari pasu* rather than being discussed in individual sub-sections.

A sustainable option seeks a balance between judicious use of resources in order to provide competitive service by destinations – on the one hand, visitors' satisfaction of their desires through ecosystem services and on the other, through experiential consumption. This requires system-led thinking because every destination component is related and each aspect contributes towards the sustainability schema. Good governance, planning and management are quite vital. Destinations, and organisations within them, should be able to integrate the synthesis of environmental, organisational and individual variables in

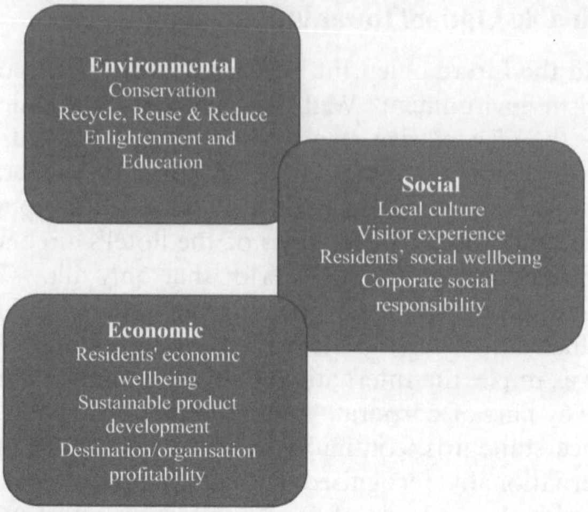

Figure 16.1 The three perspectives.

exploring the adoption of sustainable options or business models in order to manage ecosystem services and destinations effectively post pandemic. Implications exist as tourists continue to demand ecosystem services that offer clean and ethically sourced linen on beds and gift shops in hotel receptions that have been fully sterilised. These put pressure on both operational and natural resources and may even necessitate excessive use of operational resources and materials that may be harmful to the environment, for example, chemicals for sterilisation. To this end, judicious resource use cannot be over-emphasised.

Tourism accrues benefits to stakeholders – desirable experience to tourists, viable business to organisations, civic pride to host community, public value to governments and means of livelihood to practitioners. For a sustainable option to be adopted the interest of stakeholders must be balanced ensuring each group derives desired benefit from engaging in tourism. One of the common theories used in explaining stakeholders' engagement with tourism is Social Exchange Theory (SET), which is premised on exchange of resources among individuals and groups in a collaborative situation. SET postulates that behaviours can be theorised as the product of cost–benefit analyses by a network of people who interact within a society and with the environment. People interact with the understanding that their engagement/involvement will bring benefits and the more benefits accruable the more encouraged they will be to participate. On the contrary, if it is perceived that the engagement is not profitable, an actor is likely not to engage; and according to Jonason and Middleton (2015), such an act could produce a short-term orientation in favour of immediate, assured rewards capable of eliciting socially aversive behaviours. In this sense and in adopting a sustainable business model, balancing the accruable benefits and cost to stakeholders becomes very crucial. The SET forms part of the basis for delineating the mutual dependency of the three key building blocks mentioned previously.

Based on the foregoing, whilst the bundling of products currently seen in the all-inclusive tourism ecosystem bubble might be a vehicle to economic sustainability, there is yet a need to ensure that cultural ecosystems reflect the culture of the host community. Sustainable product development that takes into cognisant the local culture and does not exclude the local resident is desirable. This type of product development is vital to destination/organisation economic viability, local socio-cultural dynamics enhancement and preservation of the ecological and biodiversity of the local environment. The products in some tourist ghettos and all-inclusive models seem ideal for some destinations however all destinations are not equally endowed with ecosystem services and resources. Also, the model of luxury tourism seen in some southern European countries is definitely not replicable in other parts of the world. Destinations in countries such as Italy and Spain have taken advantage of funds made available for European recovery to reposition themselves, targeting more affluent visitors who have high spending power and are able to stay longer. Ideally, this is the desire of many destinations however resources are not equally distributed. One lesson that is needed to be learnt here is that the role of government and funding bodies is crucial to the resurgence and resilience of the sector particularly if destinations are to recover after the devastating effects of the pandemic.

For the foreseeable future, visitors will prefer less crowded destinations for obvious health reasons. A decrease in infection rate because of vaccination has necessitated the easing of lockdowns and social distancing measures in many countries such as the United Kingdom; however, the emergence of new variants of the coronavirus such as the Delta variant continues to pose challenges for policy makers, organisations and citizens. Managers responsible for product development in destinations with limited ecosystem services and less funding assistance have to be innovative by adopting "mass produced yet 'customised' strategy". This strategy is exemplified in Krippendorf's (1987) analysis. Product development should adopt a "mass production yet 'customised' strategy" with safety and security measures at the core of the development. There is a need to educate tourists to be more sustainable beyond the reuse and sharing of facilities and materials commonly practised in the industry. Even though tourists are conversant with some basic reuse-recycle policies, for example, hotel guests reusing towels, post Covid-19, the options are likely to be stretched further. The notion of "leisure-casual" consumer further emphasises the issue of sustainability awareness among stakeholders and the importance of organisations managing sustainability issues and developing appropriate culture in the development of sustainability mindset (Oriade et al., 2021).

The creation of such a sanitised hospitality bubble where guests must show proof of vaccinations and where consumable items have been hermetically sealed during production is just one way in which this brave new world of the leisure environment may appear as we emerge post Covid-19. An environment where the leisure casuals are directed, educated and informed of the modus operandi of experience, where suppliers, manufacturers and ultimately service providers in the form of staff are compliant with good practice and food service delivery standards. This is an ecosystem where human interaction is clean and safe, where the requirements of good hygiene and compliance are shared and agreed, supported and practiced by all. An ecosystem service that offers clean and ethically sourced linen on your bed, a gift shop in the hotel reception that has been fully sterilised and food on your table that is fully sourced from accredited and approved local suppliers.

Such good operational practices do not necessarily need to be the preserve of the corporate chains, but once developed and piloted could be exported to smaller-scale family-/locally run enterprises within the vicinity. Such a concept also needs to take into account host population's needs and those who are seasonally employed and might need to travel some distances to work and might unknowingly spread infections due to employment mobility. A key part of such a development would need to consider improved measures for accommodating host population employees close to employment venues that are also close to their home base so that long distances associated with employment migration can be mitigated. Other such benefits associated with host employment should consider value-added benefits for host employees. Those rewards benefit the locals and make them feel enfranchised as opposed to the traditional disenfranchised and antagonistic nature of the social confines that host employability can potentially cause. Such benefit mechanisms for the host employees and their family dependents might include health care provision, educational scholarships, financial hardship funds and poverty alleviation measures, all potentially coming from the reengineering and better design of the traditional all-inclusive holiday destination.

Even though all-inclusive tourism bubbles should strive to be self-sustaining ecosystems that thrive with limited external interaction, Schofield et al. (2018) and Oriade, Broad and Gelder (2020) advocate that collaboration among destinations and organisations within them will drive creativity, innovations and competitiveness. Collaboration and cooperative service delivery are important in delivering sustainability. For example, destination partnering with airlines that use Sustainable Aviation Fuel (SAF) reduces greenhouse gas emissions across the aviation industry. A partnership such as this will emphasise the seriousness of destinations not only in furthering economic gain and staging a come-back after the pandemic, but shows commitment to deliver a global sustainable tourism industry. Collaboration also aids process innovation which enhances efficiency in terms of cost reduction of production and service delivery, and enhancement of quality. Collaboration and cooperation also become crucial in avoiding the criticism of the all-inclusive strategy for being exploitative of local tourism and hospitality operator and host community who are prevented from benefiting economically and socially, respectively, from visitors. Consideration must be given to how local social and economic ecosystem services will be marketed and distributed safely and hygienically to visitors.

Whilst it is questionable whether technology can actually substitute the friendly manner humans deliver service, the adoption of technology cannot be overemphasised in managing perceived risk and enhancing service delivery where social distancing may remain for a foreseeable future (Dale et al., 2021). Deployment of technology in managing pre-trip experience is tried and tested; however, operators would need to enhance process innovation by deploying technology more effectively on site, during visitors' stay (Gretzel et al., 2015). The use of technology in the accommodation sector is well established but operators will need to move away from "gimmicky" robots that say "hello" to more sophisticated, fully functional robots and artificial intelligence that can anticipate, respond to and understand customers' needs. Guests in hotels pre-Covid have been embracing technologies and there is the possibility of guests' expectations growing further in expecting their room to be a "home away from home" with extra safety measures in place. Other forms of regular technologies should not be discarded because these have proved expedient in recent times. For instance, to minimise traffic during rush hours in restaurants and/or visitor attractions, Mobile Applications may be designed to manage booking, zoning and redirection. In addition, the use of technology in marketing needs to be reconceived. Real-time marketing should be embraced with the use of an assortment of technological apparatus. Real-time marketing continues to change the way the organisations maintain relationships with the customers by engaging on the basis of "pull marketing" advertising where captivating contents are posted via social technologies and are deployed to initiate and maintain dialogue with tourists. Destinations left out of this trend may find it difficult to compete and retain their visitors.

16.8 Conclusion

The chapter has reviewed the role of ecosystem services in a post-pandemic environment. Specifically exploring the role of the all-inclusive tourism ecosystem bubble, the adoption of this by the "leisure-casual" consumer and its future place as a sustainable option. Ultimately, the Covid-19 pandemic period has seen many deaths resulting. The travel/service sector must ensure that its house is in order and appropriately designed mechanisms both at strategic and operational levels are in place to combat any such reoccurrences. The global and economic wealth of society has much to do with the ability of the travel and service sector to operate unhindered and at full capacity. The manner by which the travel and service sector is better able to prevent such a future spread is fundamental to social well-being, shared economic prosperity and global peace. At the global level, the collective financial receipts and contribution that the service industries make to each government's exchequer easily run into trillions, not including the paid taxation of profits from large multinationals that go to fund health care provision. The role of the service sector in ensuring a return to normality and its financial contribution, to social harmony and disease prevention goes a long way from the initial premise of people holidaying in the sun. Once the Covid-19 pandemic has passed and there is movement towards a more prosperous and confident social mindset, it must be ensured that ongoing operational practices and suitably designed social management rules and new practices are adhered to facilitate disease prevention.

It is of key importance that the global service sector is better prepared to deal with any uncertainty that might impact its business model. Not only does the service sector contribute hugely in terms of employment and social development, it also has the potential to better serve community development and host living standards. Indeed, a simple trick that the hospitality/service sector has been missing for some years or has not properly developed its ability to build social capital in terms of host infrastructure development and well-being. One such model would not be different to that of the coffee growers in Kenya, who benefit from housing, commercial enterprise, fair trade and educational scholarship opportunities. The future will be challenging, but we now have a once-in-a-lifetime opportunity to better design such a business model that will benefit both hosts, employers and tourists.

References

Arora, V. (2021) *India's Covid Nightmare Leaves Its Travel Industry Reeling*. Skift Research, 6th May 2021.

Atieno, L. and Njoroge, J.M. (2018) The ecotourism metaphor and environmental sustainability in Kenya. *Tourism and Hospitality Research* 18(1), 49–60.

BBC News (2021) Covid-19: Dutch sign up for test holiday on Greek island. Retrieved from https://www.bbc.com/news/amp/world-europe-56528112

Czúcz, B., Arany, I., Potschin-Young, M., Bereczki, K., Kertész, M., Kiss, M., et al. (2018) Where concepts meet the real world: A systematic review of ecosystem service indicators and their classification using CICES. *Ecosystem Services* 29, 145–157. DOI: 10.1016/j.ecoser.2017.11.018

Dale, C., Robinson, N. and Sheikh, F. (2021) Tourist safety and security post COVID-19: Global perspectives. In Gowreesunkar, V., Maingi, S., Roy, H. and Micera, R. (Eds.), *Tourism Destination Management in a Post-Pandemic Context*. Emerald Publishing Limited, London: UK.

De Groot, R.S., Wilson, M.A. and Boumans, R.M. (2002) A typology for the classification, description and valuation of ecosystem functions, goods and services. *Ecological Economics* 41(3), 393–408.

De Kadt, E. (1984) *Tourism: Passport to Development?* New York: Oxford University Press.

Díaz-Sánchez, J.P. and Obaco, M. (2020) The effects of coronavirus (COVID-19) on expected tourism revenues for natural preservation. The case of the Galapagos Islands. *Journal of Policy Research in Tourism, Leisure and Events*. DOI: 10.1080/19407963.2020.1813149

Drius, M., Bongiorni, L., Depellegrin, D., Menegon, S., Pugnetti, A. and Stifter, S. (2019) Tackling challenges for Mediterranean sustainable coastal tourism: An ecosystem service perspective. *Science of the Total Environment* 652, 1302–1317.

El-Said, O. and Aziz, H. (2021) Virtual tours a means to an end: An analysis of virtual tours' role in tourism recovery post COVID-19. *Journal of Travel Research*, 1–21. DOI: 10.1177/0047287521997567

Fotiadis, A., Polyzos, S. and Huan, T.-C. (2021) The good, the bad and the ugly on COVID-19 tourism recovery. *Annals of Tourism Research* 87. DOI: 10.1016/j.annals.2020.103117

Girma, L.L. (2021) *Why Tourism Needs to Step Up and Push for Vaccine Equity*. Skift Research, 4th May 2021.

Gowreesunkar, V., Maingi, S., Roy, H. and Micera, R. (2021) *Tourism Destination Management in a Post-Pandemic Context: Global Issues and Destination Management Solutions*. Emerald Publishing Limited, London: UK.

Gretzel, U., Werthner, H., Koo, C. and Lamsfus, C. (2015) Conceptual foundations for understanding smart tourism ecosystems. *Computers in Human Behavior* 50, 558–563.

Haines-Young, R. and Potschin-Young, M. (2018). Revision of the common international classification for ecosystem services (CICES V5. 1): A policy brief. *One Ecosystem, 3*, e27108. https://doi.org/10.3897/oneeco.3.e27108

Hall, M.C., Scott, D. and Gossling, S. (2020) Pandemics, transformations and tourism: Be careful what you wish for, tourism geographies. *Tourism Geographies*. DOI: 10.1080/14616688.2020.1759131

Haywood, K.M. (2020). A post-COVID future: Tourism community re-imagined and enabled. *Tourism Geographies*. DOI: 10.1080/14616688.2020.1762120

Huang, A. and Farboudi Jahromi, M. (2021) Resilience building in service firms during and post COVID-19. *The Service Industries Journal* 41(1–2), 138–167. DOI: 10.1080/02642069.2020.1862092

Ioannides, D., Roslmaier, M. and van der Zee, E. (2019) Airbnb as an instigator of "tourism bubble" expansion in Utrecht's Lombok neighbourhood. *Tourism Geographies* 21(5), 822–840. DOI: 10.1080/14616688.2018.1454505

Jaung, W., Carrasco, R. and Bae, J.S. (2019) Integration of ecosystem services as public values within election promises: Evidence from the 2018 local elections in Korea. *Ecosystem Services* 40, 101038.

Jonason, P.K. and Middleton, J.P. (2015) Dark triad: The "dark side" of human peersonality. In Wright, J.D. (Ed.), *International Encyclopedia of the Social & Behavioral Sciences* (2nd edition, pp. 671–675). Elsevier, Oxford, UK.

Kim, J., Park, J., Lee, J., Kim, S., Gonzalez-Jimenez, H., Lee, J., Choi, Y.K., Lee, J.C., Jang, S., Franklin, D., Spence, M.T. and Marshall, R. (2021) COVID-19 and extremeness aversion: The role of safety seeking in travel decision making. *Journal of Travel Research*, 1–18. DOI: 10.1177/00472875211008252journals.sagepub.com/home/jtr

Klien, D. (2021) Italian Mafia's Cut in Tourism Sector is Over $2.6 Billion. Retrieved from https://www.occrp.org/en/daily/14298-report-italian-mafia-s-cut-in-tourism-sector-is-over-2-6-billion

Krippendorf, J. (1987) *The Holidaymakers: Understanding the Impact of Leisure and Travel*. Oxford: Butterworth-Heinemann.

Lange, G.M. (2015) Tourism in Zanzibar: Incentives for sustainable management of the coastal environment. *Ecosystem Services* 11, 5–11.

Lashua, B., Johnson, C. and Parry, D. (2021) Leisure in the time of coronavirus: A rapid response. *Leisure Sciences* 43(1–2), 6–11.

Laterra, P., Nahuelhual, L., Gluch, M., Sirimarco, X., Bravo, G. and Monjeau, A. (2019) How are jobs and ecosystem services linked at the local scale? *Ecosystem Services* 38, 207–218.

Lee, Y.-J., Kim, J., Jang, S., Ash, K. and Yang, E. (2021) Tourism and economic resilience. *Annals of Tourism Research* 87.

Matzdorf, B. and Meyer, C. (2014) The relevance of the ecosystem services framework for developed countries' environmental policies: A comparative case study of the US and EU. *Land Use Policy* 38, 509–521.

McVeigh, T. (2014) All-inclusive boom leaves local workers and tour operators out in the cold. *The Observer*, 8 March 2014.

Milwood, P.A. and Crick, A.P. (2021) Culinary tourism and post-pandemic travel: Ecosystem responses to an external shock. *Journal of Tourism, Heritage & Services Marketing* 7(1), 23–32.

Oriade, A., Broad, R. and Gelder, S. (2020) Alternative use of farmlands as tourism and leisure resources: Diversification, innovations and competitiveness. *International Journal of Management Practice* 13(5), 565–586.

Oriade, A., Osinaike, A., Aduhene, K. and Wang, Y. (2021) Sustainability awareness, management practices and organisational culture in hotels: Evidence from developing countries. *International Journal of Hospitality Management* 92, 102699.

Outeiro, L., Rodrigues, J.G., Damásio, L.M.A. and Lopes, P.F.M. (2019) Is it just about the money? A spatial-economic approach to assess ecosystem service tradeoffs in a marine protected area in Brazil. *Ecosystem Services* 38. DOI: 10.1016/j.ecoser.2019.100959

Pueyo-Ros, J. (2018) The role of tourism in the ecosystem services framework. *Land* 7, 111.

Ren, L., Li, J., Li, C. and Dang, P. (2021) Can ecotourism contribute to ecosystem? Evidence from local residents' ecological behaviors. *Science of the Total Environment* 757, 143814.

Robinson, N. and Dale, C. (2003) The battle of the two SARS: Hong Kong's hospitality and tourism provision since 1997. *Hospitality Review* 5(3), 26–33.

Schofield, P., Crowther, P., Jago, L., Heeley, J. and Taylor, S. (2018) Collaborative innovation: Catalyst for a destination's event success. *International Journal of Contemporary Hospitality Management* 30(6), 2499–2516.

Schwartz, D.B. (2019) How America's Ugly History of Segregation Changed the Mean. Retrieved from https://time.com/5684505/ghetto-word-history/

Sengel, U. (2021) COVID-19 and "New Normal" tourism: Reconstructing tourism. *Journal of Tourism & Development* 35, 217–226.

Suneson, G. (2020, 21 March 21) Industries hit hardest by coronavirus in the US include retail, transportation, and travel. USA Today. https://www.usatoday.com/story/money/2020/03/20/us-industriesbeing-devastated-by-the-coronavirus-travel-hotels-food/111431804/

Tourism Concern (2014) All-inclusies. Retrieved from https://www.tourismconcern.org.uk/all-inclusives/

Ulrich, K. (2015) The perceived impacts of all-inclusive package holidays on host destinations. Retrieved from www.tourismconcern.co.uk

UNWTO (2021) Tourist numbers down 83% but confidence slowly rising. Retrieved from https://www.unwto.org/news/tourist-numbers-down-83-but-confidence-slowly-rising

Wan, L., Chan, E. and Luo, X. (2021) ROBOTS COME to RESCUE: How to reduce perceived risk of infectious disease in Covid19-stricken consumers? *Annals of Tourism Research* 88.

WTTC (2021) Economic Impact Reports. Retrieved from https://wttc.org/Research/Economic-Impact

Zhou, D., Dejnirattisai, W., Supasa, P., Liu, C., Mentzer, A.J., Ginn, H.M., Zhao, Y., Duyvesteyn, H.M., Tuekprakhon, A., Nutalai, R. and Wang, B. (2021) Evidence of escape of SARS-CoV-2 variant B. 1.351 from natural and vaccine-induced sera. *Cell* 184(9), 2348–2361. DOI: 10.1016/j.cell.2021.02.037

Part VI

Global Case Studies

17 The Contribution of Whale Watching to Conservation in Marine-Protected Areas

The Example of Bahía De San Antonio, Patagonia

Guadalupe Sarti, Matilde Encabo, and Raúl González

17.1 Introduction

17.1.1 Tourism in Conservation Territories within the Post-Pandemic Context

The COVID-19 pandemic has caused different impacts on Marine Protected Areas (MPAs). The majority of those regions have suffered from an illegal removal of natural resources, a reduction of visitors, and the loss of income for both the areas themselves and the economies of entities linked to them (Mitchell & Phillips 2021). The effects on societies dependent on tourism have involved a nearly total shock, with the recuperation of that sector now requiring a prolonged length of time (Mooney & Zegarra 2020). The COVID-19 worldwide crisis has forced different societies to reflect on the relationship between man and the natural environment and how to best offset anthropic impacts (Mitchell & Phillips 2021). With respect to tourism and recreational activities, the post-pandemic context involves persons who search for experiences in natural territories (Spenceley *et al.* 2021; WTO[1] 2021). The United Nations (2020) accordingly predicted that ecotourism *per se* would increase rapidly within that context, thus exerting greater pressure on the protected sites and their surroundings if they failed to plan and conduct the management of those areas adequately.

The MPAs are a form of territorial ordering and thus a fundamental tool for the conservation and restoration of marine ecosystems along with a reconstruction of the biodiversity of the oceans (UICN[2]-CMAP[3] 2018, p.1). Globally, 7.74% of marine waters fall within protected areas (ONU 2021). Marine and coastal ecosystems are continually impacted negatively by human activities, such as overfishing, maritime transport, tourism, pollution, and climate change (Notarbartolo di Sciara & Hoyt 2020). The modern obsession with consumption and production has created progressive environmental disruptions (Silveira 2011; Gudynas 2012) that have been even more evident in the COVID-19 pandemic (Feola *et al.* 2021). The tourism and recreation industries have an equal responsibility in these developments (Encabo *et al.* 2014, 2016;

DOI: 10.4324/b23145-23

Spenceley *et al.* 2021), just as does tourism, which is "based on the increase and satisfaction" of the visitors (Encabo *et al.* 2014, p.5), massive tourism, nature tourism, or poorly managed ecotourism (Lück & Aquino 2021). In view of these considerations—and with a mind to the predicted intensification of ecotourism and recreational activities within the post-pandemic scenario—the commercial entities involved must consider in their operational schemes not only those effects resulting from their own inaction but also any other related influences, such as climate change and the issue of biodiversity (Spenceley *et al.* 2021, p.111); in that manner combining the principles of sustainable development in order to promote green-favoring attitudes (Gössling & Hall 2006; Spenceley *et al.* 2021). One way to achieve those ends is to take advantage of the experience of the tourism and recreation industries in order to educate citizens on the essential role of the MPA (Waithaka *et al.* 2021).

Therefore, the approach adopted by this investigation is the conceptual model entitled Recreation and Tourism in Conservation (RaTiC), a rationale that transcends the value inherent in recreation and tourism *per se* with respect to both the satisfaction of the visitors and their increased numbers, participating through a clear understanding of the essential and imperative responsibility implied in the enjoyment of natural environments and their fundamental biodiversity (Encabo *et al.* 2016). The model focuses especially on the visitors owing to the increase in their presence in natural regions worldwide but specifies that the responsibility of conserving natural biodiversity is shared, part and parcel, by all the actors involved in recreation and tourism (Encabo *et al.* 2016, p.18).

17.1.2 Whale Watching and Conservation of Marine and Coastal Ecosystems

The marine environments on which cetaceans depend are threatened by human activities in different ways that affect the marine mammal's health and well-being, and even their life—namely, incidental captures, entanglement in cables and fishing nets, collisions with passing boats, and environmental contamination, among others (Vernazzani *et al.* 2017; ICB[4] 2021).

The southern right whale *Eubalaena australis*, heretofore the SRW, is a migratory species that moves annually between zones of feeding and reproduction (Arias 2019). Different international, national, and regional protective measures for the SRW have ensured that the whale populations continue to grow so as to achieve the classification of a species of "lesser concern" in the Species Red List of the International Union for Conservation of Nature (IUCN; Cooke & Zerbini 2018) as with the mammals of Argentina in danger of extinction (D'Agostino *et al.* 2019). Nevertheless, in the example of whale watching (heretofore WW), if a noncompliance with the proper norms transpires, that inappropriate activity can generate changes in the respiration rate of the animals, their behavioral state, their acoustic communication, their patterns of movement, and their habitat use (Arias 2019, p.5). Likewise, recent studies in the Valdés-Peninsula Nature Reserve (Patagonia) have reported a significant increase during recent decades in the unintentional enmeshing of SRWs, as well as their ingestion of plastic residues (ICB 2021; Sironi 2021).

The WW implies the observation of any cetacean whatsoever in its natural setting by means of some viewing place—be it, for example, on board a ship, from an airplane, or on the coast (Zeppel & Muloin 2014, p.110). That activity has, in fact, become a major sector of tourism directed at viewing nature (Cárdenas *et al.* 2021, p.1). In agreement with other authors (Ballantyne, Packer & Hughes 2009; Kessler, Harcourt & Bradford 2014; Bertella 2018), WW can fall within the rubric-designated *wildlife tourism*, which implies the interaction of visitors with wild species of flora and fauna in their natural habitat (WTO[5]). That touristic modality involves some 7% of all tourism worldwide and exhibits an annual growth rate of about 3% (ONU 2020, p.16). Accordingly, WW under the modality of wildlife tourism involves key principles of ecotourism: for example, that form of tourism should educate the participants regarding subjects associated with natural systems (Wearing *et al.* 2014) and enhance those persons' sensitivity with respect to the essentiality of conserving the natural and cultural goods and services of wildlife (UNWTO[6]). In like manner, wildlife tourism can generate significant experiences in the visitors and foment and strengthen their sense of the ethical imperative in supporting conservation by informing those persons' acquaintances, attitudes, beliefs, and behaviors (Zeppel 2008; Zeppel & Muloin 2014; Sarti *et al.* 2019; Cárdenas *et al.* 2021). To accomplish those ends, the work of the tourists' guides is fundamental (Hoyt 2001, 2005), in that they link the conservation of the whales to the protection of the natural state of the marine habitat (Andersen & Miller 2006) and transmit a multidisciplinary message (Andersen & Miller 2006). Moreover, wildlife tourism with a strong educational focus can contribute favorably to the willingness of the visitors to pay or motivate other personal actions on their part that promote wildlife conservation (Zeppel & Muloin 2014, p. 116).

17.1.3 Study Case

This chapter represents a study on WW in the Bahía de San Antonio Marine Protected Area (BSAMPA) within the San Matías gulf (Río Negro province, Patagonia, Argentina). WW began here in the year 2012 with the SRW and from that time on has been monitored by a scientific program and regulated through specific management directives for every season (Arias 2019). WW in the BSAMPA had been the objective of previous studies (Arias *et al.* 2016; Sarti *et al.* 2016) that had contributed to a better understanding of the status of conservation and the population pattern of the SRW in that region plus the characteristics of the tourist needs. From a touristic and recreational perspective, however, the contribution of WW to the objectives of the conservation of the BSAMPA is unknown. This lack of knowledge is with respect to understanding whether or not this form of WW generates meaningful experiences in the visitors and if the latter duly perceive the relationship between recreation and tourism and the essential conservation of the whales and their habitat.

This chapter centers around an evaluation of the recreation-tourist experiences in WW in the BSAMPA, with the final objective aiming at understanding the relationship between the visitors, the SRWs, and the MPA. The knowledge generated will make a contribution to the management of this MPA, and the results obtained could aid in fine-tuning the present directive actions for that tourist activity as well as provide a precedent for the direction of the general tourism of watching marine fauna within protected territories.

17.2 Methodology

17.2.1 Study Area

The BSAMPA (Figure 17.1) was created in 1993 and has an area of about 812 km², of which 25% corresponds to the continental portion and the remaining part, the marine sector. The reserve is managed by a plan appropriate for a so-called *multiple-use reserve*, promoting the harmonic co-participation among productive human activities along with the maintenance of the natural environment, its biodiversity, and its ecosystem services.

Figure 17.1 Location of the Bahía de San Antonio Marine Protected Area (BSAMPA) in the Province of Río Negro, Patagonia, Argentina. The San Antonio Este port and Las Grutas are indicated.

WW in this territory takes place each year between the months of August and October. The excursion trips used by the tourist services are semi-rigid boats with a maximum capacity of ten passengers permitted per trip. Two points are used for boarding, one in the San Antonio Este port and the other in Las Grutas (*cf.* Figure 17.1) at a distance of 60 km from each other. The field tasks (questionnaires, observations, and field notes) were centered at Las Grutas, the site constituting the departure point of the greater number of boats and passengers during each tourist season.

17.2.2 Recreation and Tourism in a Conservation Conceptual Model

The approach adopted by this investigation is the conceptual model entitled Recreation and Tourism in Conservation (RaTiC), a rationale that transcends the value inherent in recreation and tourism *per se* with respect to both the satisfaction of the visitors and their increased numbers, participating through a clear understanding of the essential and imperative responsibility implied in the enjoyment of natural environments and their fundamental biodiversity (Encabo *et al.* 2016). The model focuses especially on the visitors owing to the increase in their presence in natural regions worldwide but specifies that the responsibility of conserving natural biodiversity is shared, part and parcel, by all the actors involved in recreation and tourism (Encabo *et al.* 2016, p.18). The model is defined by three key principles: (1) both tourism and recreation in natural environments depend on and are coresponsible for the conservation of biodiversity. A consideration of the maintenance of biodiversity must be integrated into the organization of recreational and tourist use, including facilities, equipment, and infrastructure (Encabo *et al.* 2016, p.14). (2) Actors related to directing the acquaintance and recreation-touristic use of the natural environment are obliged to guarantee the maintenance of biodiversity. These actors must be aware of and familiar with the regulations regarding nature conservation and duly comply with those norms. (3) Visitors who enjoy that environment have the responsibility to conserve biodiversity. The knowledge provided to visitors is an opportunity to strengthen their awareness of environmental values.

17.2.3 Data Collection: Questionnaires and Participant Observations on Board

During the years 2018 and 2019, surveys were conducted via an individual, self-administered semi-structured questionnaire that was presented and explained to the visitors more than 18 years of age participating in the activity, requesting their consent to complete it. The sampling was performed two times: the first part of the questionnaire was completed before the trip and the second part after the experience. The design of the questionnaire (30 questions) enabled the acquisition of the demographic profile of the participants along with their recreation-touristic background—*i.e.*, their experiences, motivations, acquaintances, beliefs, attitudes, and behaviors in relation to the conservation of the SRW and its natural habitat. In particular, the questionnaire requested the visitor's personal opinion on the experience of WW and, in the second part, the learning obtained, the animals observed, the acquaintance with MPAs and the SRWs in the area, and the behavior of the participants during the excursion. The information received during the excursion as well as the degree of satisfaction with the diverse aspects of the activity of WW along with the degree of satisfaction with the total experience was qualified by the visitors by means of a Likert scale ranging from 1 = inadequate and unsatisfied to 5 = totally adequate and satisfied. Finally, the survey of 2019 was modified slightly from the version in 2018 by incorporating questions associated with the interest on the part of the visitors in contributing to the conservation of the MPS and its habitat.

As a complement to those surveys, the recreation-tourism services conducted other surveys on board through direct observation. In those instances, they registered information on the number of visitors per trip; on the presence and/or absence of a guide; on the subjects on board discussed by the guide, if present, and/or the driver/skipper of the trip touching on SRW conservation and the BSAMPA; on the questions, commentaries, and behavior of the visitors; on the presence and/or absence of other trips making use of the same viewing waters; and on the maneuvers performed by the boat to gain a vantage point within the proximity of the animals.

17.2.4 Procedure and Data Analysis

The responses to the questionnaires and the records of the sheets of on-board observations were analyzed by descriptive statistics. Graphs of the distribution of variable frequencies with percentage accumulations and, in some instances, measurements of the central tendencies (*i.e.*, the means, medians, and modes) were constructed, and the standard deviations and variances were calculated in order to ascertain in which values of agreement the responses were concentrated. The relationship between *learning obtained by the visits* as well as *presence of a guide on the excursion* and *intention of the visitors to collaborate with the conservation of the SRW and their habitat* were evaluated by the chi-squared test. In order to determine the relationship between the *level of* agreement *by the visitors to each variable linked to WW* and the *level of agreement by the visitors to the entire experience*, the Pearson coefficient-of-correlation test was applied. The level of agreement by the visitors to each one of the intervening variables in the experience of WW was evaluated from the Likert scale. Therefore, for the three items that called for responses in that scale—namely, to indicate *if the information received in the excursion was adequate*, the *degree of agreement with the different variables linked to WW*, and the *degree of overall satisfaction with the entire experience*—a reliability test was performed from the alpha coefficient of Cronbach (Hernández, Fernández & Baptista 2014).

17.3 Results

17.3.1 The Obligation of Actors Related to Directing the Acquaintance and Recreational-Touristic Use of the Natural Environment to Guarantee the Maintenance of Biodiversity

- Previous acquaintance with the experience

The majority (48.5%) of the visitors expressed that information received from different communication media before the WW hadn't brought them any knowledge of the SRW or its habitat, though 27% had been informed by other persons, with 24.4% having acquired information via the *Internet*. Other means of gaining information were tourist packages (8.8%), graphics media (8.4%), and government tourism offices (8%). The rest of the information came from other media (*e.g.*, radio) and tourist accommodation sites. Of the 33.9% that reported having received previous information, 40.2% specified that information received was related to the biologic characteristics of the species. Other aspects of the visitors' previous information—for example, on conservation and other species of marine fauna—all together amounted to less than 5%.

Of the visitors, 31% said that they became acquainted with the norms for performing the WW and that they related those to the conservation of the SRW (41%) or to the provision of the WW service. The majority (78%) considered that some type of precaution is necessary in order to conduct activities in nature. Nevertheless, nearly half (47%) didn't know of any to mention. Among those visitors who responded, the majority of the responses (29%) were associated with the animals and their surroundings or with aspects linked to the safety of the service (14%) and to the knowledge of and compliance with the regulations (6%).

- Learning

One-half of the visitors surveyed indicated having learned "something" during their WW experience. A majority (62%) mentioned having learned about aspects of the biology and ecology of the SRWs—for example, their feeding habits, size, weight, natural history, behavior, types of groups, and habitats—and to a lesser degree (18%) about other species.

Of the visitors, 47% replied affirmatively to having been informed that they would be coming across in the BSAMPA, but the majority of that group (71%) did not know the exact name. During the surveys on board, the number of drivers/skippers and of those acting as guides who provided practically no information on the BSAMPA were also registered.

Nearly half of those surveyed (48%) did not know how to respond to the name of the species of whale, while the rest answered with incorrect names. In this manner was documented the finding that, after the experience of WW, the majority of the visitors were still ignorant of the name of the BSAMPA and the species of whale they had observed there.

The information supplied on board by the persons officiating as guides contained the following information at the indicated frequency: aspects of the biology and ecology of the SRW (79%), conservation of the whale and its habitat (12%), mention or identification of other species (5%), aspects of the tour service (3%), and the BSAMPA (1%). In essence, the information on the subjects that was obtained was determined primarily by the questions raised by the visitors and not by any structured guide-presentation plan.

Of the visitors who indicated having learned something during the WW experience, half of those were accompanied by a driver/skipper that acted as a guide. Nevertheless, in general, no relationship was established between the learning on the excursion and the presence of a guide during the trip.

17.3.2 The Responsibility of Visitors Who Enjoy Natural Environments to Conserve Biodiversity

- Demographic data

A total of 408 self-administered questionnaires were obtained from 51 excursion trips. Almost 100% of those visitors were Argentines, with 42.8% coming from the Río-Negro province; while the majority (60.8%) were within the age range of 24–54 years, with the greatest percentage (23.4%) being between ages 35 and 44 years. More than half (63.1%) were women (Table 17.1).

- Motivation and expectations

Of the majority of the visitors surveyed, the principal motive for coming to the BSAMPA was to participate in WW (Figure 17.2).

Table 17.1 Demographic data of the visitors to whale watching

Demographic Data	*% (n)*
Sex	
Male	36.86% (136)
Female	63.14% (233)
Age	
18–24	13.25% (51)
24–34	19.74% (76)
35–44	23.38% (90)
45–54	17.66% (68)
55–64	13.25% (51)
65 or more	12.73% (49)
Nationality	
Argentines	97.71% (384)
Foreigners	2.29% (9)

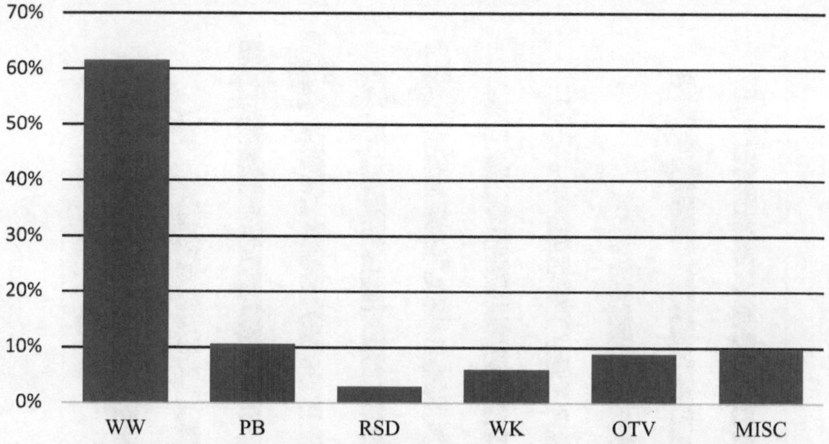

Figure 17.2 Visitors who perform whale watching in the Bahía de San Antonio Marine Protected Area (BSAMPA). References: WW, whale watchers using excursions; PB, passers-by; RSD, local residents; WK, local workers, OTV, out-of-town visitors of relatives or friends; MISC, other miscellaneous watchers.

The factors of WW most highly valued by the visitors are grouped into five responses: (1) the observation of the SRWs in their natural habitat, (2) the contact with nature, (3) the close observation of marine fauna, (4) the information and learning involved, and (5) the respect for nature.

The visitors' favorite moments in the experience were the close observation of the SRWs and their behavior (49%), the experience in its totality (16%), and the observation of the different species of marine fauna (13%). In contrast, the work of the guide (1.83%), as well as the information received on the excursion (1.68%), was only scantily recognized by the participants.

- Satisfaction with a conservation experience with whales

In order to learn the degree of satisfaction of the visitors with respect to the WW experience in the BSAMPA, the survey asked them to qualify the different factors of the experience from the standpoint of the degree to which WW fulfilled several of their possible expectations.

The majority of the visitors expressed being totally satisfied with each one of the evaluated aspects (Figure 17.3), with a greater variation in that degree of satisfaction: the number of available WW trips and seeing many whales and different species. As to the latter, the majority of the visitors surveyed (71.3%) acknowledged having observed a variety of species, including whales, dolphins, sea lions, and birds, but the guides had made very little mention of species other than the SRWs. The factors that elicited the greatest degree of satisfaction were the commitment of the driver regarding the care of the environment, seeing at least one whale, and seeing whales in a manner respectful of their territory.

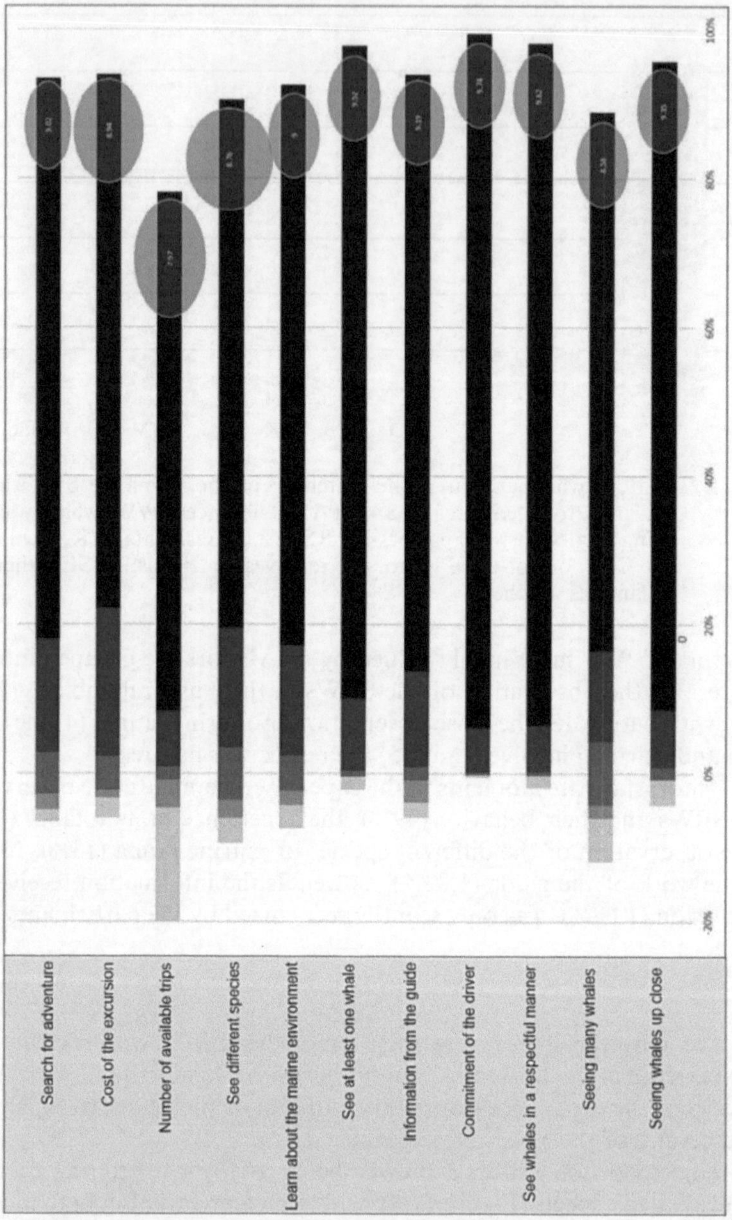

Figure 17.3 Visitor's degree of satisfaction with the different factors of the experience of whale watching. The percent of visitors that partially or fully conform regarding whale watching lie to the right of the vertical zero line; values to the left of that line represent visitors who were disappointed to different degrees. Categories of satisfaction are shown in white, black, or in gray scales. Circled numbers on each bar denote the mean value of each variable.

Table 17.2 Relationship between the possible expectations in whale watching and the complete satisfaction with the experience (*cf.* Figure 17.3)

Possible Expectations in Whale Watching	R^2
Seeing different species	98.8%
Seeing many whales	98.1%
Seeing whales up close	97.8%
Learning about the marine environment	97.6%
Search for adventure	97.6%
Commitment of the driver/skipper	96.8%
Information from the guide	96.8%
Seeing at least one whale	96.7%
Seeing whales in a respectful manner	96.6%
Cost of the excursion	96.5%
Number of available trips	92.9%

Regarding the degree of satisfaction with the experiences, the value obtained by the survey was 71%, with the majority of the visitors (61.7%) expressing total satisfaction. The possible factors that evoked the greatest degree of total satisfaction with the WW experience were seeing different species and seeing many whales, whereas the factors that least affected that satisfaction were the cost of the excursion and the number of trips available for WW (Table 17.2).

Upon consultation regarding an interest in contributing to the SRW and its habitat, fewer than half of the visitors suggested their willingness to do so as a result of what they had learned during the experience (24%) or through their own personal attitude and respect for the cause (18%). Finally, a relationship between the learning experience in the excursion and the interest in collaborating in the conservation of the SRWs and their habitat was verified statistically (χ^2 [1, n = 366] = 31.46; $p < 0.05$).

17.4 Discussion and Conclusions

Tourism and recreation in natural environments both depend on and are coresponsible for the conservation of biodiversity. Although WW in AMPBSA has the potential to contribute to the site's conservation, that capability has not yet been fully exploited. To promote conservation experiences in natural areas, visitors must know the regulations regarding nature conservation and comply with those norms. Nevertheless, most visitants do not receive information about this MPA. Therefore, upon consideration that this territory is defined, regulated, and managed in order to conserve, among other species, the SRWs as an environment in which they both reproduce and rest during their migration, the BSAMPA is offering spaces within this protected area where the use by public tourism is permitted through prioritizing "the harmonic coexistence of productive human activities and the ongoing maintenance of natural environments along with their native resources" (Decree 398/14, p.52). Accordingly, part of the WW regulations established as an ulterior objective a contribution to the consolidation of an environmental consciousness that would promote the harmonic integration of man and nature. Thus, as currently performed, WW would not be delivering completely the concern regarding a responsibility toward the natural system of the BSAMPA.

The lack of specialization in those who act as guides results in answers to visitors' questions that are often imprecise. For the purpose of this investigation, a *guide* is considered to be whatever person who—upon request, or through the agreement of the tourist company's operators—provides information during the excursion but is not the driver/skipper of the trip. When such persons have been absent from the trip, the driver/skipper of the boat has been the one who has acted as a guide by default. Thus, the absence of specialized guides can sometimes confuse visitors when a speech is not prepared with precise information. Likewise, there is a scarcity of information provided on aspects of the conservation of the SRWs and the regulations for responsible WW, by emphasizing only the biology-oriented information on the species. Nonetheless, for their part, a majority of the visitors expressed a complete satisfaction with the information received during the trip. This shortcoming in the information bestowed on the visitors could be associated with the possibility that the guides adjust the information communicated to the people according to what they perceive as their necessities and expectations on the basis of the profile of tourists and recreationists seeking natural environments.

We should mention that the norms of WW in the BSAMPA establish that all the boat trips include a specialized guide—*i.e.*, with a formal training in subjects pertaining to ecology, biology, tourism, culture, and conservation, among others, associated with not only the SRW but also the BSAMPA in general including other species that inhabit the reserve. The actors who are responsible for carrying out the activity don't comply with this regulation so, they don't contribute to guaranteeing biodiversity conservation. For this reason, action should be taken to acquire or train specialized guides along with the creation of complementary material to read. The comprehensive training of guides will ensure that they respond to visitors' questions in a manner that is both responsible, complete, and correct and will also contribute to incorporating into those tours a mention of the environmental value of the territory with respect to, for example, birds and dolphins, in that manner enriching the experience of the visitors and ensuring that their satisfaction extends beyond an appreciation of only the whales. The training of those guides is essential for generating experiences that, besides being agreeable and causing a greater satisfaction in the visitors, will be truly educational in providing information that not only is diverse and fascinating but also will create an environmental consciousness in favor of conservation.

The majority of the visitors do not have previous experience in the activity, and WW was the principal motive for visiting the BSAMPA. The majority are Argentines coming from nearby locations and representing intermediate age groups. These results coincide with the average age both of the observers of whales and of the visitors to protected areas worldwide. A major percentage of the visitors expressed an interest in contributing to the conservation of the SRWs and their habitat although the majority of the surveyed people did not know how to do so. To consider promoting those kinds of positive attitudes in the planning of the WW trips is an opportunity for enabling each visitor to become an active promoter of the best practices for the well-being of the MPA. Those who act as guides or direct the activity could also indicate how to promote the conservation of the SRWs and the BSAMPA in a concrete manner. Among those aspects most greatly valued before performing the WW *per se* are the acquaintance with and learning about the marine fauna in general, the contact with nature, and the observation of that fauna in its natural habitat, experiences that support a propensity for observing wild cetaceans in a non-consumeristic manner and the tendency to do so on a worldwide level associated with the rise in ecotourism.

The majority of the visitors who evaluated WW were totally satisfied with the entire experience as well as with each one of the elements inquired. Different possible expectations regarding WW influenced the visitor's satisfaction with the activity such as the possibility of observing different species during the trip. Likewise, the quantity of trips available is the condition that least influenced the participants' overall satisfaction. This result of excursions could be associated with the non-acquaintance with the norms of responsible WW. That non-acquaintance of the WW norms on their part evidences the necessity to deal with this aspect in the planning, management, and diffusion of the activity, in which scheme other actors would intervene, such as the local conservation authority.

WW as a generator of significant experiences in nature conservation is incompletely exploited. The experiences are significant as a result of the knowledge communicated to the visitors about the biology and ecology of the whales, but the awareness regarding the conservation of the species and the SRW's habitat has been insufficient. The results also demonstrate that the majority of visitors fail to perceive the relationship between tourism-recreation and the conservation of whales and their habitat. Although the responsibility of improving this relationship depends on several actors, the environmental-recreational tourism guide's performance in that capacity is key. Finally, WW in BSAMPA still has no set management plan but for any such approach in the future the weaknesses and problems manifested by this study should be material for an analysis aimed at the establishment of remedial objectives as well as measures to monitor and verify compliance. The present study underscores the essential role of WW as a tool for promoting the conservation of the species and the territory of the BSAMPA through education and the necessity to improve the tools of communication in the activity by the appropriate training of those acting as guides during the excursions.

Notes

1 World Tourism Organization.
2 International Union for Conservation of Nature.
3 World Commission on Protected Areas.
4 Instituto de Conservación de Ballenas.

5 https://www.unwto.org/asia/unwto-chimelong-why-wildlife.
6 https://www.unwto.org/es/desarrollo-sostenible/ecoturismo-areas-protegidas.

References

Andersen, M. and Miller, M. 2006. Onboard marine environmental education: Whale watching in the San Juan Islands, Washington. *Tourism in Marine Environments*, Vol. 2, No. 2, pp. 111–118. DOI: 10.3727/154427306779436327

Arias, M. 2019. Distribución, comportamiento y evaluación del impacto de las embarcaciones turísticas sobre la ballena franca austral Eubalaenaaustralis en el Golfo San Matías. PhD Thesis. Departamento de Ecología, Genética y Evolución, Facultad de Ciencias Exactas y Naturales, Universidad de Buenos Aires. (Accessed: June 2021).

Arias, M., Sarti, G., Sylwan, C., Svendsen, G., Romero, A., Crespo, E. and Gonzalez, R. 2016. Bases para el desarrollo sustentable del turismo de avistamientos de ballena franca austral en el golfo San Matías. Informe técnico final. Consejo Federal de Inversiones. Provincia de Río Negro.

Ballantyne, R., Packer, J. and Hughes, K. 2009. Tourists' support for conservation messages and sustainable management practices in wildlife tourism experiences. *Tourism Management*, Vol. 30, N o. 10, pp. 658–664. DOI: 10.1016/j.tourman.2008.11.003

Bertella, G. 2018. Sustainability in wildlife tourism: Challenging the assumptions and imagining alternatives. *Tourism Review*. DOI: 10.1108/TR-11-2017-0166

Cárdenas, S., Gabela-Flores, M.V., Amrein, A., Surrey, K., Gerber, L.R. and Guzmán, H.M. 2021. Tourist knowledge, pro-conservation intentions, and tourist concern for the impacts of whale-watching in Las Perlas Archipelago, Panama. *Frontiers Marine Science*, Vol. 8, 627348. DOI: 10.3389/fmars.2021.627348

Cooke, J.G. and Zerbini, A.N. 2018. Eubalaena australis. The IUCN Red List of Threatened Species 2018. e.T8153A50354147. DOI: 10.2305/IUCN.UK.2018-1.RLTS.T8153A50354147.en (Accessed: July 2021).

D'Agostino, V.C., Mandiola, A., Bastida, R., Giardino, G., García, N.A., Romero, M.A. and Coscarella, M.A. 2019. Eubalaena australis. En: SAyDS–SAREM (eds.) *Categorización 2019 de los mamíferos de Argentina según su riesgo de extinción*. Lista Roja de los mamíferos de Argentina. Versión digital: http://cma.sarem.org.ar

Encabo, M., Sanchez, S., Mastrocola, Y., Vazquez, M. and Paz Barreto, D., 2014. Area Natural Protegida Parque Universitario Provincia del Monte como espacio recreativo – Campus Universidad Nacional del Comahue – Neuquen – Argentina. VI Congreso Latinoamericano de Investigación Turística. Construyendo conocimiento en turismo: diversas miradas sobre un campo complejo. Neuquén, Argentina.

Encabo, M., Sanchez, S., Torre, G., Paz Barreto, D., Andrés, J., Mastrocola, Y., Vazquez, M. and Cánepa L. 2016. Responsible use of biodiversity: Revisiting the "Recreación y Turismo en Conservación" (RyTeC) model. In M.G. Torre, (Ed.) *Anuario de Estudios en Turismo – Investigación y Extensión*. Facultad de Turismo – Universidad Nacional del Comahue Neuquén – Argentina, pp. 8–20.

Feola, G., Büscher, B., Fischer, A. and Koster, M. 2021. How not to go 'back to normal' after COVID-19: Planning for post neoliberal developme. Posted at EADI blog *Debating Development Research*. Available at: http://www.developmentresearch.eu/?p=935#more-935 (Accessed: July 2021).

Gössling, S. and Hall, C., 2006. *Tourism and Global Environmental Change Ecological, social, economic and political interrelationships*. Routledge, NY, USA.

Gudynas, E. 2012. Debates sobre el desarrollo y sus alternativas en América Latina: Una breve guía heterodoxa in *Más allá del desarrollo. Grupo Permanente de Trabajo sobre Alternativas al Desarrollo*. 1a ed. Ciudad de México, Mexico. pp. 21–53.

Hernández Sampieri, R., Fernández Collado, C. and Baptista Lucio, P. 2014. *Metodología de la Investigación*. 6ta ed. McGRAW-HILL / Interamericana Editores, S.A. de C.V. Impreso en México, pp. 4–600.

Hoyt, E. 2001. *Whale Watching 2001: Worldwide tourism numbers, expenditures, and expanding socioeconomic benefits*. International Fund for Animal Welfare, Yarmouth Port, MA, USA, pp. i–vi; 1–158.

Hoyt, E. 2005. Sustainable ecotourism on Atlantic islands, with special reference to whale watching, marine protected areas and sanctuaries for cetaceans. *Biology & Environment Proceedings of the Royal Irish Academy*, Vol. 105B, No. 3, pp. 141–154. DOI: 10.3318/BIOE.2005.105

ICB. 2021. Impactos de redes y sogas de la actividad pesquera sobre las ballenas francas de Península Valdés. Informe técnico. Available at: https://ballenas.org.ar/ (Accessed: July 2021).

Kessler, M., Harcourt, R. and Bradford, W. 2014. Will whale watchers sacrifice personal experience to minimize harm to whales?. *Tourism in Marine Environments*, Vol. 10, No. 1–2. DOI: 10.3727/154427314X14056884441662

Lück, M. and Aquino, R. 2021. Domestic nature-based tourism and wellbeing – A roadmap for the new normal?. In *Tourist Health, Safety and Wellbeing in the New Normal*, pp. 1–33.

Mitchell, B. and Phillips, A. 2021. A global tragedy in search of answers: editors' introduction in IUCN WCPA. *PARKS. The International Journal of Protected Areas and Conservation*, Vol. 27, No. 3, Gland, Switzerland: IUCN, pp. 7–12. DOI: 10.2305/IUCN.CH.2021PARKS-27SI.en

Mooney, H. and Zegarra, M. 2020. COVID-19: Shock sin precedentes sobre el turismo en América Latina y el Caribe. Resumen de políticas del BID N°339. Banco Interamericano de Desarrollo, p. 20.

Notarbartolo di Sciara, G. and Hoyt, E. 2020. Healing the wounds of marine mammals by protecting their habitat. *Ethics in Science and Environmental Politics*, Vol. 20, pp. 15–23. DOI: 10.3354/esep00190

Organización de las Naciones Unidas. 2020. Informe de politicas: La COVID-19 y la transformación del turismo. pp. 1–30

Organización de las Naciones Unidas. 2021. Available at https://www.unep.org/es/noticias-y-reportajes/comunicado-de-prensa/el-mundo-cumple-su-meta-de-areas-terrestres-protegidas (Accessed: June 2021).

Sarti, G., Arias, M., Sylwan, C. and Gonzalez, R. 2016. Perfil de la demanda y factores condicionantes de la satisfacción de los usuarios en el turismo de avistaje de ballenas francas (eubalaena australis) en Rio Negro, Argentina. Presentación en el XI Congreso de la Sociedad Latinoamericana de Especialistas en mamíferos Acuáticos RT17. Valparaiso, Chile.

Sarti, G., González, R. and Encabo, M. 2019. Turismo en conservación. Avistaje de ballenas. Area Natural Protegida Bahía de San Antonio – Argentina. Póster. In *III Congreso de Áreas Protegidas de Latinoamérica y El Caribe*. Lima, Perú.

Silveira, M. 2011. Territorio y ciudadanía: reflexiones en tiempos de globalización. Informes de investigación y ensayos inéditos, Vol. 11 N°3. Facultad de Educación-Universidad de Antioquia. Medellín, Colombia.

Sironi, M. 2021. ¿Cómo afectan los enmalles y el plástico a las ballenas francas de península Valdés? Presentación oral en Instagram. Instituto de Conservación de ballenas.

Spenceley, A., McCool, S., Newsome, D., Báez, A., Barborak, J., Blye, C., Bricker, K., Cahyadi, H., Corrigan, K., Halpenny, E., Hvenegaard, G., King, D., Leung, Y., Mandić, A., Naidoo, R., Rüede, D., Sano, J., Sarhan, M., Santamaria, V., Sousa, T. and Zschiegne, A. 2021. Tourism in protected and conserved areas amid the COVID-19 pandemic. In IUCN WCPA. *PARKS. The International Journal of Protected Areas and Conservation*, Vol. 27, No. 3, Gland, Switzerland: IUCN, pp. 103–118. DOI: 10.2305/IUCN.CH.2021PARKS-27SI.en

UICN-CMAP. 2018. Estándares Globales de Conservación de la UICN aplicables a las Áreas Marinas Protegidas (AMP). Medidas efectivas de conservación a través de las AMPs, para asegurar la salud y el desarrollo sostenible de los océanos. Versión 1.0. Gland, Suiza. 4p.

Vernazzani, B., Burkhardt-Holmb, P., Cabrera, E., Iñíguez, M., Luna, F., Parsons, E., Ritter, F., Rodríguez-Fonseca, J., Sironi, M. and Stachowitsch, M. 2017. Management and conservation at the International Whaling Commission: A dichotomy sandwiched within a shifting baseline. *Marine Policy*, Vol. 83, pp. 164–171.

Waithaka, J., Dudley, N., Álvarez, M., Mora, S., Chapman, S., Figgis, P., Fitzsimons, J., Gallon, S., Gray, T., Kim, M., Pasha, M., Perkin, S., Boixeda, P., Sierra, C., Valverde, A. and Wong, M. 2021. Impacts of COVID-19 on protected and conserved areas: A global overview and regional perspectives. *PARKS. The International Journal of Protected Areas and Conservation*, Vol. 27 (3), Gland, Switzerland: IUCN, pp. 41–56. DOI: 10.2305/IUCN.CH.2021PARKS-27SI.en

Wearing, S., Cunningham, P., Schweinsberg, S. and Jobberns C. 2014. Whale watching as ecotourism: How sustainable is it? *Cosmopolitan Civil Societies Journal*, Vol.5, No. 3. DOI: 10.5130/ccs.v6i1.3714

World Tourism Organization (UNWTO). 2021. COVID-19 and tourism 2020: A year in review. *UNWTO*. Available at https://www.unwto.org/covid-19-and-tourism-2020 (Accessed: June 2021).

Zeppel, H. 2008. Education and conservation benefits of marine wildlife tours: Developing free-choice learning experiences. *The Journal of Environmental Education*, Vol. 39, No. 3, pp. 3–17.

Zeppel, H. and Muloin, S. 2014. Green messengers or nature's spectacle: Understanding visitor experiences of wild cetacean tours. In J. Highman, L. Bejder and R. Williams (Eds.), *Whale-watching: Sustainable tourism and ecological management*, pp. 100–127.

18 Marketing of Tourism Ecosystem Services in Times of Uncertainty

Lessons from a Health Tourism Destination in Poland

Adrian Lubowiecki-Vikuk, Jacek Borzyszkowski, and Mirosław Marczak

18.1 Introduction

Before the coronavirus pandemic (COVID-19) paralysed the world economy, tourism had been developing very dynamically over the past decades (Nientied, 2021). The coronavirus pandemic has completely changed the situation on the tourism market (Vegnuti, 2020) to the extent that, in many tourist destinations, a new phenomenon became the topic of discussion, i.e. "no tourism" (Koh, 2020). Therefore, changes on a global scale require the tourism business to operate under uncertain and ambiguous market conditions, which can be understood as a crisis. Kotler and Caslione (2009) suggested that in a turbulent period – and the COVID-19 pandemic should be considered as such – every product needs to be modified and innovative solutions need to be relied on. An example of this approach was given by Apostolopoulos *et al.* (2021), who suggested combining a medical tourism product with the production of generic medicines.

It is Destination Management Organisations (DMOs), among others, that are responsible for the promotion of tourist products. More than ever, they should focus on eliminating false messages and disinformation as to the threats and risks perceived among tourists, and thus on appropriate crisis communication and the selection of content generated by users with an impact on the image of the brand of a tourist destination (Sigala, 2020). The image of a tourist destination may change over time. It is important to bear in mind the effects of the COVID-19 pandemic on the formation of this image (Zenker & Kock, 2020). The COVID-19 pandemic has "created" a negative image of travelling, including the context of public health (Godovykh & Ridderstaat, 2020); this also applies to the mental health of workers employed in the tourism industry. It is worrying that workers in the tourism and hotel industries have displayed racism by blaming other countries for the outbreak of the pandemic (Park, Kim & Kim, 2020). The evolving nature of their emotional states will be the main determinant of their attitudes and future behaviour towards "returning" customers. It also seems that some attributes of the image, especially medical attributes, will be crucial to future travel destinations. Those tourist destinations that may have been held

DOI: 10.4324/b23145-24

responsible for the outbreak of the COVID-19 pandemic must fight to rebuild their damaged image.

To strengthen the brand of the destination, one should concentrate on the visual identification of a health tourism destination (HTD) using the brand design. What is fundamental is the creation of a coherent vision of the brand while maintaining the integrity and values which encourage one to purchase (McWilliam & Dumas, 1997) and choose a specific destination for health purposes. It is also essential to take into account the development of ecosystem services, especially considering the nature of tourism, i.e. environmental aspects (e.g. Hancock, 1993). All of this requires specific organisational conditions and the involvement of the representatives of DMOs in building the HTD brand as one that is open to cooperation and development of entrepreneurship based on the provision of health tourism services. Among the responsibilities of these organisations, the implementation of a joint strategy of communication in specific foreign markets is emphasised (Polish Tourism Organisation, 2016).

Based on the available literature, there has not been a great deal of research to date that has focused on the need to communicate the value of health through the DMO visual identity system. It has not been necessary to do so. However, increasing awareness of the need to protect the ecosystem is of growing importance. In addition, the context of the COVID-19 pandemic is a special circumstance which necessitates an even closer look at the activities of DMOs and a review thereof in terms of promoting health tourism about the marketing of ecosystem services. The authors' ambition, therefore, is not to make generalisations, which usually fill a gap in theoretical research. A clear attempt has been made to illustrate issues related to the circumstances in which the analysed phenomena occur. The purpose of the chapter is to identify DMO activities in building the brand of a tourist destination based on health and safety while considering the importance of ecosystem services. To achieve this aim, it was necessary to answer the following research question (RQ):

> RQ: How do DMO activities aimed at creating the image and brand translate into visual identification in the context of ecosystem services?

18.2 Literature Review

18.2.1 Brand and Image: The Context of a Tourist Destination

Even though the concept of the brand of an area (territory) is a relatively new issue, it has raised several doubts and controversies from the very beginning. Among scholars and practitioners, there are generally two types of views as to the sense and possibilities of the creation of so-called territorial brands. The first concept assumes that destinations can be the subject of branding and management in a similar fashion as consumer goods and services (e.g. Boisen *et al.*, 2011; Florek & Janiszewska, 2015; Marczak, 2018b). The second concept questions the entire process.

Supporters of the former concept emphasise, among other things, the universal nature of the mechanisms used in brand building and management and the possibility of using these in the process of creating a territorial brand (a tourist reception area). However, some authors (e.g. Ashworth, 2010), while expressing some moderate scepticism in this respect, emphasise that marketing specialists are too ready to assume that territories simply constitute extended products. Meanwhile, the number and diversity of the elements that they are composed of is so great that territories should be referred to as "mega products" (Marczak, 2018a). The authors of this study also accept this approach, and for the purposes of this analysis they accept an assumption that identifies the territory of a given country with the so-called mega product of a tourist destination, among other things. Therefore, taking this assumption into account, and considering the perception of the brand as a certain network of associations functioning in consumers' minds, the brand of a destination (an area, a territory) can be defined as a network of associations (a combination of various associations) in the minds of consumers based on a visual, verbal and behavioural expression of the destination, which is embodied by objectives, communication, values, as well as the general culture of the stakeholders of the destination and its design (Zenker *et al.*, 2017).

The source literature also contains a concept based on views that question the sense and possibilities of building territorial brands. Some authors (e.g. Dolnicar & Mazanec, 1998) provide a chief argument in this respect, namely the fact that territories are not easily eligible for branding because, in the case of destinations, they are usually "fuzzy" and dispersed. In particular, the nature of destination branding raises doubts which, according to Olins (2004), are always complex, controversial and disputable, mainly because it must focus on creating a brand management system focused on the identity of a particular destination (Freire, 2005), considering several historical, political, religious and cultural elements. Furthermore, considering the political nature of the destination brand management process and the lack of clear leadership in the process, branding is subject to cycles and fluctuations (Florek & Janiszewska, 2015). According to Kavaratzis (2008), another threat resulting from this process is the risk of partnerships between the public sector (e.g. central government, local government) and the private sector (tourism industry), which initiate the branding process without consulting the local community, which is not appropriate because destinations are a social construct formed from a selection of spatial elements by humans (Boisen *et al.*, 2011). Compared to consumer products, an area as a product is therefore an open system, one that is historically determined, constantly changing and evolving with the relationships occurring between people and the environment.

Regarding tourist destinations, the brand is to be considered in light of all the elements that define it. There are three basic groups of these elements in the literature:

- elements directly related to the construction and structure of the brand, including but not limited to brand identity (Pereira *et al.*, 2019), brand

competencies (Kim *et al.*, 2017), brand personality (Hanna & Rowley, 2019; Kusumawati, 2019), and brand *USP* (Henthorne *et al.*, 2016; Marczak, 2019; Sigwele *et al.*, 2018),
- elements satisfying consumer needs in the form of various types of benefits: rational (functional), symbolic, and emotional (e.g. Kladou *et al.*, 2016; Pike & Page, 2014),
- elements that define the main assumptions and the tools of the so-called "brand communication" system that covers any exchange of information with the market (including a visual identification system that covers elements such as the name, the symbol – logo, the promotional slogan, and colours) (e.g. De Las Heras-Pedrosa *et al.*, 2020).

Precise identification of the three elements of the tourist destination brand will enable the effective implementation of one of the most important objectives of modern DMOs using the strength and the market position of the brand, in the form of making efforts to obtain the desired image. Apart from building brand awareness (branding), this is the second important element of the source of its strength (Basera & Baipai, 2020). At present, strong brands are primarily those enjoying both a high level of awareness and a strong, beneficial and unique image. Achieving such an effect is only ensured by reliable preparation of the strategic guidelines of the brand concept and a detailed description of its identity, key competencies that distinguish a product or a service, determination of positioning, indication of the brand value, as well as its vision and mission. The final stage of the creation of guidelines related to the brand is the formation of its verbal aspect, i.e. the brand name, and its non-verbal aspect (i.e. its visual aspect), which consists of elements such as the sign, the logo, the arrangement and shape of the letters, or the colour scheme used (Barisic & Blazevic, 2014).

As mentioned previously, the term "branding" is inseparable from the concept of the brand. According to Marland *et al.* (2017), we must understand branding as a proactive act of bringing a brand to life through concerted and concentrated efforts from all key stakeholders. Branding is perhaps one of the strongest marketing tools available for experts in destination marketing (Morgan *et al.*, 2011). A common strategy is that every country has a unique culture, landscape and cultural heritage, and particular places are exposed as locations friendly to tourists and visitors in the context of human hospitality and a high standard of customer service and convenient facilities (Marczak, 2018b). As a result, the need to create the image of a place with a unique identity that will distinguish a given destination from the competition is becoming more urgent (Kozak & Mazurek, 2011). Thus, the purpose of place marketing is to increase the attractiveness of the place, so that branding not only involves occasional promotional activities but a holistic development process that affects the perception of the place as well (Ma *et al.*, 2019).

The analysis presented in this chapter covers the image of health tourism in Poland. Assuming that the tourist reception area is to be considered a mega tourist product, it should also be assumed that the image of health tourism forms part of a comprehensive brand image of a tourist destination. In the source literature, the term "brand image of the tourist reception area" is quite frequently identified with the term "the image of the tourist reception area". The relations between these concepts were aptly defined by, among others, Nawrocka (2013), who stated that in the case of a given tourist reception area building a single branded product, the image of that area will be identified with the brand image. However, when many so-called branded products (e.g. health tourism) are created in each area for different groups of buyers, the positive image of the area becomes a kind of "umbrella" for them (offering various products and services under one brand: a brand umbrella), and each of these products has an influence on the overall image of the destination among its recipients. Therefore, understanding the tourist reception area as an autonomous mega tourist product entitles one to identify the brand image of a given tourist reception area with its comprehensive tourist image.

18.2.2 Investigation of DMO Activities in Building the Brand of a Tourist Destination Based on Health and Safety

Today's DMOs have evolved from place promotion agencies dating back to the mid-19th century into professional tourism organisations with a wide range of marketing and management responsibilities (Pike & Page, 2014). The objectives of DMOs can now be regarded as particularly diversified (Borzyszkowski & Lubowiecki-Vikuk, 2019). DMOs are frequently assigned a broad range of activities, such as destination branding, lobbying, strategy formulation, quality assurance, crisis management and policy-making (Pearce, 2015).

In general, the literature emphasises the importance of promotion or, more broadly, of destination marketing in activities undertaken by DMOs. For example, the significance of marketing activities conducted by DMOs was emphasised by Morgan *et al.* (2011). From the perspective of the activities pursued by DMOs, the brand is of special significance. DMOs serve as a guardian of the destination brand. They are responsible for the development, coordination and implementation of the destination network brand, working to induce images in the minds of consumers of destination experiences (Marzano & Scott, 2009). This results from the problem of the knowledge of the brand in the context of the tourist reception area. Investigations carried out on the German market served to confirm this problem (Steinecke, 2001). It became evident that the knowledge of brands from the individual sectors of the tourism economy is highly diversified, and it is as follows: tour operators – 51% of all the indications, airlines (36%), hotels (5%), things used during the trip (4%), companies that offer car rental (2%), and destinations (2%).

In the case of activities aimed at branding and brand development, the creation of the system of visual identification is of primary significance. The graphic sign, i.e. the logo, forms the basis of such a system. It performs numerous functions in the scope of the creation and strengthening of brands

(Blain *et al.*, 2005). As stated by Keller (1993), every time the marketer creates a certain name, a logo or a symbol for a new product (from the technical perspective), a brand is simultaneously created. The use of graphical signs is popular among DMOs regardless of the type of organisation and the level which they represent. The logo must be visually appealing and aesthetically pleasing while conveying the identity of the destination (Bonnardel *et al.*, 2020). However, a certain diversification can be seen in the scope of the graphic elements used. This can be shown using the example of national logotypes (Raftowicz-Filipkiewicz, 2008), which are created within the following categories:

- thematic (referring to the specificity of the brand), for example, the logo of Spain with a sun,
- symbolic (e.g. an image of plants or animals), for example, the logo of Holland (currently the Netherlands) including a tulip,
- inspired by heraldry (referring to coats of arms to emphasise the connection between the brand and a defined region or state), for example, the logo of Malta with the cross of the knights of Malta,
- inspired by lettering (logotypes, monograms) or digits (referring to the symbolism of letters or digits in each alphabet),
- inspired by the name of a brand, a flag or abstract signs of the brand.

From the perspective of the issues raised in this chapter, the problem of the formation of the HTD image is important for further consideration. By accepting the general assumptions from Manhas *et al.* (2016) concerning the tourist destination brand image and taking the specificity of health tourism into consideration, we may accept that the image of an HTD brand depends on the specificity of a destination and the innovations used therein, the social and demographical profile of a health tourist and their experiences, as well as marketing activities, including those of people who are responsible for its creation (Dryglas & Salamaga, 2018). The following is to be considered an HTD brand design:

- name and logo: adapted to the specificity of an HTD taking the colours (Séraphin *et al.*, 2016), topographics and the graphical language into consideration. The logo, which is appreciated chiefly from the practical perspective (e.g. Blain *et al.*, 2005; Lee *et al.*, 2012), is "the most relevant tangible artifact, resulting from intentional branding by DMOs" (Beritelli & Laesser, 2018), whereas its colour and saturation play an important role in the creation of the brand (Ghaderi *et al.*, 2015),
- advertising slogan: simple to pronounce, easy to remember, stands out compared to competitive destinations, and emphasises the HTD concept. Galí *et al.* (2017) observed that the slogan should possess an exclusive appeal.

A review of the literature demonstrates quite clearly that the formation of the HTD image by DMOs is relatively little practised. However, one needs to remember that identification with an HTD or the absence thereof does not prejudge the effectiveness of the actions undertaken by DMOs. Some organisations may build their identity based on other tourism products (Borzyszkowski & Lubowiecki-Vikuk, 2019). These products can directly refer to the use of natural resources (renewable and non-renewable) in the HTD branding strategy. It can also be assumed that the HTD brand should represent the values of cultural services, which is one of the ecosystem service categories. We especially have ecological values and socio-cultural values in mind. Given this, the HTD brand attributes should represent the elements of the natural environment and human well-being in a broad sense. It is important to agree with the findings of Giannopoulos, Piha and Skourtis (2021), who proposed a stepwise process of strategic imperatives for co-branding in a destination context. This process refers to the multi-directional flows of brand meaning at different levels of the so-called tourism ecosystem, and thus interprets stakeholders' efforts to co-create sustainable brands that gain relevance in the global tourism market.

18.2.3 Health Tourism as a Component of the Ecosystem

The service ecosystem lens implies that innovation is grounded in a novel and useful way to integrate resources to co-create and realise value within a service ecosystem (Edvardsson & Tronvoll, 2013; Lusch & Nambisan, 2015). The concept of ecosystem services is interpreted differently in the literature. Most commonly it refers to the set of creations and ecosystem functions that are useful to society and to the economy. It can be assumed that it is "a relatively self-contained, self-adjusting system of resource integrating actors connected by shared institutional arrangements and mutual value creation through service exchange" (Lusch & Vargo, 2014, p. 23). Examples of innovation within ecosystem services can include cultural services (Machnik, 2019, p. 6) and redefined methods of engaging the public in health care and promoting healthy lifestyles. Tourism and its health function correspond to this. It is health tourism that gives intangible benefits obtained from ecosystems, such as cognitive value, recreational value, reflective value, aesthetic experience, or spiritual enrichment (e.g. Costanza *et al.*, 1997; Dominati, Patterson & Mackay, 2010). Health tourism is thus an excellent example of the composition of ecosystem services, and its promotion by DMOs seems crucial, especially in the post-COVID-19 era.

Health tourism constitutes an important product of the tourism market (Dryglas, 2018; Wen *et al.*, 2021). It includes those types of tourism where health aspects are the main motivation for the journey and that contribute to improving one's physical, mental and/or spiritual health through medical and health activities. They increase the ability of individuals to satisfy their needs and to function better in their environment and society (World Tourism Organisation & European Travel Commission, 2018). What is meant here is the therapeutic tourism product, the wellness tourism product, and the medical tourism product (Dryglas, 2018). The development of health tourism will probably not only help to mitigate the effects of the coronavirus pandemic and to regenerate the health of society, but it may also provide an opportunity to promote those tourist destinations which are in decline, or which currently are not appreciated enough (Navarrete & Shaw, 2021).

Health tourism in terms of a tourist product is most often realised in an HTD, frequently understood as a health region. The concept of a health region is presented as a tool for increasing the attractiveness of tourist destinations (Białk-Wolf *et al.*, 2016; Dryglas & Lubowiecki-Vikuk, 2019b). With reference to the classic division of the marketing mix, in this case the health tourism product is generally not redefined based on sensations, experiences or values as in the case of experience marketing (Lubowiecki-Vikuk, 2021). Health is the highest value of a human being. Thus, this product is a permanent marketing instrument, which should be looked at from the perspective not only of a company and business but also the perspective of a destination and the environment. The environment is uncertain, which is linked to a random factor, namely the COVID-19 pandemic.

18.2.4 The Role of DMOs in Creating Ecosystem Services: The Importance of Organisations in Tourism Development in the Post-COVID-19 Era

Rather than tourism being a single type of business, it is a diverse ecosystem of businesses where each business type depends on the others for its survival and well-being (Tourism Alliance, 2021). In this regard, it is recognised that DMOs are important players in the tourism ecosystem and are able to share a common vision and their experiences (Giannopoulos *et al.*, 2021). DMOs can and should be expected to engage all stakeholders in the sustainable development of tourism at the destination (World Tourism Organisation, 2004). The role of DMOs goes beyond the classic framework of ecosystem activities. They are organisations that should be guided by a set of principles, dealing with the coordination and effective distribution of tourism services (Antionos *et al.*, 2020). With the rise of sustainable development and sustainable tourism, new approaches to destination management are required. Destination managers, developers and DMOs now need to include sustainable tourism development in their strategies and policies (Welford & Ytterhus, 2004). Especially in recent years, the expectations of DMOs have increased. They are the ones who should raise environmental awareness to encourage travellers to take responsible action. Therefore, DMOs should adopt the appropriate approach of interactive communication, implementing the strategies of responsible use of tourism assets (Sultan *et al.*, 2021).

The considerations presented here can serve as a starting point for trying to define the role of DMOs in tourism recovery and creating ecosystem services after a pandemic. It is these organisations that are expected to work particularly hard in the largest-ever crisis in global tourism. It is important to remember that DMOs are not perfect organisations. In many cases, they are burdened with bureaucratic problems. The current, as well as the future, post-pandemic situation will certainly accelerate many changes in the structure and functions of DMOs (Vargas, 2020). These organisations will be responsible for the return to "normalcy" soon. Their activities will rely heavily on supporting the development of ecosystem services, such as promoting products and experiences related to nature, health and well-being (European Travel Commission, 2020). In many cases, this will require DMOs to significantly change their operating philosophy, if only in terms of a better understanding of the domestic market (Lück & Seeler, 2021). This should result not only in marketing but, among other things, in an improvement in the quality of tourism products.

18.3 Methodology

This chapter takes a two-pronged methodological approach. First, a traditional narrative review was conducted, involving an arbitrary mapping of existing intellectual output. To this end, secondary source search methods were used to analyse the constituent body of knowledge on branding and image creation and DMO branding competencies (manifested through logo design) concerning ecosystem services. The authors moved away from analysing only papers published in Web of Science indexed journals and included the "useful in many ways" (Briner & Walshe, 2014, p. 417) nature of a traditional review. This type of review makes an essential contribution to the discussion of interdisciplinary topics in which ecosystem services are embedded (Millennium Ecosystem Assessment, 2005). Loosening the rigor, in this case, allowed the review to capture a broader perspective, with different coverage, including regional/local, in a language other than English, was reviewed, and important reports published by public institutions and organisations were studied. The criterion for selecting a particular publication for review was strictly related to the issue addressed, which was verified by the authors of this chapter. Secondly, case studies were used here, referring to the activities of national, regional, or local Polish DMOs. The chosen method, although criticised, is currently experiencing a renaissance in management sciences. Its aim is, among others, practical orientation (more broadly executive research) referring to a better understanding of reality. The authors, through a purposeful and arbitrary selection of analysed cases according to the methodological recommendations of Dul and Hak (2008), sought to understand the circumstances and decision-making paths of specific decision-makers.

18.4 Results and Discussion

DMOs in Poland have a relatively short history. It is accepted that these organisations may be discussed considering the three-stage system in tourism, which was introduced in 2000 on the basis of the Polish Tourism Organisation Act (i.e. the Act of 25 June 1999 on the Polish Tourism Organisation, 1999). This Act distinguished three types of organisations, namely the Polish Tourism Organisation (PTO), regional tourism organisations (RTO) and local tourism organisations (LTO). Generally, the PTO is responsible for tourism promotion and the creation of the tourism image of the country as a whole, while RTOs (of which there are 16) and LTOs (approximately 130) are responsible for similar tasks implemented at lower levels of the administrative division of the country (Borzyszkowski, 2015). DMOs in Poland at the national, regional and local levels see the opportunity to look at tourism, health and the active lifestyle of society, which is perceived as an opportunity to change the position of the country's brand and its image (Lubowiecki-Vikuk & Basińska-Zych, 2011). The image of HTDs is composed of medical and non-medical attributes (Dryglas & Lubowiecki-Vikuk, 2019a).

The idea to create a logo of the Polish Economy Brand arose from human capital, i.e. Poles themselves who are internationally recognised as qualified professionals. The logo is made up of two elements: (1) the shape in the form of the letter "P": as in Polska (Poland) and (2) a network of complementary people (Lubowiecki-Vikuk, 2021, p. 157). The addition here is a more general term for the "medical tourism" economy sector, which is "Health". Adding an image slogan allows one to attach more importance to what can be achieved, thanks to the name or symbol itself. White and red are the leading colours which refer to the national colours of Poland. The typography is related to the natural proportion of signs, owing to which the font is universal. The graphic language, in turn, permits any interpretation of the graphic sign by giving it a recognisable style that makes references to modernity and tradition.

In relation to the logo, the following advertising slogan was accepted: "Poland: your health destination". This action, however, lacks consistency. Instead of the Polish name of "Polska", the English name of "Poland" was used; apart from this, its typography departs from the one that was previously agreed. The producers' intentions are subject to assumptions, yet the idea was to be distinct from other brands, and thus to avoid the (frequent) phonetic error of identifying the word "Poland" with "Holland". Apart from this, Kladou *et al.* (2016) pointed out that the name of the destination is of greater significance than a logo or a slogan. We may assume that this may be similar in relation to HTDs. An official website with the logo and slogan accepted was prepared. The keywords for this destination were used in the internet domain. The website is both in English and Polish, which serves to emphasise the awareness of the occurrence of national health tourism. Apart from basic information, the website contains a search engine according to two criteria: medical services and healthcare providers. This action was supplemented with the use of the logo and slogan in the promotional video.

The development of Poland's destination branding as an HTD is based primarily on the participation of the interested medical entities and medicine-related entities, including tourist entities in foreign economic missions and at health tourism fairs. Furthermore, the decor of the exhibition stands is related to Poland's national colours and includes the previously accepted logo. Originally, training sessions and press conferences were organised, promotional materials were distributed, and articles were prepared for foreign magazines. Public relations activities were conducted. At present, the focus is on TV programmes that promote Polish pro-health services and the organisation of study visits for foreign journalists. According to a study by Lubowiecki-Vikuk (2021, p. 157), the logo "Poland: your health destination" is recognised by more than half of domestic and foreign medical tourists undertaking treatment in Poland, especially by those aged 21–30 and 31–40. Those aged 50 and above do not yet recognise the logo. This result provides an impetus to seriously engage DMOs in the responsible promotion and health care for people in this growing demographic category.

Świętokrzyskie Province is distinctive when compared to other Polish regions in the context of HTD. First, it is worth noting that the Regional Tourism organisation of Świętokrzyskie Province possesses a "catalogue of visual identification for the tourist logotype of Świętokrzyskie Province". This document describes the idea and distinguishing features of the tourist logotype of the region in detail.

It is not only the graphic elements included in the logotype that indicate an attempt to expose the tourism potential of Świętokrzyskie Province but above all their colouring. As emphasised by the authors of the logotype, "the range of colours has been selected to create a coherent and harmonious entirety and, at the same time, to illustrate the wealth and attractiveness of the region". It is worth noting the importance of the colour red which, as intended, is to depict "life, vital forces and beauty". Green, in turn, is supposed to connote "youth, freshness, fertility". The elements refer to a large extent to health aspects and, considering the essence of the activities pursued by the Regional Tourism Organisation of Świętokrzyskie Province, to health tourism itself. This corresponds directly to the provisions of the "Strategy for the Development of Tourism in Świętokrzyskie Province for the years 2014–2020", where the health tourism product is clearly distinguished as one of the priorities in the development of the tourist function in the region.

There are more cases of HTD image building at a local level than in regional structures. The experiences of individual municipalitics, cities and even local tourism organisations are interesting. One of the most interesting examples is certainly "Roztocze Vitality by Nature", the brand of the Roztocze region, based on the marketing strategy developed by the Roztocze Local Tourism Organisation.

The concept of the brand assumes several actions aimed at increasing the recognition of the region as a place that is attractive for tourists and that fulfils the promise of healthy, active leisure pursuits in nature (Lokalna Organizacja Turystyczna Roztocze, 2020). Thus, there is a clear reference to aspects of health tourism. The additional aspects distinguished include specialist tourism and nature tourism.

As a rule, the logotype is characterised by considerable minimalism and aesthetics, and it refers to the innovativeness of actions and a modern approach to the management of the brand of the region. In addition, the dynamic form of the logotype and appropriately selected colours are to symbolise the positive and energetic nature of the destination. In turn, the slogan "Vitality by Nature" is memorable for its approachable and easy manner.

Additionally, its purpose is to build specific associations with the region. According to the assumptions,

> (…) a concise and concrete message possesses a multifaceted meaning, whose leading idea is to emphasise the beneficial influence of the natural values of the region, which enable one to obtain an internal balance and harmony between the body, the soul and the mind. The region, while being based on its unique natural resources, stimulates the vitality of the organism and, using spa treatment methods and the properties of the Roztoczańskie spa springs, further strengthens the image of this place built on health values.
>
> (Synergia, 2020)

The colouring of the logotype and the graphic elements are also important; they are closely related to aspects of health tourism. The graphic symbol represents a person in motion, one who is active and satisfied. Orange, in turn, depicts energy, while green points to harmony and balance, a regeneration of vitality and associations with nature.

An overview of the examples presented in this chapter demonstrates that Polish DMOs (or, more precisely, a specific part of these) use HTD-related elements that directly relate to the value of ecosystem services. Admittedly, this phenomenon is certainly not versatile in its nature but individual cases serve to confirm that actions in this area are practised. This confirms previous research, which demonstrates that actions in HTDs in other countries are undertaken at diversified levels (Borzyszkowski & Lubowiecki-Vikuk, 2019).

We believe that the considerations presented herein are particularly important in view of the current situation regarding COVID-19. There is no doubt that the pandemic has resulted in huge losses in global tourism. It must be assumed that a reconstruction of tourism will be a lengthy process and, in most cases, will require a change of mindset in terms of tourism development. DMOs will play a vital role in this. Regardless of the forms and areas of activity presented, it is they that should take over the greatest possible responsibility in "bringing tourism back to normalcy". The considerations contained in this chapter are also important for yet another reason. Opinions are increasingly often being voiced that it is health tourism, including activities in the HTD area, that should play an important role both in the current period of the pandemic and in the subsequent period (the post-COVID-19 stage) (e.g. Stackpole, Ziemba & Johnson, 2021). Practical action is already being taken in this area, including in the country which is the subject of analysis. In Poland, since the very beginning of the pandemic, great hopes have been placed on stimulating health tourism. To a large extent, this is due to the role of domestic tourism, which practically dominated the tourist traffic in Poland in the first months of the COVID-19 pandemic. In addition to this, the development of health tourism in this period could result from the needs of Polish society in terms of restoring one's health or taking care of one's health. What is more, the PTO accepted an assumption that health tourism should be a particularly strongly emphasised product of Polish tourism (Gasior, 2020).

Interesting initiatives are taken at lower levels. One of these is the project of the "Silesian Package for Tourism" (the regional level), which offers comprehensive support to the tourism industry, which is struggling with the effects of the pandemic. The programme is valued at a total of PLN 3.5 million (approximately USD 1 million) and consists of five key pillars: Social Campaigns, Support for Tourist Events, Support for Organised Tourism, Support for Tourist Brands and Products, and Support for Tourist Infrastructure. It is worth noting that Śląskie Province has placed a strong emphasis on the issue of health tourism, for example by encouraging people to go out and enjoy the tourist attractions of Śląskie Province, gastronomy, and trails. The campaign also promoted the tourist attractiveness of the subregions of Śląskie Province by means of proposals prepared concerning bicycle trips and one-day slow-moving tours, such as "Śląskie Outside the City" or "Śląskie for Health". Additionally, the creation of an offer promoting the potential of the province in medical tourism and therapeutic tourism was accepted (Śląskie wspiera turystykę, 2020).

The direction of this type of marketing activities is to be considered appropriate but not sufficient. This is, for example, because the effects of COVID-19 are noticeable in practically all countries and regions and, what is particularly important, in most of the sectors of the tourist economy. COVID-19 has also had negative consequences for health tourism in a broad sense of the term, as well as for individual sectors, such as medical tourism. During the peak months of the pandemic in Europe, over 28 million scheduled surgeries were cancelled, a significant proportion of which concerned medical tourists. It is worth adding that the countries in Europe particularly affected by the COVID-19 pandemic are, to a large extent, those which are significant or even the most important players in the medical tourism market, such as Spain or France (Medical Tourism Magazine, 2020). Therefore, the actions we mention in this chapter not only include exemplary undertakings aimed at supporting or even saving tourism but, importantly, also health tourism itself. Considering the crisis and the health-related needs of society, we see an opportunity for DMOs to implement socially responsible marketing. Such marketing decisions are treated as holistic solutions to current and future ethical, environmental, legal, public, social and cultural values and problems that are in the interest of society, i.e. tourists, residents and other stakeholders in the health tourism market (e.g. Kasim, 2008).

We are aware of certain limitations of this study. First, we based our analysis on the study of selected examples of DMOs operating in one country. This characteristic should be extended to include practices from other countries. In addition, the considerations in this chapter are based on an analysis of the situation during the pandemic. As we know, at this stage (the second half of 2021), the situation in many countries is quite difficult, considering, among other things, the high numbers of new infections and deaths, as well as the many restrictions resulting in limitations in tourist traffic.

18.5 Conclusion

The considerations in this chapter have theoretical and practical implications. First, they complement and enrich the existing scientific resources in relation to the issue of health tourism and HTDs. Certainly, an important value of these considerations is an attempt to correlate these issues with changes resulting from the pandemic. Thus, this chapter also enriches the scientific literature resources in relation to socio-economic problems in the COVID-19 era.

An added value of the chapter is its practical implications. Although the analysis is based on selected examples from one country only, certain trends in activities in relation to contemporary DMOs in other destinations can be indicated. Our deliberations form part of an open discussion about the directions of DMO activities in crisis conditions. Representatives of these organisations should do much more than they are currently doing to present an appropriate offer to tourists and, even now, they should consider what kind of tourist products customers will be expecting after the COVID-19 pandemic. The pandemic should not be a time of stagnation for them but a time of intensive and serious efforts. To answer the research question, building the image of an HTD by means of visual identification is the right direction to be followed for this moment. Providing a safe and healthy place for leisure can be crucial, and betting on ecosystem services continues to present challenges. It is obvious that especially domestic tourist traffic in tourist destinations after the pandemic will be observed, which requires the perspective of time (Zenker & Kock, 2020). The COVID-19 pandemic has brought tourism to an almost complete halt. We express the view that uncertainty and ambiguity do not release us from thinking about the future, and above all about the health of society, the development of the tourism business and respect for the natural environment. Marketing activities should consider various segments and appropriately selected marketing communication channels or a mix of these, for example, social media for Generation Z, the Internet for Generation Y, and TV for the so-called Silver Generations (Fraiz Brea, Falcó & Vila, 2020, pp. 565–566).

The cases of DMOs described herein do not exhaust our arguments. They only serve as an example that can be used in the benchmarking of other DMOs. The directions of further research should focus on a continuous analysis of activities and undertakings in the scope of the creation of an HTD offered by individual DMOs. It is also worth considering what attributes of the HTD image will be desirable among tourists after the end of the COVID-19 pandemic and what elements of visual identification will be considered in the decision making and consumer behaviour on the health tourism market in relation to social responsibility for oneself, the society and the environment. Finally, one needs to study how DMOs will fit into marketing activities with respect to socially responsible business related to health tourism.

References

Antionos, G., Lamprini, P., & Skourtis, G. (2020). Destination branding and co-creation: A service ecosystem perspective. *Journal of Product and Brand Management*. DOI: 10.1108/JPBM-08-2019-2504.

Apostolopoulos, N., Makris, I., Liargovas, P., Apostolopoulos, S., & Varelas, S. (2021). Building national branding strategy in medical tourism and production of generic medicines: National branding and health. In V. Pistikou, A. Masouras, & M. Komodromos (Eds.), *Handbook of Research on Future Policies and Strategies for Nation Branding* (pp. 309–323). USA: IGI Global. Pennsylvania

Ashworth, G.J. (2010). *Towards Effective Place Brand Management: Branding European Cities and Regions*. London: Edward Elgar Publishing Ltd.

Barisic, P., & Blazevic, Z. (2014). Visual identity components of tourist destination. *International Science Index*, *7*(5), 875–879.

Basera, V., & Baipai, R. (2020). The online marketing strategies of Destination Marketing Organisations (DMOs), South Africa Tourism (SAT) Southern Africa benchmark. *International Journal of Tourism and Hotel Management*, *2*(1), 6–12.

Beritelli, .P, & Laesser, C. (2018). Destination logo recognition and implications for intentional destination branding by DMOs: A case for saving money. *Journal of Destination Marketing & Management*, *8*, 1–13.

Białk-Wolf, A., Pechlaner, H., Nordhorn, C., & Zacher, D. (2016). Awareness of health issues in the Pomeranian region as a precondition for developing a health region. *Studia Periegetica*, *2*(16), 45–58.

Blain, C., Levy, S.E., & Ritchie, J.B. (2005). Destination branding, insights and practices from destination management organizations. *Journal of Travel Research*, *43*(4), 328–338.

Boisen, M., Terlouw, K., & van Gorp, B. (2011). The selective nature of place branding and the layering of spatial identities. *Journal of Place Management and Development*, 4(2), 135–147.

Bonnardel, V., Séraphin, H., Gowreesunkar, V., & Ambaye, M. (2020). Empirical evaluation of the new Haiti DMO logo: Visual aesthetics, identity and communication implications. *Journal of Destination Marketing & Management*, *15*, Article 100393.

Borzyszkowski, J. (2015). *Organizacje zarządzające obszarami recepcji turystycznej. Istota, funkcjonowanie, kierunki zmian*. Koszalin: Wydawnictw Uczelniane Politechniki Koszalińskiej.

Borzyszkowski, J., & Lubowiecki-Vikuk, A. (2019). Destination management organizations and health tourism visual identification in Central and Eastern Europe. *European Research Studies Journal*, *22*(4), 241–261.

Briner, R.B., & Walshe, N.D. (2014). From passively received wisdom to actively constructed knowledge: Teaching systematic review skills as a foundation of evidence-based management, *The Academy of Management Learning and Education*, *13*(3), 415–432.

Costanza, R., D'Arge, R., de Groot, R., Farber, S., Grasso, M., Hannon, B., Limburg, K., Naeem, S., O'Neill, R.V., Paruelo, J., Raskin, R.G., Sutton, P., & van den Belt, M. (1997). The value of the world's ecosystem services and natural capital. *Nature*, *387*, 253–260.

De Las Heras-Pedrosa, C., Millan-Celis, E., Iglesias-Sánchez, P.P., & Jambrino-Maldonado, C. (2020). Importance of social media in the image formation of tourist destinations from the stakeholders' perspective. *Sustainability*, *12*, Article 4092, 1–27.

Dolnicar, S., & Mazanec, J.A. (1998). Destination marketing: Reinventing the wheel or conceptual progress. In *Destination Marketing* (pp. 55–87). Reports of the 48th AIEST Congress, Marakech.

Dominati, E., Patterson, M., & Mackay, A. (2010). A framework for classifying and quantifying the natural capital and ecosystem services of soils. *Ecological Economics*, *69*(9), 1858–1868.

Dryglas, D. (2018). *Designing a Health Tourism Product Structure Model in the Process of Marketing Management*. Warsaw: PWN.

Dryglas, D., & Lubowiecki-Vikuk, A. (2019a). Image of Poland as perceived by German and British medical tourists. *Tourism Review*, *74*(4), 861–871.

Dryglas, D., & Lubowiecki-Vikuk, A. (2019b). The attractiveness of Poland as a medical tourism destination from the perspective of German and British consumers. *Entrepreneurial Business and Economics Review*, *7*(2), 45–62.

Dryglas, D., & Salamaga, M. (2018). Segmentation by push motives in health tourism destinations: A case study of Polish spa resorts. *Journal of Destination Marketing & Management*, *9*, 234–246.

Dul, J., & Hak, T. (2008). *Case Study Methodology in Business Research*. London: Butterworth-Heinemann.

Edvardsson, B., & Tronvoll, B. (2013). A new conceptualization of service innovation grounded in S-D logic and service systems. *International Journal of Quality & Service Sciences*, *5*(1), 19–31.

European Travel Commission. (2020). *COVID-19 Recovery Strategies for National Tourism Organisations*. Brussels.

Florek, M., & Janiszewska, K. (2015). Brand and its identity as a source of added value for a metropolitan area. *Studia Oeconomica Posnaniensia*, *3*(8), 49–66.

Fraiz Brea, J.A., Falcó, E.T., & Vila, N.A. (2020). El turismo de bienestar y mindful travel en el nuevo escenario pos-COVID-19. In M.S. Cruz, R.H. Martín, & N.P. Fumero (Eds.), *Turismo pos-COVID-19: Reflexiones, retos y oportunidades* (pp. 559–568). Canarias: Universidad de La Laguna.

Freire, J. (2005) Geo-branding, are we talking nonsense? A theoretical reflection on brands applied to places. *Place Branding*, *1*(4), 347–362.

Galí, N., Camprubí, R., & Donaire, J.A. (2017). Analysing tourism slogans in top tourism destinations. *Journal of Destination Marketing & Management*, *6*(3), 243–251.

Gasior, A. (2020). Uzdrowiska polskie: jeśli przetrwamy, szybko się odrodzimy. Retrieved from https://www.waszaturystyka.pl/uzdrowiska-polskie-jesli-przetrwamy-szybko-sie-odrodzimy. Accessed October 24, 2020.

Ghaderi, M., Ruiz, F., & Agell, N. (2015). Understanding the impact of brand colour on brand image: A preference disaggregation approach. *Pattern Recognition Letters*, *67*(1), 11–18.

Giannopoulos, A., Piha, L., & Skourtis, G. (2021). Destination branding and co-creation: A service ecosystem perspective. *Journal of Product & Brand Management*, *30*(1), 148–166.

Godovykh, M., & Ridderstaat, J. (2020). Health outcomes of tourism development: A longitudinal study of the impact of tourism arrivals on residents' health. *Journal of Destination Marketing & Management*, *17*, Article 100462.

Hancock, T. (1993). Health, human development and the community ecosystem: Three ecological models. *Health Promotion International*, *8*(1), 41–46.

Hanna, S., & Rowley, J. (2019). The projected destination brand personalities of European capital cities and their positioning. *Journal of Marketing Management*, *35*(11–12), 1135–1158.

Henthorne, T.L., George, B., & Miller, M.M. (2016). Unique selling propositions and destination branding: A longitudinal perspective on the Caribbean tourism in transition. *Tourism*, *64*(3), 261–275.

Kasim, A. (2008). Socially responsible hospitality and tourism marketing. In H. Oh, & A. Pizam (Eds.), *Handbook of Hospitality Marketing Management* (pp. 32–58). Oxford: Butterworth-Heinemann.

Kavaratzis, M. (2008). *From City Marketing to City Branding: An Interdisciplinary Analysis with Reference to Amsterdam*. Budapest and Athens: University of Groningen.

Keller, K.L. (1993). Conceptualizing, measuring, and managing customer-based brand equity. *Journal of Marketing*, *57*, 1–22.

Kim, W., Malek, K., Kim, N., & Kim, S.H.J. (2017). Destination personality, destination image, and intent to recommend: The role of gender, age, cultural background, and prior experiences. *Sustainability*, *10*(87), 2–18.

Kladou, S., Kavaratzis, M., Rigopoulou, I., & Salonika, E. (2016). The role of brand elements in destination branding. *Journal of Destination Marketing & Management*, *6*(4), 426–435.

Koh, E. (2020). The end of over-tourism? Opportunities in a post-Covid-19 world. *International Journal of Tourism Cities*, *6*(4), 1015–1023.

Kotler, P., & Caslione, J.A. (2009). *Chaotics: The Business of Managing and Marketing in the Age of Turbulence Hardcover*. New York: AMACOM.

Kozak, M., & Mazurek, M. (2011). Destination branding: Brand equity, brand identity, brand extensions and co-branding. *Folia Turistica*, *25*, 93–111.

Kusumawati, A. (2019). Exploring and measuring city brand personality in small cities. *Advances in Social Science, Education and Humanities Research*, *343*, 300–306.

Lee, S., Rodriguez, L., & Sar, S. (2012). The influence of logo design on country image and willingness to visit: A study of country logos for tourism. *Public Relations Review*, *38*(4), 584–591.

Lokalna Organizacja Turystyczna Roztocze. (2020). Retrieved from http://roztocze-wita.pl/marka–roztocze/. Accessed September 29, 2020.

Lubowiecki-Vikuk, A. (2021). *Postrzeganie kompozycji marketingowej w podmiotach podaży turystyki medycznej przez konsumentów*. Warszawa: Oficyna Wydawnicza SGH.

Lubowiecki-Vikuk, A., & Basińska-Zych, A. (2011). Sport and tourism as elements of place branding. A case study on Poland. *Journal of Tourism Challenges and Trends*, *IV*(2), 33–52.

Lück, M., & Seeler, S. (2021). Understanding domestic tourists to support COVID-19 recovery strategies – The case of Aotearoa New Zealand. *Journal Of Responsible Tourism Management*, *1*(2). DOI: 10.47263/JRTM.01-02-02.

Lusch, R.F., & Nambisan, S. (2015). Service innovation: A service-dominant logic perspective. *MIS Quarterly: Management Information Systems*, *39*(1), 155–175.

Lusch, R.F., & Vargo, S.L. (2014). *Service-Dominant Logic: Premises, Perspectives, Possibilities*. Cambridge: Cambridge University Press.

Ma, W., Schraven, D., de Bruijne, M., de Jong, M., & Lu, H. (2019). Tracing the origins of place branding research: A bibliometric study of concepts in use (1980–2018). *Sustainability*, 11, 2999.

Machnik, A. (2019). Natural capital and ecological ecosystem services: Methods of measuring socio-economic value of nature. In W. Leal Filho, A. Azul, L. Brandli, P. Özuyar, & T. Wall (Eds.), *Responsible Consumption and Production. Encyclopedia of the UN Sustainable Development Goals* (pp. 1–13). Cham: Springer.

Manhas, P.S., Manrai, L.A., & Manrai, A.K. (2016). Role of tourist destination development in building its brand image: A conceptual model. *Journal of Economics, Finance and Administrative Science*, *21*(40), 25–29.

Marczak, M. (2018a). *Zarządzanie marką obszaru recepcji turystycznej przez narodowe organizacje turystyczne*. Warszawa: CeDeWu.

Marczak, M. (2018b). Branding as an essential element of the destination management process using the example of selected states. *Management Sciences*, *23*(2), 29–40.

Marczak, M. (2019). Attempt at assessment of the effectiveness of activities undertaken by national tourism organisations aimed at the creation of the country's tourism brand. *Folia Turistica*, *51*, 9–34.

Marland, A., Lewis, J.P., & Flanagan, T. (2017). Governance in the age of digital media and branding. *Governance*, *30*, 125–141.

Marzano, G., & Scott, N. (2009). Power in destination branding. *Annals of Tourism Research*, *36*(2), 247–267.

McWilliam, G., & Dumas, A. (1997). Using metaphors in new brand design. *Journal of Marketing Management*, *13*(4), 265–284.

Medical Tourism Magazine. (2020). COVID-19 and the Future of Medical Tourism across Europe. Retrieved from https://www.magazine.medicaltourism.com/article/covid-19-and-the-future-of-medical-tourism-across-europe. Accessed October 22, 2020.

Millennium Ecosystem Assessment. (2005). *Ecosystems and Human Well-being: Biodiversity Synthesis*. Washington, DC: World Resources Institute.

Morgan, N., Pritchard, A., & Pride, R. (2011). *Destination Brands: Managing Place Reputation*. Oxford: Butterworth-Heinemann.

Navarrete, A.P., & Shaw, G. (2021). Spa tourism opportunities as strategic sector in aiding recovery from Covid-19: The Spanish model. *Tourism and Hospitality Research*, *21*(2), 245–250.

Nawrocka, E. (2013). *Wizerunek obszaru recepcji turystycznej. Podstawy konceptualizacji i czynniki jego kreowania*. Wrocław: Uniwersytet Ekonomiczny we Wrocławiu.

Nientied, P. (2021). Rotterdam and the question of new urban tourism. *International Journal of Tourism Cities*, *7*(2), 344–360.

Olins, W. (2004). *Branding the Nation. In On Br@nd*. London: Thames and Hudson.

Park, E., Kim, W.-H., & Kim, S.-B. (2020). Tracking tourism and hospitality employees' real-time perceptions and emotions in an online community during the COVID-19 pandemic. *Current Issues in Tourism*. DOI: 10.1080/13683500.2020.1823336.

Pearce, D.G. (2015). Destination management in New Zealand: Structures and functions. *Journal of Destination Marketing & Management*, *4*(1), 1–12.

Pereira, L.A., Limberger, P.F., da Silva Flores, L.C., & de Lima Pereira, M. (2019). An empirical investigation of destination branding: The case of the city of Rio de Janeiro, Brazil. *Sustainability*, *11*(90), 1–17.

Pike, S., & Page, S.J. (2014). Destination Marketing Organizations and destination marketing: A narrative analysis of the literature. *Progress in Tourism Management*, *41*, 202–227.

Polish Tourism Organisation. (2016). *Market Analysis – Health Service Promotion*. Retrieved from http://pot.gov.pl/6-4fundusze-ue/l/program-promocji-uslug-prozdrowotnych/koncepcja-programu-promocji-uslug-prozdrowotnych. Accessed September 29, 2020.

Raftowicz-Filipkiewicz, M. (2008). National branding in European Union. *Studenckie Prace Prawnicze, Administratywistyczne i Ekonomiczne*, *5*, 55–63.

Séraphin, H., Ambaye, M., Gowreesunkar, V., & Bonnardel, V. (2016). A marketing research tool for destination marketing organizations' logo design. *Journal of Business Research*, *69*(11), 5022–5027.

Sigala, M. (2020). Tourism and COVID-19: Impacts and implications for advancing and resetting industry and research. *Journal of Business Research*, *117*, 312–321.

Sigwele, L., Prinsloo, J.J., & Pelser, T.G. (2018). Strategies for branding the city of Gaborone as a tourist destination. *African Journal of Hospitality, Tourism and Leisure*, *7*(2), 1–19.

Śląskie wspiera turystykę. (2020). Retrieved from https://www.slaskie.pl/content/slaskie-wspiera-turystyke. Accessed October 28, 2020.

Stackpole, I., Ziemba, E., & Johnson T. (2021). Looking around the corner: COVID-19 shocks and market dynamics in US medical tourism. *International Journal of Health Planning and Management*, *36*(5), 1407–1416.

Steinecke, A. (2001). *Key-note Presentation at the Conference on Cultural Tourism at the ITB*. Berlin: ITB.

Sultan, M.T., Sharmin, F., Badulescu, A., Stiubea, E., & Xue, K. (2021). Travelers' responsible environmental behavior towards sustainable coastal tourism: An empirical investigation on social media user-generated content. *Sustainability*, *13*(56). DOI: 10.3390/su13010056.

Synergia. (2020). *Roztocze Witalność z Natury. Księga Znaku*. Synergia, sp. z o.o. Retrieved from http://roztoczewita.pl/wp-content/uploads/2017/04/SIW_Roztocze_witalnosc_z_natury.pdf. Accessed October 15, 2020.

The Act of 25 June 1999 on the Polish Tourism Organisation. (1999). DzU no. 62, 689.

Tourism Alliance. (2021). *Tourism Alliance DMO Review Submission*. London: Tourism Alliance.

Vargas, A. (2020). Covid-19 crisis: A new model of tourism governance for a new time. *Worldwide Hospitality and Tourism Themes*, *12*(6), 691–699. DOI: 10.1108/WHATT-07-2020-0066.

Vegnuti, R. (2020). Cinque Terre, Italy – A case of place branding: From opportunity to problem for tourism. *Worldwide Hospitality and Tourism Themes*, *12*(4), 471–483.

Welford, R., & Ytterhus, B. (2004). Sustainable development and tourism destination management: A case study of the Lillehammer region, Norway. *International Journal of Sustainable Development & World Ecology*, *11*(4), 410–422.

Wen, J., Kozak, M., Yang, S., & Liu, F. (2021). COVID-19: Potential effects on Chinese citizens' lifestyle and travel. *Tourism Review*, *76*(1), 74–87.

World Tourism Organisation. (2004). *Indicators of Sustainable Development for Tourism Destinations: A Guidebook*. Madrid: UNWTO.

World Tourism Organisation, & European Travel Commission. (2018). *Exploring Health Tourism – Executive Summary*. Madrid: UNWTO.

Zenker, S., & Kock, F. (2020). The coronavirus pandemic – A critical discussion of a tourism research agenda. *Tourism Management*, *81*, 104164.

Zenker, S., Braun, E., & Petersen, S. (2017). Branding the destination versus branding the place: The effects of brand complexity and identification for residents and visitors. *Tourism Management*, *58*, 15–27.

19 Understanding the Cultural Ecosystem Service of Heritage Tourism

The Case of Jatiluwih Heritage Tourism

I Nengah Subadra

19.1 Introduction

Heritage sites have been long designated as tourist attractions in Bali, Indonesia, in addition to the richly Balinese cultures, such as arts, religious rites and way of life, and splendidly natural resources, such as beaches, lakes, rivers, valleys, hills and mountains, which currently ruled under regional regulation of Bali Province Number 5 of 2020 regarding Standards of Balinese Cultural Tourism Operations (Government of Bali Province, 2020). This regulation is served as the legal reference for developing tourism in Bali.

Jatiluwih Village is one of the agricultural sites in Penebel District, Tabanan Regency, Bali. The 22.23-kilometre square village is sited approximately 700 metres above sea level and is reserved for one of Bali's central rice productions to supply the staple food demands on the island. The rice cultivations in this village have significant roles for the village's food security, which is directly consumed by the local farmers, and for Bali's food security, which is sold by the farmers to other villages, districts or regencies as their earnings. The earning pattern of the locals in the certain region belongs to a cultural element as it is learnable, dynamic and exchangeable (Ardika & Subadra, 2018). The primary function of Jatiluwih is a rice cultivation centre that belongs to the Subak Landscape of Catur Angga Batukaru area, wherein its developments and preservations involve stakeholders including farmers, *Subak* (irrigation system) association, traditional village and government (Subadra & Nadra, 2006). They play a great role in sustaining Jatiluwih rice terrace landscape and its embedded universal cultural values which have been designated as a world heritage site in Bali since 2012 (UNESCO, 2012).

Cultural ecosystem services (CES) encourage a more thorough and inclusive discussion on sustainability. This research revisits and extends the application of sustainable tourism at Subak Jatiluwih Heritage Tourism Site (SJHTS) undertaken by Subadra and Nadra (2006), which was only focused on the analysis of natural, economic and social-cultural impacts of tourism shared within the stakeholders as evidence that the site applies sustainable tourism concept and extends the study of Ryan et al. (2011) discussing the impacts of heritage tourism on local communities and their perceptions of

DOI: 10.4324/b23145-25

the designation of their region as an international heritage site. Additionally, this research also widens the discussions on cultural ecosystem services, which is mainly discussed from the tourists' perspectives, particularly what aspects were experienced by the tourists (Lindhjem et al., 2015) and how tourists gained value from the visited tourism sites (Sanna & Eja, 2015). This current research seeks to understand the hosts' perspectives, including local communities, heritage site managers and tour guides, and the guests' perspectives, particularly the tourists who have been to the site to analyse and understand the notion of CES at the SJHTS before and post-Covid-19 pandemic (Milcu et al., 2013). The discussion of these two different rapports contributed to the understanding of CES issues arising at SJHTS as the novelty of this research.

19.2 Theoretical Background

Tourism is closely related to cultural ecosystem services as it involves the mobilisation of people travelling to certain destinations to escape from their routines to gain enjoyment through sightseeing and engagement with the tourist attractions set authentically or staged for tourism in the visited destinations (Poria et al., 2003; Subadra 2015). The designations of natural and cultural resources as tourism sites are beneficial to both local people residing in the destination and tourists who visit the region (Subadra, 2019). The extended immaterial benefits which local people and tourists receive from these resources are defined as cultural ecosystem services. These benefits include recreation, pictorial enjoyment, physical and mental health benefits and even spiritual delight consumed by the tourists onsite (Jacobides et al., 2018; Milcu et al., 2013). Such impressive experiences are essential in creating a greater sense of place, encouraging more people to visit the site as well as to vigorously increase the qualities of the locals' welfare and tourists' awareness and respect of other cultures. In other words, CES is beneficial to both local communities who own and prepare the heritage-based tourist attractions and tourists who consume such perishable tourism products, which bind them in a mutual relationship wherein the local people gain cash while the tourists enrich their experiences.

Local people play great roles in tourism development. Citing the work of Koontz and O'Donnell (1972), Subadra (2015) argued that local communities involve in any tourism development stages, which include planning, organising, staffing, directing and controlling. The locals' involvement is aimed at allowing local people to gain the economic value of the tourism developed in their regions through direct and indirect employments (Byrd, 2007; Subadra, 2015). Additionally, Okazaki (2008) argued that locals' involvement aids in minimising the negative impacts of tourism developments and also enhancing positive impacts for better value. In other words, community involvement in tourism encourages local people not only to gain value from tourism development but also to raise their awareness and respect for the religious and cultural values designated for tourism.

19.2.1 Heritage Tourism and Cultural Ecosystem Services

Heritage sites have been long designated as tourism attractions worldwide, especially in cultural tourism destinations. These include the sites stipulated by the national authorities and the international body The United Nations Educational, Scientific and Cultural Organization (UNESCO), which authorises to nominate and stipulate the names and lists of the world heritage sites, for instance, The Great Wall (China), Taj Mahal (India), Borobudur Temple (Indonesia), Venice (Italy), Ancient Thebes (Egypt), Tower of London (the United Kingdom), and Sydney Opera House (Australia). The outstanding values of the heritages are believed to bring peace to the world through three major global collaborations, including education, sciences and culture, which are solely packed and shared in tourism (UNESCO, 2021).

This initiated the rise of heritage tourism, which promotes and shares the sites' significance and values with tourists to increase their knowledge and meet their holiday demands by immediate visits and onsite engagements which served as authentic tour experiences (Subadra, 2015). Tourists visiting heritage tourism sites do not merely gaze at the attractions but, more importantly, experience them to immediately feel such newly visited destinations, which serves as memorable experiences (Urry, 2002). As Poria et al. (2003) claimed, the motivation of tourists visiting heritage sites is to experience and learn the great history behind them and obtain a profound understanding of their significances. Additionally, the attachment of sites and attractions also motivates tourists to visit heritage tourism sites (Kang, 2019; Nanjangud & Reijnders, 2021). In the case of Bali tourism, for instance, learning and engaging with culture-related attractions, including heritages, were considered by the tourists to be the foremost motivation visiting Bali, in addition to exploring the beautiful nature and hunting for unique Balinese souvenirs to take home (Subadra, 2015).

The notion of heritage tourism has been long scientifically discussed among academics who have seen heritage tourism from different perspectives, including heritage tourism and social media (Falk & Hagsten, 2020), heritage attachment as visiting motivation (Kang, 2019), tourists' engagements and locals' resistances in heritage tourism sites (Ardika & Subadra, 2018), heritage tourism branding (Poria et al., 2011), impacts of heritage tourism (Ryan et al., 2011), tourists' perceptions upon heritage sites and their tour behaviours (Poria et al., 2003) and tour guides' interpretations on heritage site (Poria et al., 2009).

Furthermore, the ecosystem is an ancient word of 1869 introduced by Ernest Haeckel, which was originally and only used in biology and which describes the relationships and dependencies of its existing resource and inhabitants within a certain environment (Begon et al., 2006; Miller & Spoolman, 2009). Begon et al. (2006) define ecology as "the scientific study of the interactions between organisms and their environment." This includes which organism or individual lives together and how the living environment and non-living environment interact with one another (Miller & Spoolman,

2009). The terminology of ecology is adapted in other fields of studies to describe the same notion of what, how, why and when such a relationship occurs (Jacobides et al., 2018; Miller & Spoolman, 2009). As Jacobides et al. (2018) noted, the notion of the ecosystem was first adopted in business perspectives in 2013 to refer to the relationships of the commerce's stakeholders and their dependencies on one another. These include enterprises that produce and sell goods and services, which are investigated from distinctive angles such as company innovations (Adner, 2017; Teece, 2017), business strategy (Augier & Teece, 2018); company's competitiveness, collaboration and challenges (Jacobides et al., 2018), which are all related to ecosystem and value creation and increase. Additionally, scholars have also extended and embraced a diverse range of concepts and theories to define ecosystem from the perspective of service trades like tourism, from which then arose the concept of cultural ecosystem service (CES), which is defined as "non-material benefits that people obtain from ecosystems through spiritual enrichment, cognitive development, reflection, recreation and aesthetic experience" (Sarukhán & Whyte, 2005). Such benefits cover both tangible and intangible possessions obtained from immediate engagements made within certain sites, including cultural heritages, which increase people's awareness and understanding of the site's significances and values that existed and were believed by the local communities (Milcu et al., 2013; Sarukhán & Whyte, 2005). The cultural value is constructed to increase the human's life dimensions which are utilised for their developments within their ecosystem (Fish et al., 2016).

The notion of value created through CES is dynamic and still debatable among the scholars since it is not just simply what value is gained by the stakeholders (Sarukhán & Whyte, 2005) but more importantly about the process of value creation and engagements of the stakeholder within the ecosystem (Chan et al., 2012; Fish et al., 2016). Thus, Chan et al. (2012) extended the definition of CES as an interactive "processes and entities that people actively create and express through interactions with ecosystems." This suggests that the discussion on CES may be prolonged further on the process of how the value of such interaction is created. Almost the same notion, Fish et al. (2016) defined CES as the "ecosystems' contribution to the non-material benefits (capabilities and experiences) that arise from human–ecosystem relationships." This adapts the CES definition of Sarukhán and Whyte (2005) and Chan et al. (2012), which focus on the added value of the products gained through interactions among the stakeholders within the ecosystem.

These wide ranges of research areas are challenging tourism and cultural academics to research on CES to understand the notions and values embedded behind (Milcu et al., 2013) and also its significance in the policy decision-making process (Fish et al., 2016). Sarukhán and Whyte (2005) had identified the values gained by people upon their visits to certain destinations from the visitors' perspectives, including spiritual and religious values, cultural values, educational values, aesthetic values, social values, recreational values and environmental value yet identified how such values created and gained from the perspectives of site managements, intermediaries and other stakeholders contributing the value creations and transfers before, amid and post Covid-19 pandemic, as discussed in this current the scientific paper.

CES increases the value of cultural resources used which are shared among the interested parties upon their joint dependencies and roles played in the system. Tourism, for instance, is a multi-sector trade of service that use environmental and cultural resources available in the destination and involves a number of stakeholders to operate (Byrd, 2007; Subadra, 2021a). Tourism does not only generate value in the form of economic benefits for the stakeholders but also for the natural and cultural resources designated as tourism sites, which allow tourists to sightsee, learn and experience the sites upon the purchase of entrance tickets (Subadra, 2019). This corresponds to the notion of triple bottom line "people, planet and profit" introduced by Fisk (2010), wherein tourism business shall be generating profit for the people (the entrepreneurs, employees and tourists), contributing to environmental sustainability and gaining profit for the tourism enterprises to enlarge and expand the business scales. This concept has been long practised in tourism development in Bali with a different term named *Tri Hita Karana* – the three sources of happiness wherein human beings live with each other horizontally, worship their God and Goddesses vertically, and respect their environments to live in a certain ecosystem for a happy life – and served as the basis for tourism development in Bali which was drawn up in the current Regional Regulation on Balinese Cultural Tourism (Dharmayudha & Cantika, 1999; Regional Regulation Number 5 of 2020; Subadra, 2015, 2021b). This stipulates the cultural resources designated as tourism sites to be beneficial for the local people and government, contribute to environmental sustainability, respect the religious values embedded with the rites conducted by the local people for the almighty God and also offer opportunities for the tourists to gaze, learn and engage with the cultures to enrich their holiday experiences, which may be gained either directly by the tourists or through tour guide's interpretations.

Tour guide plays a great role in interpreting the heritage site to understand its outstanding significance (Ababneh, 2017). The knowledge generated from the immediate engagements and tour guide' interpretations during the visit becomes essential values, in addition to onsite enjoyments of the surrounding heritage atmospheres, which make the visit more experienced and memorable. It is served as a novelty as the tourists can learn heritage more deeply for their educational advancement. Thus, the tourists' experiences and value gained by the tourists depend on the interactions between the heritage sites and tourists (Fish et al., 2016) and also interpretations made by the intermediary bridging the sites and tourists (Ababneh, 2017; Holloway, 2006; Kenter et al., 2016).

The tourist flows to heritage tourism sites initiate the rise of creativity made by the management and local people to offer more engagement alternatives and enrich experiences for the tourists and even increase their holiday value upon the purchased heritage tour packages (Richards & Wilson, 2007). This corresponds to the study of Smit et al. (2020) finding that offering unforgettable visiting experiences to the tourists is highly essential for tourism destinations as the so-called core tourism business attracts new tourists and maintains the existing ones. The tangible heritage products which creatively emerged with hospitality services add value both for the tourists and heritage sites. To some extent, the creativity of heritage tourism allows tourists to escape from their routines, enjoy the uniqueness of the newly visited heritage, engage with the heritage-related activities, learn the significance of the heritage site, to being recognised as tourists who have a high awareness and respect on humankind and culture, and to relax refreshing their minds while enjoying the surrounding atmosphere (Richards, 2020). These five heritage tourism experiences add value for the tourists, who spent money purchasing heritage tour programmes. On the other hand, the heritage tourism management challenges to being creative, providing typical attractions and services which are distinct from others to generate value in the form of profits gained from the tourists who interact and experience the attractions in the sites. In other words, the core notion of CES in heritage tourism is the values of the heritage itself, both its primary and outstanding value as a heritage site and the additional value which is created and staged for tourism (Fish et al., 2016; UNWTO, 2020). Those two-side values of the culture embedded with heritage tourism have led and encouraged the stakeholders that involve in heritage tourism to sustain its ecosystem (Chan & Satterfield, 2015).

Furthermore, sustaining and managing heritage tourism is always challenging, both before and amid the Covid-19 pandemic. The global Covid-19 outbreaks have shut down world tourism businesses and have taken two years to restart and even more for recovering fully. In the case of Bali, tourism management compulsorily adopted a health disaster mitigation system ruled by the government to reopen tourism and awaken community resilience (Subadra & Hughes, 2021). Subadra (2021b) claims that supplementary amenities related to health safety such as washbasin, hand sanitiser and thermo-gun scanner are installed in the destination and, also, immediately involve state (policemen and soldiers) and *pelacang* – the traditional Balinese security officers to manage the tourists' flows to tourism sites during the pandemic. These are the dynamics of tourism management, wherein the amenities and stakeholders are adopted under the current situations and government policies regarding tourism operational rules amid and post-pandemic to keep tourism open and allow tourists to visit tourism destinations (Subadra, 2021a; Subadra & Hughes, 2021). In other words, the Covid-19 pandemic has shifted conventional management into a new heritage tourism management, including tourism policy, site management, tourists' behaviours, and local people's behaviours hosting tourist visitation amid and post-pandemic.

The tourist visits also allow local people to establish tourism-related enterprises around the heritage sites to gain benefits from tourism. This confirms that heritage tourism contributes to the rise of entrepreneurship acquisitions, wherein local people see opportunities upon the tourist visits to their region and are creative to run businesses that support the tourism development and meet onsite tourists' demands (Subadra, 2021a). The creativities of the local people include both creation of supporting tourism businesses and establishment of supplementary tourism attractions.

Richards (2020) claims creative attractions related to heritage tourism provided for the tourists affect the image of the visited destinations, which has a significant role in marketing. Destination image is the tourists' perspectives of the products and services of a certain destination (Kotler & Keller, 2016). The notion of destination image has been long discussed by tourism scholars, which relate the tourist destinations and tourists' visiting intentions before their visits and tourists' satisfactions post visitations and experiences (Gallarza et al., 2002; Liu et al., 2015). Subadra (2019) argues that the image of the heritage site is one of the factors which motivates tourists to visit the site, particularly for the repeating tourists since they had already gained the "value" of their vacations through immediate experiences. The satisfaction gained by the tourists does not only initiate tourists to revisit the site, which is referred as to loyalty but also potentially share their experiences to fellow tourists to come and witness the tourist attractions, which is known as word-of-mouth marketing and which targets families, colleagues and friends (Kotler & Keller, 2016; Xu et al., 2020).

WOM plays a great role in impacting tourist behaviours in deciding destinations to visit and is considered as the most effective marketing distribution channel since the information and promotions made based on the experiences of the fellow tourists post visitations through a personal website, blog, online travel reviews and other types of social media communications (Gong et al., 2018; Kim et al., 2019; Litvin et al., 2018; Xu et al., 2020; Yan et al., 2018). In other words, value is served as the key success of tourism marketing to maintain the relationship between producer – the local people in the host destination who prepared tourism products and services and consumers – the tourists who purchase and experience the tourism attractions and amenities (Kotler & Keller, 2016). Moreover, in the specific case of heritage tourism marketing, Falk and Hagsten (2020) found that social media like Instagram plays a great role in promoting the heritage tourism world widely, where the tourists post and share the location and characteristics of the visited heritage on their wall, which can be instantly seen and commented by their online friends and families.

In conclusion, CES in tourism is not only significant to the stakeholders involved in the tourism business environment such as local people, tourism site management, related-tourism enterprises, government and tourists but also vital to tourism policymaking process, cultural sustainability and tourism destination marketing.

19.2.2 Research Method

This research adopts the case study method, which is one of the qualitative research paradigms (Creswell, 2014; Denzin & Lincoln, 2018) and was undertaken at the Jatiluwih Heritage Tourism Site of Penebel District, Tabanan Regency, Bali – Indonesia between January and August 2021. The case study is aimed at gaining an in-depth, multi-angled understanding of a complex issue that factually occurs within the community (Crowe et al., 2011;

Table 19.1 Research informants

No.	*Informants*	*Abbreviation*	*Total*
1	Operational Manager	OM	1
2	Water Irrigation Chief	WIC	1
3	Local Farmers	LF	2
4	Tour Guide	TG	1
5	Domestic Tourists	DT	5
6	International Tourists	IT	5
Grand Total			**15**

Schwandt & Gates, 2018). It used a single case study to investigate the specific problem in the existing context of the cultural ecosystem services in JHTS using primary data collected from multi-sources including field observations to observe the current condition of the tourism heritage site and capture related photos to support the research; face-to-face in-depth interviews using interview question guides for each informant; and secondary data published online in tourism-related media (Baxter & Jack, 2008; Patton, 2015; Schwandt & Gates, 2018; Yin, 2018). It involved 15 informants who were knowledgeable about the case under study (Dul & Hak, 2008) (Table 19.1). The multiplied data were triangulated to understand the researched case more deeply by looking at such circumstances in its real-world setting from diverse angles (Schwandt & Gates, 2018; Yin, 2018).

19.3 Finding and Discussion

19.3.1 Cultural Ecosystem Services Development in Heritage Tourism

Subak Jatiluwih is an agricultural area of 303 hectares located in Penebel Sub-District, Tabanan Regency, Bali Province – Indonesia and belongs to the Subak Catur Angga Batukaru, which accounts for 17376.1 hectares of rice paddy grown up to supply the staple food of the local people and also others residing on Bali Island (Figure 19.1). This agrarian region gives lives to 1,043 farmers of the total population of 2,780 people residing in Jatiluwih Administrative Village (UNESCO, 2012; Jatiluwih Village Profile, 2021).

Some rice varieties are planted within the farming site of this largest rice-paddy field, including white rice, red rice and organic rice. The harvested grains are also produced as white flour and red rice tea, which are popular among the local Balinese people. A farmer stated:

> We mainly grow rice paddy here … many rice variants. We consume them … sell for daily necessities, rituals, tradition … support children's educations… Farming is our culture. Thus, we all commit to preserving this agrarian site where cultures exist and practise.
>
> (LF-1, 01/07/2021)

Figure 19.1 Subak Jatiluwih – The UNESCO Heritage Site in Bali. Photograph by the author.

The high values attached to the agricultural sites of Jatiluwih are served as the principal reasons for farmers to preserve the site as it is a part of their culture. Jatiluwih is used as the earning place where they earn their livings and also as a medium to practise their beliefs and daily cultures. These two significances have been used by the local communities to sustain the stunning agronomic areas of Jatiluwih which has cultural value. The opportunities to practise their beliefs worshipping God while engaging with agricultural activities in Jatiluwih have enriched the farmers' spiritual senses and loves in addition to the grains harvested.

> I keep working when no tourists coming here like today … just plant rice paddy as normal even though there is no tourist seeing me working… I earn a living from farming … to preserve my traditions and cultures… Subak and rituals.
>
> (LF-1, 01/07/2021)

The intentions and commitments of the local people preserving their agricultural regions are due to their immediate attachments towards the sites which have been long inherited to present generation (Kang, 2019); and there are not any pressures from other interests like tourism. The local people of Jatiluwih play their natural roles as farmers consistently even if Bali tourism was paused from April to July 2020 and the tourism operation in Jatiluwih is now temporarily closed as ruled in Emergency Order of Community Activity Restrictions (Government Regulation, 2021; Subadra, 2021a; Subadra & Hughes, 2021). In other words, agricultural life continuously exists even if no tourism activities running there. The presence of tourism is the supplementary "bonus" for Jatiluwih as it has great scenery of rice terrace lining from the top hill down to the streams and also offers fresh and unpolluted air to breathe as it is located in the rural area which attracts tourists to visit and experience the site. This builds a cultural ecosystem that consists of the local community, government, traditional villages, tourism management, travel agencies and tourists that involve and interact with one to another in the process of staging Jatiluwih as a heritage tourism site (Chan et al., 2012; Fish et al., 2016) to gain economic benefits and non-economic reimbursements in the form of relaxation, spiritual delight and cognitive advancement (Sarukhán & Whyte, 2005; Subadra & Nadra, 2006).

Jatiluwih has now been officially designated as a tourism site for 23 years since the issue of the decree of the Regent of Tabanan Regency Number 470 of 1998. This ruling approves tourists entering the SJHTS with entry tickets. Contemporarily, an adult foreign tourist is charged with IDR 40,000 (equals to £ 2) and child is charged IDR 30,000 (equals to £ 1.5); and an adult domestic tourist is charged IDR 15,000 (equals to £ 0.7) and child is charged with IDR 5,000 (equals to £ 0.2) (OM, 14/08/2021). The site of Jatiluwih has primarily functioned for agricultural activities; thereby, all attractions are authentic without being purposely staged for tourism. Consequently, tourists may gaze and experience different views depending on the times they visit. They may see the rice plantation phase, the growth phase or the harvesting time which all have different sets of rites performed by the farmers on the field.

The popularity of SJHTS attracts thousands of national and international tourists each year (Table 19.2). In general, the number of tourists visiting the site increased before the Covid-19 pandemic and declined amid the pandemic.

This great number of tourist visits has provided opportunities for the local people to gain benefits from tourism and initiated the establishments of tourism-related businesses such as restaurants, snack bars, hotels, villas and homestays along the main road, which use the SJHTS as their front views to allow tourists gazing at this beautiful Bali landscape (Figure 19.2).

SJHTS is managed by the Jatiluwih Tourist Attraction Management Agency, which was set up by the government of Tabanan on 13 February

Table 19.2 Tourists visiting SJHTS between 2016 and 2021

Year	*Domestic Tourists*		*Foreign Tourists*		*Total*
	Adult	*Child*	*Adult*	*Child*	
2016	28,026	215	179,890	5,378	213,509
2017	39,537	148	205,145	6,143	250,975
2018	47,768	331	220,786	7,021	276,906
2019	66,554	1,153	239,328	7,407	314,443
2020	53,339	1,587	38,053	764	93,743

Figure 19.2 Tourism businesses established within the Subak Jatiluwih Heritage Tourism Site. Photograph by the author.

2014 in collaboration with the administrative and traditional villages of Jatiluwih. This agency is managed by an operational manager who is responsible for running the tourism operations including site management, human resource management, admission collections, and profit distributions. This agency maintains relationships among the stakeholders that bond in the ecosystem through profit shares which are aimed at sustaining the existing cultures staged for tourism. The fund collected from the entry fees is allocated by site management for tourism operation, while the profits are shared with the governments of Tabanan and Jatiluwih Village, which account for 45% and 55%, respectively. The manager confirmed:

> … all tourists are charged as ordered by the law … we share the money proportionally … to fund the operations … state revenue … village incomes … to host regular rituals.
>
> (OM, 14/07/2021)

This suggests that the designation of agricultural sites as SJHTS has been economically beneficial to local people who involve directly in the tourism operational management wherein they are formally employed and earn a monthly salary. Additionally, the villages and associations also received a certain amount of money shared from the profits generated from SJHTS, including Desa Dinas (administrative village) of Jatiluwih, two Desa Adat (traditional villages) of Jatiluwih and Gunungsari and three Subak (traditional irrigation system associations) of Jatiluwih, Abian Jatiluwih and Gunungsari, which makes up 15%, 33%, 22%, 26%, 2% and 2% in any order (OM, 14/08/2021). Some of the shared money is used to support religious rites in the said administrative and traditional villages. These direct and indirect economic impacts generated from SJHTS have been consistently and proportionally shared among those stakeholders since the author researched tourism sustainability in 2006 (Subadra & Nadra, 2006). The collaboration of the stakeholders which bonds in a tourism ecosystem has been economically beneficial to the respective parties (Byrd, 2007; Subadra, 2015) and the funds gained from these tourism collaborations are aimed at conducting religious rites to enrich their spiritual and sustain their cultures (Sarukhán & Whyte, 2005).

In addition, during the Covid-19 pandemic, the agency plays a great role in accelerating the tourism recovery of SJHTS while complying with health disaster mitigations: adapting the current government policy relating to tourism openings and closures, which dynamically changed according to the number of Covid-19 cases, preparing standard health protocol infrastructures and measures and educating local people and tourists to keep washing hands regularly, wearing face-mask and practise social distancing when visiting SJHTS amid and post Covid-19 pandemic for self-preventions and protections.

The splendour of the rice terrace and the uniqueness of the Jatiluwih which commits to preserving the Subak met the requirements and qualifications for world heritage nomination which was finally designated by UNESCO as one of the world heritages on 29 June 2012 under property name "Cultural Landscape of Bali Province: the Subak System as a Manifestation of the *Tri Hita Karana* Philosophy" due to its outstanding value attaching to its ecosystem, which consistently respects and practises the native Hindu Balinese philosophy of *Tri Hita Karana* as life guidance since the 9th century (UNESCO, 2012).

> There exist a great variety of Landscapes that are representative of the different regions of the world. Combined works of nature and humankind, they express a long and intimate relationship between peoples and their natural environment.
>
> (UNESCO, 2012)

Community obligations are bond in traditional village association called "Desa Adat," wherein all members are obliged to apply the indigenous philosophy of *Tri Hita Karana* – the three relationships leading to happiness including *parhyangan* (relationship between human and God), *pawongan* (relationship between human and beings) and *palemahan* (relationship between human and environment) (Dharmayudha & Cantika, 1999; Subadra, 2015). These three relationships can be simply understood and termed as GPE (God, People and Environment). The notion *parhyangan* is related to *Widhi Sradha* (believing in God) which is one of the five principal beliefs of

Hinduism in Bali, wherein God is believed to inhabit any space of the island and connected with humans and other creatures living in an ecosystem of a certain environment. In the case of SJHTS, farmers worship *Bhatara Sri* (the God of prosperity), which gives them lives where they can grow rice paddy along with their lands within the areas of Subak Jatiluwih. The farmers establish shrines in their properties to show their respects to God by serving offerings during the seedling, planting and harvesting. They believe that God's power plays a great role in their agricultural activities. A farmer argued:

> It is a part of our belief … anything exists due to His power … we can only work hard and do our best … the harvesting depends on how the God protects and bless the plantations… It works very well so far as we worship the God from the rice field.
>
> (LF-2, 20/06/2021)

> There are 15 rituals conducted by farmers until the rice ready for consumption … farmers do the same rites … during the water arrival, cultivation process, paddy growth, harvesting, storing, cooking … parts of our cultures and lives … practised until today.
>
> (WIC, 14/07/2021)

The preservation of Jatiluwih agricultural areas made by the local communities is initiated by internal interest, wherein they are motivated to gain personal incomes and search for religious and spiritual delights, and external interest wherein they are bound in community compulsory obligations in both traditional village and Subak associations to preserve their cultures and environment continuously as the implementations of *pawongan* and *palemahan*. As a Bali-based water distribution system, Subak plays a great role in the preservation of SJHTS, wherein farmers and their fellows collaborate to manage and share the water equitably as well as decide the definite schedules of cultivating, seedling, planting and harvesting. Both internal and external interests prove that CES co-exist within the world heritage site of Jatiluwih.

19.3.2 Cultural Ecosystem Services Experience in Heritage Tourism

Subak Jatiluwih heritage tourism site presents a wide range of tourism-related activities, which allow tourists to have shallow experience by gazing at agricultural landscape view and also deep experiences by engaging with the site and learning the significance of local cultures. The image of SJHTS has been positioned in the tourists' minds as the centre of rice cultivations and Subak, where they can sightsee a large farm and learn its unique irrigation system which blended with Balinese Hindu rituals that differentiate with the water distribution systems, applies beyond Bali. Domestic and foreign tourists argued:

> UNESCO site … we could really enjoy the green view … was unique. It shows the Balinese traditional agricultural system that applies multilevel irrigation for rice fields that we couldn't find at the other place.
>
> (DT-2, 17/10/2020)

> It really is beautiful … a landscape of rice field terraces … the only place on earth that has 3 annual rice harvests! … incredibly well-maintained rice terraces.
>
> (IT-1, 26/03/2020)

The understanding of the traditional Balinese irrigation system "Subak," which sustains the local culture and environment is the evidence of cognitive development generated through CES at Jatiluwih; in addition to the aesthetic and authentic experiences of the surrounding views and atmosphere gained during the visit (Sarukhán & Whyte, 2005). The engagements and attachments of the place increase the sense of love towards SJHTS and make their trips more impressive and memorable. Domestic and international tourists claimed:

> Beautiful day ever … came to Jatiluwih rice field … what a beautiful day, clean air and quiet … very inspiring place and we'll never forget this moment.
>
> (IT-2, 05/12/2019)

> Huge, scenic at every turn … did the shorter walk … is scenic and picturesque at any angle … is majestic to walk around and look as far as your eyes … see the rice terraces … enjoy nature and be grateful for the abundance.
>
> (IT-3, 11/12/2019)

> …well-known as the central production of red rice… witness the rice production process … had a satisfactory experience … recommended for tourists.
>
> (DT-3, 7/23/2021)

The value gained by the tourists from their owned gazes and immediate interactions with the Jatiluwih heritage site either in the form of recreations, relaxations or knowledge advancements fulfils the tourists' vacation demands, which led to satisfaction. Such satisfactory experiences are valuable for destination marketing, which builds trust and loyalty to tourists either intending to revisit SJHTS or recommending fellow tourists to visit the site (Kotler & Keller, 2016). Foreign tourists shared their experiences:

> The Jatiluwih rice terraces are massive and absolutely stunning! …the fields were green and lush … a rare and precious thing in Bali … spent an evening and a sunny morning at the terraces … just sweet and helpful locals …Be sure to stay the night at Padi Bali's inexpensive bungalows. The owner and his family are warm and welcoming … see the sunrise right from your front door.
>
> (IT-4, 30/03/2020)

> We had stopped by a small & narrow stream which is part of the Subak. We had put our hands in the water & found it to be extremely cold &

> refreshing. I instantly felt like swimming! We were able to appreciate the scenic views, the breeze, fresh air & peacefulness… We highly recommend a visit to Jatiluwih.
>
> (IT-5, 26/03/2020)

Experiencing atmosphere and witnessing authentic agricultural attractions make the heritage tour more memorable as tourists gained the tour core value during the trip. Additionally, the image of Jatiluwih as the heritage landscape of Bali is deeply positioned in the tourists' minds, which is further publicly shared online and used as travel references for other tourists. Thus, CES is significant to the marketing strategy of this heritage tourism. The success of positioning the image of Subak Jatiluwih as a world heritage site has also been supported by the role of tour guides who escort tourists to the site. A foreign tourist argued:

> The most iconic rice paddies in Bali … Jatiluwih is the biggest and most iconic rice paddies that you can find in Bali. It's so breath-taking and you can enjoy this place for hours. In order to have a better experience you may get a tour guide to bring you to the best spot available here.
>
> (IT-5, 30/04/2021)

The role of the tour guide is highly needed to show the directions and to guide the tourists to the site where they can engage with the site more deeply. The tour guide's interpretations are also essential in widening the tourists' knowledge of cultural heritage sites as evidence of cognitive development attainment in CES (Sarukhán & Whyte, 2005). Moreover, the interpretations made by the tour guide also build loyalty for the tourists to revisit SJHTS. A tour guide claimed:

> I often guide repeating tourists to Jatiluwih … they have been to Jatiluwih many times. Moreover, since legalised as world heritage site in 2012 … always eager to enjoy the scenic views and understand more about the culture… Subak… Tri Hita Karana …traditional cultivations, agricultural concept… guide them courtesy.
>
> (TG, 26/07/2021)

The competency of tour guides in mastering the knowledge and philosophy attached to the daily culture practised by the local people contributes to the knowledge development of the tourists. Additionally, the guiding skills and attitudes applied during the tours are essential in assuring the tourists gain enjoyment and relaxation, which served the value of their heritage tour to SJHTS.

19.4 Conclusion

The designation of Jatiluwih Subak and landscape as a world heritage tourism site needs to be researched to understand how CES is created by the tourism management and experienced by stakeholders and tourists in addition to its primary functions as rice productions and earning sources in Bali. CES is naturally created by the local people and authentically staged for tourism by the destination management in collaboration with stakeholders, including local farmers, Subak associations, traditional and administrative villages and government (Chan et al., 2012; Fish et al., 2016). CES have been existing at SJHTS and both local people and tourists have gained from CES. The locals primarily use the site for agriculture purposes to fulfil their daily needs and support regional rice supplies as the material benefits obtained from the site. Additionally, it also offers the non-material benefit of spiritual enhancements through religious rites practised in their agricultural activities wherein they can worship *Bhatara Sri* – the God of prosperity while preserving their cultures. In other words, the development of CES was embedded with the locals' daily cultures and authentically performed as cultural tourism attractions, which allow local people to gain economic benefits from tourism and sustain their respective cultures while practising their beliefs for spiritual delights. Furthermore, tourists visiting SJHTS do not only gaze at the scenic agrarian landscape to refine their minds through recreational activities in the green and fresh atmosphere but more importantly to engage with the agricultural activities such as planting and harvesting rice paddy and experience tourist activities such as trekking, cycling and photo-taking as well as tasting the local cuisine served in local restaurants. Thus, the process of enjoyment and "being there" are the most essential value gained by the tourists visiting SJHTS, which are collaboratively managed by the respective stakeholders that termed the process "cultural ecosystem services" (Chan et al., 2012; Fish et al., 2016; Sarukhán & Whyte, 2005). This suggests that the non-economic benefits of SJHTS, which are managed in collaboration with local stakeholders, are valuable for both indigenous residents and tourists. This research only focuses on how CES is created by the management and gained by the tourists as the supplementary value in addition to the financial benefits in SJHTS. The finding can be used as a recommendation and reference for other Bali tourism site managements adapting such collaboration patterns which have successful offered satisfactory and impressive experiences visiting SJHTS that initiated tourists either sharing their experiences online through social media platforms or intended to revisit the site (Chan et al., 2012; Fish et al., 2016; Gong et al., 2018; Kim et al., 2019; Litvin et al., 2018; Xu et al., 2020; Yan et al., 2018). This needs to be researched further confirming whether CES affect the tourists' loyalties towards SJHTS using the tourist behaviour approach.

References

Ababneh, A. (2017) Tour guides and heritage interpretation: Guides' interpretation of the past at the archaeological site of Jarash, Jordan. *Journal of Heritage Tourism*. DOI: 10.1080/1743873X.2017.1321003

Adner, R. (2017) Ecosystem as structure: An actionable construct for strategy. *Journal of Management*, Vol. 43, No. 1, pp. 39–58.

Ardika, I.W. & Subadra, I. N. (2018) *Warisan Budaya Dunia: Pura Taman Ayun dan Pura Tirtha Empul sebagai Daya Tarik Wisata di Bali*. Denpasar: Pustaka Larasan.

Augier, M. & Teece, D.J. (2018) *The Palgrave Encyclopedia of Strategic Management*. London: Palgrave Macmillan.

Baxter, P. & Jack, S. (2008) Qualitative case study methodology: Study design and implementation for Novice researchers. *The Qualitative Report*, Vol. 13, No. 4, pp. 544–559.

Begon, M., Townsend, C.R. & Harper, J.L. (2006) *ECOLOGY: From Individuals to Ecosystems*. Oxford: Blackwell Publishing Ltd.

Byrd, E.T. (2007) Stakeholders in sustainable tourism development and their roles: Applying stakeholder theory to sustainable tourism development. *Tourism Review*, Vol. 62, No. 2, pp. 6–13. DOI: 10.1108/16605370780000309

Chan, K.M.A. & Satterfield, T. (2015) Managing cultural ecosystem services for sustainability, in M. Potschin, R. Haines-Young, R. Fish & R.K. Turner (Eds.). *Routledge Handbook of Ecosystem Services*. Oxon: Routledge.

Chan, K.M.A., Satterfield, T. & Goldstein, J. (2012) Rethinking ecosystem services to better address and navigate cultural values. *Ecological Economics*, Vol. 74, pp. 8–18. DOI: 10.1016/j.ecolecon.2011.11.011

Creswell, J.W. (2014) *Research Design: Qualitative, Quantitative, and Mixed Methods Approach*. California: SAGE Publications, Inc.

Crowe, S., Cresswell, K., Robertson, A., Huby, G., Avery, A. & Sheikh, A. (2011) The case study approach. *BMC Medical Research Methodology* Vol. 11, No. 1, pp. 1–9.

Denzin, N. K., & Lincoln, Y. S. (Eds.). (2018). *The SAGE Handbook of Qualitative Research*. Los Angeles: SAGE

Dharmayudha, I.M.S. & Cantika, I.W.K. (1999) *Filsafat Adat Bali*. Denpasar: Upada Sastra.

Dul, J. & Hak, T. (2008) *Case Study Methodology in Business Research*. Oxford: Elsevier Ltd.

Falk, M.T. & Hagsten, E. (2020) Visitor flows to World Heritage Sites in the era of Instagram. *Journal of Sustainable Tourism*. DOI: 10.1080/09669582.2020.1858305

Fish, R., Church, A. & Winter, M. (2016) Conceptualising cultural ecosystem services: A novel framework for research and critical engagement. *Ecosystem Services*, Vol. 21, pp. 208–217.

Fisk, P. (2010) *People, Planet, Profit: How to Embrace Sustainability for Innovation and Business Growth*. London: Kogan Page Limited.

Gallarza, M.G., Saura, I.G. & Garcıa, H.C. (2002) Destination image towards a conceptual framework. *Annals of Tourism Research*, Vol. 29, pp. 56–78.

Gong, S., Li, Q., Zhao, P. & Ren, Z.W. (2018) Marketing communication in the digital age: Online advertising, online word of mouth and mobile game sales. *Nankai Business Review*, Vol. 21, pp. 28–42.

Government of Bali Province (2020) Regulation Number 5: Standards of Balinese Cultural Tourism Operations.

Government of Bali Province (2021) Circular Number 12 of 2021: Order of Community Activity Restrictions Level 4 Corona Virus Disease 2019 in New Life Norm in Bali Province.

Holloway, J.C. (2006) *The Business of Tourism* (7th ed.). Essex: Pearson Education Limited.

Jacobides, M.G., Cennamo, C. & Gawer, A. (2018) Towards a theory of ecosystems. *Strategic Management Journal* Vol. 39, No. 8, pp. 2255–2276.

Jatiluwih Village Profile (2021) Demography based on the population distribution. Available at https://jatiluwih.desa.id/first/wilayah. Accessed on 3 July 2021.

Kang, S.K. (2019) Place attachment, image, and support for marijuana tourism in Colorado. *SAGE Open*, pp. 1–11. DOI: 10.1177/2158244019852482

Kenter, J.O., Reed, M.S., Irvine, K.N., O'Brien, E., Bryce, R., Christie, M., Cooper, N., Hockley, N., Fazey, I., Orchard-Webb, J., Ravenscroft, N., Raymond, C.M., Tett, P. & Watson, V. (2016) Shared values and deliberative valuation: Future directions. *Ecosystem Services*, Vol. 21, 358–371 DOI: 1016/j.ecoser.2016.10.006

Kim, J.J., Nam, M. & Kim, I. (2019) The effect of trust on value on travel websites: Enhancing well-being and word-of-mouth among the elderly. *Journal of Travel & Tourism Marketing*, Vol. 36, No. 1, pp. 76–89.

Koontz, H. & O'Donnell, C. (1972) *Principles of Management: An Analysis of Managerial Functions*. New York: McGraw-Hill.

Kotler, P. & Keller, K.L. (2016) *Marketing Management* (15th ed.). Essex: Pearson Education Limited.

Lindhjem, H., Reinvang, R. & Zandersen, M. (2015) *Landscape Experiences as a Cultural Ecosystem Service in a Nordic Context: Concepts, Values and Decision-Making*. Copenhagen: Nordic Council of Ministers.

Litvin, S.W., Goldsmith, R.E. & Pan, B. (2018) A retrospective view of electronic word-of-mouth in hospitality and tourism management. *International Journal of Contemporary Hospitality Management*, Vol. 30, No. 1, pp. 313–325.

Liu, X., Li, J. & Kim, W.G. (2015) The role of travel experience in the structural relationships among tourists' perceived image, satisfaction, and behavioral intentions. *Tourism and Hospitality Research*, Vol. 17, No. 2, pp. 1–12.

Milcu, A.I., Hanspach, J., Abson, D. & Fischer, J. (2013) Cultural ecosystem services: A literature review and prospects for future research. *Ecology and Society*, Vol. 18, No. 3, pp. 44. DOI: 10.5751/ES-05790-180344

Miller, G.T. & Spoolman, S.E. (2009) *Essentials of Ecology*. Belmont: Cengage Learning.

Nanjangud, A. & Reijnders, S. (2021) 'I felt more homely over there …': analysing tourists' experience of Indianness at Bollywood Parks Dubai. *Current Issues in Tourism*. DOI: 10.1080/13683500.2021.1968804

Okazaki, E. (2008) A community-based tourism model: Its conception and use. *Journal of Sustainable Tourism*, Vol. 16, No. 5. pp. 511–529. DOI: 10.1080/09669580802159594

Patton, M.Q. (2015) *Qualitative Research and Evaluation Methods* (4th ed.). California: SAGE Publications, Inc.

Poria, Y., Biran, A. & Reichel, A. (2009) Visitors' preferences for interpretation at Heritage Sites. *Journal of Travel Research*, Vol. 48, No. 1, pp. 92–105

Poria, Y., Butler, R. & Airey, D. (2003) The core of heritage tourism. *Annals of Tourism Research*, Vol. 30, No. 1, pp. 238–254.

Poria, Y., Reichel, A. & Cohen, R. (2011) World Heritage Site – Is it an effective brand name? A case study of a Religious Heritage Site. *Journal of Travel Research*, Vol. 50, No. 5, pp. 482–495.

Richards, G. (2020) Placemaking through creative tourism, in *International Conference on Creative Tourism and Development*. Bangkok: Chulalongkorn University.

Richards, G. & Wilson, J. (2007) *Tourism, Creativity and Development*. Oxon: Routledge.

Ryan, C., Chaozhi, Z. & Zeng, D. (2011) The impacts of tourism at a UNESCO Heritage Site in China – A need for a meta-narrative? The case of the Kaiping Diaolou. *Journal of Sustainable Tourism*, Vol. 19, No. 6, pp. 747–765.

Sanna, S. & Eja, P. (2015) Recreational cultural ecosystem services: How do people describe the value? *Ecosystem Services*, Vol. 26, pp. 1–9. DOI: 10.1016/j.ecoser.2017.05.010

Sarukhán, J. & Whyte, A. (2005) *Ecosystems and Human Well-Being: Synthesis (Millennium Ecosystem Assessment)*. Washington: Island Press.

Schwandt, T.A. & Gates, E.F. (2018) Case study methodology, in N.K. Denzin & Y.S. Lincoln (Eds.). *The SAGE Handbook of Qualitative Research*. California: SAGE Publications, Inc., pp. 600–630.

Smit, B., Melissen, F., Font, X. & Gkritzali, A. (2020) Designing for experiences: A meta-ethnographic synthesis. *Current Issues in Tourism*. DOI: 10.1080/13683500.2020.1855127

Subadra, I.N. (2015) *Preserving the Sanctity of Temple Sites in Bali: Challenges from Tourism*. Lincoln (United Kingdom): PhD Thesis. University of Lincoln.

Subadra, I.N. (2019) Alleviating poverty through community-based tourism: Evidence from Batur Natural Hot Spring Water – Bali. *African Journal of Hospitality, Tourism and Leisure*, Vol. 8, No. 51. EID: 2-s2.0-85073111279.

Subadra, I.N. (2021a). Pariwisata Budaya dan Pandemi Covid-19: Memahami Kebijakan Pemerintah dan Reaksi Masyarakat Bali. *Jurnal Kajian Bali (Journal of Bali Studies)*, Vol. 11, No. 1, pp. 1–22. DOI: 10.24843/JKB.2021.v11.i01.p01

Subadra, I.N. (2021b) Destination management solution post COVID-19: Best practice from Bali – A world cultural tourism destination. In V.B.B. Gowreesunkar, S.W. Maingi, H. Roy, & R. Micera (Eds.), *Tourism Destination Management in a Post-Pandemic Context: Global Issues and Destination Management Solutions*. Bingley: Emerald Publishing Limited.

Subadra, I.N. & Hughes, H. (2021) Pandemic in paradise: Tourism pauses in Bali. *Tourism and Hospitality Research*, pp. 1–7. London: Sage Publication. DOI: 10.1177/14673584211018493

Subadra, I.N. & Nadra, N.M. (2006) Dampak Ekonomi, Sosial-Budaya, dan Lingkungan Pengembangan Desa Wisata di Jatiluwih-Tabanan. *Jurnal Manajemen dan Pariwisata*, Vol. 5, No. 1, pp. 46–64.

Teece, D. J. (2017). Profiting from innovation in the digital economy: Standards, complementary assets, and business models in the wireless world. Research Policy Tusher Centre for Management of Intellectual Capital. Available at: https://www.haas.berkeley.edu/wp-content/uploads/Tusher-Center-Working-Paper-No.-16.pdf

UNESCO (2012) Convention Concerning the Protection of the World Cultural and Natural Heritage. Retrieved from https://whc.unesco.org/archive/2012/whc12-36com-19e.pdf. Accessed on 2 May 2021.

UNESCO (2021) Sustainable Tourism: UNESCO World Heritage and Sustainable Tourism Programme. Retrieved from https://whc.unesco.org/en/tourism/. Accessed on 30 July 2021.

UNWTO 2020 UNWTO Inclusive Recovery Guide Sociocultural Impacts of COVID-19. World Tourism Organization Publications. Available at: https://www.e-unwto.org/doi/pdf/10.18111/9789284422579

Urry, J. (2002) *The Tourist Gaze*. London: SAGE Publications Ltd.

Xu, F., Niu, W., Li, S. & Bai, Y. (2020) The mechanism of word-of-mouth for tourist destinations in crisis. *SAGE Open*, pp. 1–14. DOI: 10.1177/2158244020919491

Yan, Q., Zhou, S. & Wu, S. (2018) The influences of tourists' emotions on the selection of electronic word of mouth platforms. *Tourism Management*, Vol. 66, pp. 348–363.

Yin, R. K. (2018). *Case Study Research and Applications: Design and Methods* (6th ed.). Thousand Oaks, CA: Sage.

20 Tourism and Restoration of Spring Ecosystem Services

Case Study of Latvia

Janis Bikse and Mahender Reddy Gavinolla

20.1 Introduction

Springs are ecosystems where the groundwater naturally flows to the earth's surface (Springer et al., 2015; Springer & Stevens, 2009). They are widely acknowledged as a hot spot of biodiversity (Hershler et al., 2014) that improves ecological functions (Springer et al., 2015) and also plays a significant role in the availability of clean water. Water springs are also considered an important aspect of human society for several centuries, due to their spiritual, sacred, wellness, mineral, and other benefits (Coltelli, 1996). For instance, Egyptians consider water spring-based streams and lakes as divine and sacred spaces (Altman, 2002). People in India and Indonesia believe that the water springs exhibit holy spirits and ensure the availability of clean water (Anthwal et al., 2010). In other words, water springs provide substantial cultural, regulatory, and provisioning ecosystem services.

For several ages, water springs have been regarded as components or resources for tourism due to their nature, characteristics, and socio-economic and environmental benefits (Bikse & Gavinolla, 2021). Most importantly, in the recent past, in several destinations, water springs are promoted as vital tourist attractions (Smith & Puczkó, 2016; Stevens et al., 2016), for wellness tourism, nature-based tourism, recreational tourism, sacred, spiritual, and religious tourism (Bikse & Gavinolla, 2021). In spite of these benefits, several studies have revealed that the springs are highly threatened ecosystems, leading to greater vulnerability to invasion by non-native plants and further leading to reduced floral diversity of the place (McKinney, 2008; Pyšek et al., 2010). These impacts vary from place to place, and in some cases, impairment exceeds about 90% and more (Stevens & Meretsky, 2008). Anthropogenic use of springs such as spiritual (Rood et al., 2005), industrial use in the catchment area (Simanjuntak, 2005), recreational and balneological use, and development of infrastructure and other tourist facilities led to change in the landscape and other features of the water springs.

Nevertheless, it is essential to elucidate that unplanned and irresponsible management of tourism at the water spring-based tourist attractions leads to the destruction of the very nature of the spring ecosystem and its services. In this background, several studies examined various aspects of sustainable

DOI: 10.4324/b23145-26

management of spring tourism. For example, an ecosystem assessment protocol is developed to understand the impact of recreational activities and tourism on spring ecosystem services (Stevens et al., 2016). Another study focused on sustainable management of SES signified the need for restraining human commotion from tourism activities at the spring tourism destinations (Nielson et al., 2019). However, research on cultural ecosystem services in connection with spring tourism and its implications on the restoration of spring ecosystem services is scant, in this regard, in order to understand the impacts and implications of tourism on various spring ecosystem services. With this foundation, the study aims to examine the impacts and implications of tourism on the restoration of spring ecosystem services in Latvia's highly potential spring tourism destinations. Latvia is selected as a study area because the country is blessed with a considerable number of spring sites associated with significant history, culture, and tourism. The study entails several issues and challenges involved in sustainable management of the spring ecosystem and provides strategies to overcome the challenges through the sustainable practice of spring tourism destinations. The outcome of the study signifies a contribution to the body of literature and new insights to practitioners on promotion and sustainable management of spring tourism resources in Nordic and Baltic States in particular and other destinations in general.

The chapter is divided into three parts. The first part provides a conceptual framework and scientific landscape of tourism and spring ecosystem services. The second part elucidates the contextual framework of the study i.e., spring tourism in Latvia. The later part includes the methodological framework of the study, followed by the result, discussions, implications, and conclusion.

20.2 Literature Review

Tourism is the largest service industry that heavily depends on the natural and cultural attractions of the destination (Afanasiev et al., 2018; Livina & Reddy, 2017). Biophysical resources such as landforms, water forms, climate, flora, and fauna of the destination (Deng et al., 2002) and cultural resources, including the tangible and intangible heritage of the place (Jansen-Verbeke, 2007; Vaeliverronen et al., 2017). Human societies is closely connected to the springs for millennia (Cuthbert & Ashley, 2014), as they have been extremely used for drinking water, harvesting, minerals (Shrestha et al., 2018), and ambushing prey (Haynes, 2008). Spring ecosystem, as cultural, natural, and activity-based resource of tourism is well exploited and promoted for the ages throughout the world (Bikse & Gavinolla, 2021; Lee et al., 2009). Spring ecosystems are "places on the earth's surface that are influenced by the exposure, and often the flow, of groundwater" (Stevens et al., 2021). Water springs play a significant role in perpetuating the ecosystem services (Rohman et al., 2021) and they play a major function in providing fresh water and water security (Cantonati et al., 2012; Shrestha et al., 2018). Springs are the places for the hot spots of biodiversity contributing to the growth and presence of diverse species, leading to the unique landscape (Cantonati et al., 2020).

Springs are geomorphologically diverse and occur in different sizes and shapes, forms, and both in underwater and terrestrial environments (Stevens et al., 2021), as a stream, lakes, ponds, waterfalls, oceans, and glaciers (Johnson et al., 2012), and they are considered as a resource of natural attraction for the tourism (Bikse & Gavinolla, 2021; Stevens et al., 2016).

In the context of tourism, springs provide several cultural ecosystem services to the visitors. Visiting water springs for health, wellness, and medical reason and spiritual and religious points of view has been in practice for ages (Kiełczawa, 2018). For example, Indians and Japanese used hot water springs both for spiritual and medical purposes over 11,000 years ago (Fridleifsson, 2001). Pilgrims visit Prislop Monastery in Romania to manifest their faith by drinking or collecting the spring water at monasteries and throwing money is quite common (Giuşcă, 2020). Pilgrims visiting the springs with religious importance commonly touch the surface of the springs before they collect and consume the spring water (Foley, 2011). However, the contemporary use of spring ecosystem services increased immensely for tourism, recreation, leisure, health, wellness, minerals, and religious and spiritual purpose (Stevens et al., 2021). Even today, in several countries, tourists are extensively visiting water springs as part of a spiritual journey or wellness tourism point of view. Several tourists are visiting host water springs in India with a belief or faith that spring water can cure certain diseases and improve longevity (Bhutiani et al., 2016). Due to the aesthetic value, elegant landscape, ambiance, and spiritual importance, a significant number of springs in Western Kazakhstan are converted into recreational and tourist spots (Akhmedenov, 2020). This has further led to the development of infrastructure and the development of the basic amenities at the site (Absalyamova, 2019); in turn, it will lead to several negative consequences and thread to the springs. In Acropolis in Greece, people visit for springs spiritual and wellness purposes, and they are under threat because of contamination from landfills and other tourist activities (Klempe, 2015).

Another important reason is that several places developed recreational and wellness centers as the spas can offer both leisure and health benefits to the tourists (de la Cruz del Río-Rama et al., 2018). In Japan, the government is improving its health system by integrating the traditional health centers with leisure and spa in the rural areas in the vicinity of the springs. This is not only to offer recreation and wellness but also economic diversification of the destination (Tabayashi, 2010). Cultural ecosystem services of water springs include improved health, well-being, and lifestyle. People are frequently traveling to the springs to improve their health and lifestyle in several regions of the world (Tabacchi, 2010). Queensland in Australia is promoting its wellness tourism that is based on the water springs such as hydrotherapy and spa (Griggs, 2013). There is a special segment of hotels providing services for tourists in the vicinity of the springs. Hot spring hotels are proving several wellness tourism services to their customers (Chen et al., 2013). The United States and Canada are promoting the mineral springs as health and spa tourism destinations and incorporating several other forms of tourism such as nature tourism and protected area tourism as part of the tour packages (Griggs, 2013).

Springs are the most significant and productive ecosystems (Stevens & Meretsky, 2008). Springwater plays a major role in the conservation and maintenance of animals and plant habitats and is also an important cultural resource of the destination (Kumazawa, 2016). However, threads or loss of springs leads to ecological integrity and sustainability (Stevens et al., 2021). In contemporary times, extensive use or unsustainable way of using the water springs led to environmental degradation. Tourists show interest to touch and drink the water as spring water consists of minerals and the power of transforming life (Hannaford & Newton, 2008; Margry, 2008). For instance, Ganga and Yamuna rivers and several water springs are degrading due to anthropogenic activities and leading to the degradation of waters, springs, and the surrounding areas (Haberman, 2006; Nagaranjan, 1998). The introduction of invasive species to the springs and their surrounding areas led to the loss and degradation of biodiversity (Bódis et al., 2016). Thousands of springs in the Kashmir Himalayan region of India is not only attracting tourist; a significant number of people also depends on the tourism economy and employment that is based on spring tourism destinations. However, there is no legislation or protective measures to preserve and conserve the springs that are deteriorating due to tourism and related activities (Cantonati et al., 2012; Gupta & Kulkarni, 2017). Climate change is also one of the impacts that are resulting due to increased temperature, detoxification and floods, and other human activities, leading to a change in the structure and functions of several spring ecosystem services (Carpenter et al., 2011). The above literature elucidates clearly that the SES provides several benefits from the tourism point of view, however, SES is affected by activities. In this regard, several empirical studies explained the need for the restoration of SES.

Several initiatives were taken to protect and conserve the spring water ecosystem in various parts of the world as local, state, or federal parks. Sustainable management of springs requires a detailed understanding of the impacts of anthropogenic activities (Stevens et al., 2021). Several studies emphasized the sustainable use and management of spring ecosystem services and the need for restraining human commotion from tourism activities (Nielson et al., 2019). A study proposed a conceptual model for the classification of spring ecosystem services based on the function and typology. This model provides an insight into the different types of spring ecosystems and helps to understand the management and restoration of springs. However, the model has not emphasized classification particularly based on the use of springs for leisure, recreation, or spiritual and religious purposes (Stevens et al., 2021). The European Commission emphasized the protection of springs as an important ecosystem, and countries like Finland and Australia laid special emphasis on the protection of springs (Cantonati et al., 2020; The European Commission, 2013). In one of the studies, emphasis was given to the ecosystem assessment protocol that was developed for understanding the impact tourism and related activities have given the little emphasis on spring ecosystem service (Stevens et al., 2016). It is evident in the past that education and awareness play an important role in the restoration of ecosystem services

(Dou et al., 2017). Visiting natural sites and experiencing the biodiversity will enhance one's understanding, awareness, appreciation of the environmental problems, and bio-geological issues of the times (Gavinolla et al., 2021). In this regard, there is a significant number of studies that examined the positive contribution of tourism to the restoration of ecosystem services (Taff et al., 2019; Willis, 2015). For instance, religious tourists and spiritual followers visit several water springs for the sacred beliefs that will enhance their concern and appreciation over water springs, and this will further lead to cultural enrichment and educational awareness (Collins-Kreiner, 2010). However, change in the human behavior on the usage of spring water around the temples and shrines is an important aspect of the restoration of the spring ecosystem, as they are mainly degraded due to unsustainable use of springs and their surrounding areas (Kumazawa, 2016). Western Kazakhstan is encouraging tourists to visit the springs to enhance knowledge, understanding, and appreciation of nature (Akhmedenov, 2020). Springs are also used to change the behavior toward the use of springs in several places. A study found that the spring water quality of the Central area of Mito-City in Japan degraded extremely because of the use of human activities, and the Japanese government is trying to change the behavior of the visitors toward the use of springs by using tourism as a tool to environmental education and create awareness (Kumazawa, 2016). Several states in the Himalayan region and other parts of India took several initiatives to protect and preserve the springs by involving the local community, NGOs, and other stakeholders, as they are becoming more and more seasonal or drying up in several areas due to the developmental activities including tourism (Shrestha et al., 2018; Tiwari, 2000). It is evident from the above-mentioned studies that the research on the springs and their cultural ecosystem services has not received substantial attention from the academic community. It may be due to the complex and multidisciplinary nature of the (Stevens et al., 2021).

20.3 Methodology

Existing literature shows that water springs provide several ecosystem services in the context of tourism, both in positive and negative ways depending on how the spring services and tourism are managed. Therefore, it is important to study the phenomenon, to better understand the contribution of tourism to the restoration of spring ecosystem services. The methodological approach is based on multiple methods, including content analysis, participant observation, semi-structured interviews with the tourism service providers and authorities involved in the management of the springs (Séraphin et al., 2019). An integrative review is wisely applied when a particular topic is at the infant stage or an emerging area (Torraco, 2016), accordingly content analysis of online reviews, news articles, magazines, and historical documents (Gowreesunkar & Gavinolla, 2020; Livina & Reddy, 2017). Qualitative research and primary data from the respondents were collected through semi-structured interviews (Livina & Reddy, 2017; Séraphin and Gowreesunkar, 2019). Qualitative research

Table 20.1 Description of the methods

Method	*Description*
Desk research, content analysis, and literature review	Both the authors reviewed the literature in order to understand the research gap and the existing state of research on ecosystem services and their restoration in the context of tourism. Website content, images, and published literature from the magazines, news clippings, tourism literature, and related articles were reviewed by both authors.
Observation	The next stage of work is mainly carried out by the first author who is a citizen of Latvia. The author made several visits to observe the destination in terms of tourist activities, and management of tourism.

using the semi-structured questionnaire is one of the important approaches of data collection in case study research (Gavinolla, et al. 2021; Gowreesunkar & Gavinolla, 2020). This will, in turn, provide an understanding and analyze the complex phenomenon in a simple manner. Table 20.1 provides a detailed overview of the various methods applied to the study.

Next, semi-structured interviews were conducted and the description of the respondents was provided in Table 20.2. The study was carried out from February to July 2020, and the primary data was collected mainly in June and July 2020 as it was the season for tourism in Latvia. Data were collected in 12 days scattered over eight weeks, with a duration of two to three hours of survey and some cases beyond three hours. Though there are several spring destinations in Latvia, only a few and popular spots were selected for the study.

Twelve respondents participated in the semi-structured interviews. Questionnaire was formulated with the emphasis on three aspects, namely description of the spring site, stakeholder's perspective on the spring ecosystem services, and role of tourism in the restoration of springs. Data was coded and structured with meaningful interpretations.

20.4 Result

20.4.1 Water Spring Tourism in Latvia

Latvia is located on the Eastern coast of the Baltic Sea area, and Riga is the capital city and historic town inscribed as a world heritage site (Livina & Reddy, 2017). Tourism is one of the important sectors that contribute significantly to the Latvian economy (Druva-Druvaskalne & Līviņa, 2014). Latvia is one of the important tourist destinations in the Baltic state, both for domestic and international tourists (Dook, 2019). Nature-based tourism, particularly visiting the protected areas, is one of the important aspects of tourism in Latvia (Berzina & Livina, 2008), and this includes the visit to the water spring tourism spots. Latvia is actively advocating the tourism stakeholders to practice the sustainability principles in their businesses and

Table 20.2 Description of the semi-structured interviews

Respondents	*Relationship with springs*	*Description of the spring site*
1	UNESCO National Commission	The UNESCO National List includes several major natural sites, such as Kuldiga Old Town in the Primeval Valley of the River Venta and Meanders of the Upper Daugava.
2	Researcher hydrogeologist	Participated in several spring-related projects, including cross-border with Estonia and commercial drinking spring water.
3	Head of Latgale Regional Administration Nature Education Center "Rāzna"	Kraka holy springs are located in the territory of the reserve.
4	Head of a tourism company that organizes educational tourism trips	Local tourists like to visit various springs as unique natural objects.
5	Executive director of Kocēni district	Resolved access and other infrastructure issues.
6	Former Member of the European Parliament, Member of the Committee on Transport and Tourism. Currently a tourism travel designer, author of a tourism design book	Different types of springs as elements of individual tourism.
7	Member of Valmiera County Council. Traveler, nature and sports enthusiast	Surveyed and maintained many county springs.
8	Chief expert of Kuldīga municipality for work with society and state institutions, manager of innovation and tourism projects	He was an expert in the Latvian sanctuary project "CULT IDENTITY," within the framework of which in joint expeditions with natural scientists, surveyed, inventoried, and made photo fixations for the main sanctuaries of Kurzeme, incl. springs, sacred exchanges, cult stones, etc.
9	Traveler, source surveyor	Most Latvian springs.
10	Master's student, researcher of sacred springs and pilgrims	Springs near Latvian shrines and churches, mainly Aglona Basilica.
11	NATO Survival Instructor	Water springs are identified when starting field activities. For example, in the Selia region, 86 springs have been identified where water is of adequate quality.
12	Rauna Parish Executive Director	In the territory of the parish, there is the most popular spring in Latvia "Raunas Staburags," which is located in the nature reserve of the same name.

services (Atstāja et al., 2011). There are numerous water springs in Latvia, and they are well known for spiritual, health, and wellness purposes (Bikse & Gavinolla, 2021). Latvia poses a long history of water springs and water spring-based tourism activities. For example, literary evidence shows that the springs flowed from Gutmanis Cave in the early 15th century (Eniņš, 2004). Historically several thermal baths, spas, and rehabilitation centers were established in the vicinity of the springs. Several tourists, mainly from Baltic and Nordic countries, visit the water spring destinations in Latvia. Considering the potential opportunities of spring tourism, the "Baltic Health Tourism Cluster" was established to promote tourism (Smith, 2015). In contemporary times, tourists visit the springs mainly for spiritual or religious, health, and wellness purposes. Tourist activities included in spring tourism are nature tourism, walking, collecting the spring water, spiritual activities, and so on.

20.4.2 State of Tourism and Its Impact on Spring Tourism

In the first part of the study, authors tried to understand the type of water spring, purpose or usage of water spring in the tourism context, and restoration measures of spring ecosystem services in those selected spring areas of Latvia. Though there are several water springs in Latvia, the most popular springs in terms of the number of visitors were selected for the study. Table 20.3 provides detailed information on the type of water spring with examples and the state of the tourism development.

Raunas Staburags is the most visited spring in Latvia, which is a travertine cliff, formed over many centuries from the coating of moss with lime from calcium carbonate-rich spring water. Linda Zūdiņa, director of Rauna parish,

Table 20.3 Types of springs and state of the tourism and spring ecosystem restoration

Type of water springs	*Examples*	*Development and protection opportunities*
Recreation SPA	Baldone, Ķemeri	Hotels, bathing places, resorts, sanatoriums, tourism
Sacred	Aglona (Catholics), Opuļu (Orthodox), Tirzas, Galtenes u.c. (Latvian traditional)	Sacred buildings, environmental improvement, water quality monitoring, tourism
Natural wonders	Travertine (Raunas Staburags), natural arches (Liepas Lielā Ellīte), caves (Gūtmaņala), specific springs (Zušu sulfur springs)	Environmental improvement, tourism promotion
Healing	Apes Acu and Veselības, Bolēnu	Environmental improvement, water quality monitoring
Historical	Gaujiena Lion mouth spring	Cultural and historical protection of the environment

was responsible for the spring protection measures in the survey explained the situation about water spring restoration that

> not only spring tourism, but every tourism object in the territory of the municipality is important because it mainly affects the economic growth, recognizability, and attendance of the place. On the other hand, as the number of visits increases, so does the need for control and monitoring related to environmental clean-up, waste management, and prevention of environmental damage. In order to reduce the negative impact of tourism, the parish administration monitors the anthropogenic flow and maintains the appropriate infrastructure of the territory, which also includes several information stands, which indicate information regarding the natural values of the nature reserve, as well as the permitted and prohibited activities therein.

The second most popular spring in Latvia is located in Gutmanis Cave. Gutmanis Cave also has the oldest cave inscription, dated 1521 (Eniņš, 2004; Urtāns Guntis Gūtmaņa ala, 2008). Gutmanis Cave was the most famous cave in the Soviet Empire – it was visited by up to 1.5 million tourists a year (Siguldas alas Za Pobedu Kommuņizma 1973). The third best-known Liepa Lielā Ellīte (Devil's Oven) is an equally old, popular tourist attraction from water-rich springs. The descriptions tell of even older annual figures (Heervagen, 1864). Lielā Ellīte, like Gūtmanis Cave, is located near the ancient road, so it must be thought that they were already known in the 12th century (Laime, 2009).

Figure 20.1 Photo by J.Bikse. Raunas Staburags – travertine cliff, located in Rauna parish, Rauna river valley.

Figure 20.2 Photo by J.Bikse. Gutmanis cave with ancient inscriptions. In the lower right corner is a spring that gives eternal youth.

Figure 20.3 Photo by J.Bikse. Liepa Lielā Ellīte – the spring that created the cave and the three natural arches.

The fourth most popular spring is the Holy Spring of Aglona Basilica. It differs from the first three, which are significant natural objects, with their built infrastructure and pronounced sacred significance. Up to 100,000 pilgrims visit Aglona Basilica every year on August 15 to celebrate the day of the Assumption of the Virgin Mary into Heaven (Juško-Štekele, 2014). Celebration is of great importance in religious tourism, but they also have negative aspects. Sixty-three per cent of the surveyed residents of the Aglona region have indicated that waste and environmental pollution have the greatest impact on the natural environment during the festival (Šnepste, 2014). It is difficult to imagine a spring for such a large number of visitors who would not have built the infrastructure. It would quickly turn into a muddy swamp and become dangerous to health. Therefore, the infrastructure has been built here and today's pilgrims cannot even imagine the natural appearance of the spring.

Catholic churches and holy springs must be seen in a much larger area – in very large areas they are inextricably linked. Similarly, there are seasonally huge numbers of visitors to church festivals. Latvian springs are similarly equipped, both sacred springs and springs and water intakes.

Figure 20.4 Photo by J. Bikse. Aglona Roman Catholic Basilica of the Assumption of the Blessed Virgin Mary.

Figure 20.5 Photo by J. Bikse. Zāģezera spring located near the Latvian – Estonian border town Valka.

Beautifully designed springs are symbols of two of Latvia's most famous balneological resorts-Ķemeri and Baldone. The symbols of both resorts are lizards, from which a spring of sulfur water flows.

Figure 20.6 Photo by J. Bikse. Baldone lizard – a famous spring at a famous SPA resort.

20.5 Findings, Discussion, and Implications

The authors tried to understand the potential benefits of spring ecosystem services in spring tourism destinations. Undoubtedly, the positive benefits from any source are the improvement of health, especially when placebo and psycho-hygiene practices, drinking water for hiking, and nature tourists are some of the positive benefits. Others include cognitive and educational tourism. Different types of protected areas or natural areas provide different benefits at the tourism destination in Latvia (Berzina & Livina, 2008), more precisely there are several cultural ecosystem services provided by the water springs in the context of tourism (Springer & Stevens, 2009; Taff et al., 2019). This study substantiates several other studies that spring tourism provides various recreational, spiritual, nature conservation, and educational benefits to the visitors and local community (Bikse & Gavinolla, 2021; Nielson et al., 2019; Stevens et al., 2016).

The result shows that if the water springs are divided according to the type of their visit, there are common and different problems for different springs. The overall threat is groundwater pollution. The biggest polluter here is intensive agriculture with fertilizers, herbicides, and pesticides. In many cases, pollution is also identified as untreated industrial wastewater or transport. Degradation of the environment is a problem in most water springs. The environment is most often degraded by human economic activity. Overall findings show that the springs are mainly affected by the various anthropogenic activities that are not much related to tourism or recreation. However, the increased tourism flows can also be dangerous for water springs as a very sensitive part of the ecosystem. Recreation, SPA, balneological resorts envisage the use of buildings (hotels, sanatoriums) and groundwater for health, recreation, and business. It often completely changes the natural environment. Here are some examples where when building a sanatorium, a spring of sulfur water was built in the form of a lizard (Baldone, Ķemeri). It shows the original spring and the link between modern architecture and nature, of which groundwater is an important part. In recent years, near the Ķemeri sanatorium, a bog trail has been built, which leads to natural sulfur ponds. This allows people not only to use the procedures but also to see the sources of sulfur in their natural environment.

The Leju holy spring is well accessible and known since ancient times. Currently, it is visited by a relatively large number of tourists. Since the source is not confined, people can get into it. Spring observed Coli bacteria and decorative building damage.

Sacred springs have several meanings. They are most often attributed to special abilities. Visitors want to take spring water to get healthy, solve some of their problems or live long. This can contaminate the spring (Figure 20.7). In order to prevent people from having the opposite effect (diseases) when

Figure 20.7 Photo by J. Bikse. Leju holy spring. In the southwest of Latvia on the border with Lithuania.

obtaining water, organizations, municipalities, or individuals tend to build infrastructure that prevents water pollution. Special water intakes have been built in Aglona Basilica and similar Catholic shrines. Opuļi holy spring has a building similar to the Orthodox church. Special buildings have also been built for some ancient Latvian springs (Tirza holy spring, Leju holy spring). The buildings completely change the appearance of the spring but allow a large number of visitors to use the spring water safely. Not all sacred springs are built. Some ancient pre-Christian springs are far from populated areas and their attractiveness lies in the untouched nature (Krāku sacred springs). Some sacred springs are located in unique natural objects – caves, cliffs, arches, riverbanks – and their construction would damage these natural objects. It is widely understood from the literature that there are several spring tourism destinations across the globe facing more or less the same kind of issues and challenges (Foley, 2011; Giuşcă, 2020; Stevens et al., 2021). Like any other destination, Latvia is also facing the challenge of balancing the use of natural objects for tourism while protecting nature and managing the destination in a sustainable manner (Berzina & Livina, 2008).

People want to see unique natural objects. Education, control of tourist flows, invisible (shy) infrastructure, and marketing are useful here. Spring monitoring should also include pollution control. This should be taken into account in particular for springs whose water is used for human consumption or as a means of improving health. Several ancient, sacred, and water-rich springs are now closed because pollution control indicates an increased presence of harmful substances or bacteria in the water. Many Latvian springs are associated with legends that tell about their miraculous properties. There are also often events related to the surroundings of the springs – caves, rocks, boulders. The springs also tend to be part of castles, manors, church complexes, and parks. There are several approaches to the protection of sources, but they usually try to preserve the source in its historical appearance and meaning.

When insisted on the tourism and its contribution to the restoration of source ecosystem services, the respondent mentioned that the properly and purposefully targeted eco-tourism, nature-friendly types of tourism with educational projects for specific segments, an attractive and well-targeted marketing campaign, as well as, of course, well-targeted information and signs in nature and mobile applications, the right tourism products, local involvement will certainly contribute to the restoration of ecosystem services.

Further, it was mentioned that inappropriate legislation and ignorance also damage the spring services. However, in certain places, the legislation is applied seriously. For example, Krustkalni Nature Reserve Law prohibits visitors to stay in the regulated regime zone, including the Kraka springs, without the written permission of the Nature Protection Board or in the presence of the employees of the Board. This prohibition does not apply to persons residing in the territory of the reserve – but they are only a few persons. Of course, there are cases when someone goes to the source without permission – but this is a violation of regulatory enactments. However, it is not an object that anyone can visit whenever they want. It is important that there is protective legislation for the conservation and sustainable management of springs

Figure 20.8 Photo by J.Bikse. Inscription at the Kalamecu-Markuzu ravines. "What you bring, take away." It is important that people do not throw their plastic bottles, food bags, cigarette packets, and cigarettes to a beautiful natural object, but take it with them.

(Cantonati et al., 2012). Like Finland and Australia, Latvia can have protective legislation for the springs in terms of carrying capacity, regulation, usage of springs, and so on (Cantonati et al., 2020; The European Commission, 2013). Latvian policymakers and destination managers may consider the successful initiatives taken by the northeast Indian states on how they have involved major stakeholders in terms of inventory management, conservation measures, and successful rejuvenation of springs (Bhat & Pandit, 2018).

In order to inform and educate the public about the values of the nature reserve, a trail of Krakas springs has been created – started in the forest near the parking lot, it leads to the springs of Karakas. This trail can be accompanied by an employee of the office by booking a visit in time. However, together with the visits of the employees of the office, a period has been set – from May 1 to October 31. The head of the Nature Education Center Latgale Regional Administration Nature Education Center "Rāzna" (place Rēzekne and Lipuški Rāzna NP of Rēzekne region) task is to accompany visitors to the Kraka springs nature trail, guided lessons, and nature education activities or events in our supervised area (in Latgale, Selia, and Vidzeme there are about 165 SPAs under the supervision of our administration).

Several respondents mentioned the problem that beautiful and interesting water sources are located on private land. In most cases, the owner of private land does not want tourists. They are afraid that tourists will leave waste and plants will be destroyed. Similarly, the number of tourists in special protection areas is also undesirable. Using tourism as a tool to protect the nature or environment is considered an important strategy in Latvia (Berzina & Livina, 2008).

Figure 20.9 Photo by J.Bikse. Ķemeri bog trail. Not far from the balneological resort and the Ķemeri symbol - Ķirzaciņa spring, it is possible to see sulfur springs in the natural environment.

Conservation of springs through awareness and education can be important aspects of this strategy (Bikse & Gavinolla, 2021; Livina & Reddy, 2017). In this regard, almost everyone mentioned education as a solution for better interaction between the tourist and springs ecosystem. Education and a smart marketing campaign increase tourists' knowledge of the springs and teach them to treat them with care. The improvement of infrastructure was also mentioned among the solutions. The most complex here is unique natural objects, the appearance of which can be damaged by infrastructures such as bridges, paths, fences, and stairs. Building sustainable infrastructure with a low impact on the environment or eco-friendly infrastructure is an important strategy for limiting the negative impacts (Gavinolla et al., 2021; Livina & Atstaja, 2015). Prompting environmentally friendly transport services and greenways in the spring tourism destinations would reduce the carbon footprint (Tambovceva et al., 2020). In the protection of springs, it is important to distinguish between the importance of springs, their use, and the possibilities to protect them. Proper documentation, digitization, and making an inventory of different types of springs and monitoring them in a sustainable way essentially elucidate the understanding of various springs and need for the restoration, and stewards will certainly help the authorities to manage springs in a sustainable manner (AGIC, 2008; ALRIS, 1993; Bikse & Gavinolla, 2021).

20.6 Conclusion and Recommendations

This study is intended to examine the impacts and implications of tourism on the restoration of spring ecosystem services in Latvia. Study shows that

Latvia has more or less similar situation like other popular spring tourism destinations. It was found that the water springs provide several cultural ecosystem services, and they are one of the important resources of tourism in Latvia. Several people visit water springs in Latvia due to various benefits that the tourist obtain, those are mainly recreational, educational, therapeutic and health, religious or spiritual, the experience of nature, and educational benefits. However, spring tourism destinations are under threat and facing several environmental challenges due to anthropogenic activities, including tourism. The study found that the private spring-protected areas are well maintained and follow several sustainability principles such as carrying capacity, regulatory legislation, and limiting the number of visitors. While the water springs are facing several challenges with unregulated and unsustainable tourism practices, it is important to note that tourism is playing a critical role in the restoration of spring ecosystem services, including the importance of nature tourism and creating awareness on the need for sustainable use of springs through education. Several suggestions were made for the sustainable management of springs and their restoration through tourism. Hence, the policymakers should take into consideration of sustainable tourism practices for the better restoration of SES in Latvia particularly and other destinations in general.

Spring tourism destinations in Latvia should practice carrying capacity principles in order to reduce the pressure on the SES. There should be clear information such as guidelines, instructions, dos and don'ts or code of conduct at the spring tourism destinations to educate the visitors and ensure the need and importance of the restoration of SES. It is important to develop sustainable and environmentally friendly infrastructure that contributes to the restoration of SES and ensures the aesthetic value of the place.

Like any other study, this chapter exhibits several limitations. The study is mainly focused on popular spring tourism spots in Latvia and the data was collected amid Covid-19 situation, due to this reason the study results may not be generalized as each destination is different in terms of types, usage, geographical, and cultural background of the place, and respondents mainly represents service providers and managers. Although not all the stakeholders were involved in the study, the authors tried to explore impacts and implications in a better manner. However, future studies may focus on various cultural spring services for their restoration by involving all the stakeholders, including the visitor in the post-Covid-19 pandemic times.

References

Absalyamova, Y. (2019). The cult of sacred springs among the bashkir. *Shaman, 27*(1–2), 139–147.

Afanasiev, O., Afanasieva, A. V., Seraphin, H., & Gowreesunkar, V. G. (2018). A critical debate on the concept of ecological tourism: the Russian experience. In Maximiliano E Korstanje (Ed.) *Critical essays in tourism research*, Nove Science Publishers, New York. pp 129–148.

AGIC (Arizona Geographic Information Council). (2008). Springs NHD. Arizona State Land Department, Arizona Land Resources Information System, Phoenix, Arizona. Available from: https://www.gisinventory.net/GISI-8671-Hydrography.html

Akhmedenov, K. M. (2020). Assessment of the prospects of springs in western Kazakhstan for use in religious tourism. *GeoJournal of Tourism and Geosites*, *31*(3), 958–965.

ALRIS (Arizona Land Resources Information System). (1993). Springs. Arizona State Land Department, Arizona Land Re-sources Information System, Phoenix, Arizona. Available from: https://land.az.gov/sites/default/files/springs.htm

Altman, N. (2002). *The Spiritual Source of Life: Sacred Water*. New Jersey: Hidden Springs.

Anthwal, A., Gupta, N., Sharma, A., Anthwal, S., & Kim, K. H. (2010). Conserving biodiversity through traditional beliefs in sacred groves in Uttarakhand Himalaya, India. *Resources, Conservation and Recycling*, *54*(11), 962–971.

Atstāja, D., Brīvers, I., & Līviņa, A. (2011). National approaches to planning and tourism in Latvia. In Hongliang Yan, Nigel D. Morpeth (Ed.) *Planning for Tourism: Towards a Sustainable Future* Oxfordshire, UK: CABI. (pp. 222–239).

Berzina, I., & Livina, A. (2008). The model on estimating economic benefit of nature-based tourism services of territories of national parks, Latvia. In *4th International Conference on Educational Technologies* (pp. 100–105). Çorfu: Greece University.

Bhat, S. U., & Pandit, A. K. (2018). Hydrochemical characteristics of some typical freshwater springs-a case study of Kashmir Valley springs. *International Journal of Water Resources and Arid Environments*, *7*(1), 90–100.

Bhutiani, R., Khanna, D. R., Kulkarni, D. B., & Ruhela, M. (2016). Assessment of Ganga river ecosystem at Haridwar, Uttarakhand, India with reference to water quality indices. *Applied Water Science*, *6*(2), 107–113.

Bikse, Janis & Gavinolla, Mahender (2021) Water springs as a resource for nature tourism in Latvia: a tourist perspective. *Environment. Technologies. Resources. Proceedings of the International Scientific and Practical Conference*. Vol. 1, pp. 30–37.

Bódis, E., Tóth, B., & Sousa, R. (2016). Freshwater mollusk assemblages and habitat associations in the Danube River drainage, Hungary. *Aquatic Conservation*, *26*, 319–332.

Cantonati, M., Fensham, R. J., Stevens, L. E., Gerecke, R., Glazier, D. S., Goldscheider, N., Knight, R. L., Richardson, J. S., Springer, A. E., & Tockner, K. (2020). Urgent plea for global protection of springs. *Conservation Biology*, in press. DOI: 10.1111/cobi.13576

Cantonati, M., Feureder, L., Gerecke, R., Jeuttner, I., & Cox, E.J. (2012). Crenic habitats, hotpots for freshwater biodiversity conservation: toward an understanding of their ecology. *Freshwater Science*, *31*, 463–480. DOI: 10.1899/11-111.1

Carpenter, S.R., Stanley, E.H., Vander Zanden, J. (2011). State of the world's freshwater ecosystems: Physical, chemical, and biological changes. *Annual Review of Environment and Resources*, *36*, 75–99.

Chen, K. H., Liu, H. H., & Chang, F. H. (2013). Essential customer service factors and the segmentation of older visitors within wellness tourism based on hot springs hotels. *International Journal of Hospitality Management*, *35*, 122–132.

Collins-Kreiner, N. (2010). Researching pilgrimage: Continuity and transformations. *Annals of tourism research*, *37*(2), 440–456.

Coltelli, L. (1996). Leslie Marmon Silko's sacred water. *Studies in American Indian Literatures*, *8*, 21–29.

Cuthbert, M. O., & Ashley, G. M. (2014). A spring forward for hominin evolution in East Africa. *PLoS One*, *9*, e107358.

de la Cruz del Río-Rama, M., Maldonado-Erazo, C., & Álvarez-García, J. (2018) State of the art of research in the sector of thermalism, thalassotherapy and spa: A bibliometric analysis. *European Journal of Tourism Research*, *19*, 56–70.

Deng, J., King, B., & Bauer, T. (2002). Evaluating natural attractions for tourism. *Annals of tourism research*, *29*(2), 422–438.

Dook. (2019). Top tourist attractions in the Baltic States. https://www.dookinternational.com/blog/top-tourist-attractions-in-the-baltic-states/

Dou, Y., Zhen, L., De Groot, R., Du, B., & Yu, X. (2017). Assessing the importance of cultural ecosystem services in urban areas of Beijing municipality. *Ecosystem Services*, *24*, 79–90.

Druva-Druvaskalne, I., & Līviņa, A. (2014). Tourism in Latvia: From fragmented resorts of Russian empire to a national brand on international level. In C. Costa, E. Panyik, & D. Buhalis (Eds.), *European Tourism Planning and Organisation Systems* (Vol. II, pp. 118–130). Bristol/ Buffalo/ Toronto: Channel View Publications.

Eniņš, G. (2004). Alas Latvijā Rīga, Zvaigzne ABC, 76.

European Commission. (2013). Interpretation manual of European habitats, EUR 28. European Commission DG Environment Nature ENV B.3, Bruxelles/Brussel Belgium. https://ec.europa.eu/environment/nature/legislation/habitatsdirective/docs/Int_Manual_EU28.pdf

Foley, R. (2011). Performing health in place: The holy well as a therapeutic assemblage. *Health & Place*, *17*, 470–479. DOI:10.1016/j.healthplace.2010.11.014

Fridleifsson, I. B. (2001). Geothermal energy for the benefit of the people. *Renewable and Sustainable Energy Reviews*, *5*(3), 299–312.

Gavinolla, M. R., Kaushal, V., Livina, A., Swain, S. K., & Kumar, H. (2021). Sustainable consumption and production of wildlife tourism in Indian tiger reserves: A critical analysis. *Worldwide Hospitality and Tourism Themes*. *13*(1), 95–108. https://doi.org/10.1108/WHATT-08-2020-0091

Giuşcă, M. (2020). Religious tourism and pilgrimage at Prislop Monastery, Romania: Motivations, faith and perceptions. *Human Geographies*, *14*(1), 149–167.

Gowreesunkar, G. V., & Gavinolla, M. R. (2020). Urbanism and overtourism: Impacts and implications for the city of Hyderabad. In A. M. Morrison & J. A. Coca-Stefaniak (Eds.), *Routledge Handbook of Tourism Cities* (pp. 101–120). Routledge: London. doi:10.4324/9780429244605-7

Griggs, P. (2013). 'Taking the waters': mineral springs, artesian bores and health tourism in Queensland, 1870–1950. *Queensland Review*, *20*, 157–173.

Gupta A, Kulkarni H (2017) Inventory and revival of springs in Himalayas for water security. Report of the NITI Aayog Working Group as part of initiatives on Sustainable Development of Mountains of Indian Himalayan Region https://niti.gov.in/writereaddata/files/document_publication/doc1.pdf

Haberman, D. L. (2006). *River of Love in an Age of Pollution: The Yamuna River of Northern India*. Berkeley: University of California Press.

Hannaford, J., & Newton, J. (2008). Sacrifice, grief and the sacred at the contemporary 'secular' pilgrimage to Gallipoli. *Borderlands e-journal*, *7*(1). Retrieved from. http://www.borderlands.net.au/volv7no1_2008/hannafordnewton_gallipoli.htm

Haynes, C. V. (2008). Quaternary caldron springs as paleoecological archives. In L. E. Stevens, & V. J. Meretsky (Eds.), *Aridland Springs in North America: Ecology and Conservation*. University of Arizona Press, Tucson. (pp. 76–97).

Heervagen, L. (1864). Vella ceplis. Mājas viesis. 13.janv. (Nr.2) 15.-16.lpp.

Hershler, R., Liu, H. P., & Howard, J. (2014). Springsnails: A new conservation focus in western North America. *BioScience*, *64*(8), 693–700.

Jansen-Verbeke, M. (2007). Cultural resources and the tourismification of territories. The tourism research agenda: Navigating with a compass. *Acta Turistica Nova*, *1*(1), 21–41.

Johnson, R. H., DeWitt, E., Wirt, L., Manning, A. H., & Hunt, A. G. (2012). Using geochemistry to identify the source of groundwater to Montezuma Well, a natural spring in Central Arizona, USA: part 2. *Environmental Earth Sciences*, *67*(6), 1837–1853.

Juško-Štekele, A., (2014). The concept of pilgrimage in the culture of Latgale. *Via Latgalica*, (6), 20.

Kiełczawa, B. (2018). Short history of thermal healing bathing. In Jochen Bundschuh, and Barbara Tomaszewska (Eds.), *Geothermal Water Management* (pp. 303–318). CRC Press: Boca Raton, Florida.

Klempe, H. (2015). The hydrogeological and cultural background for two sacred springs, Bø, Telemark County, Norway. *Quaternary International*, *368*, 31–42.

Kumazawa, T. (2016). Human behaviour in open space around spring water in a central area of Mito-City in Japan. *Planning Malaysia*, *14*(4).

Laime, S. (2009). *Svētā pazeme*. Rīga: Zinātne, p. 474.

Lee, C. F., Ou, W. M., & Huang, H. I. (2009). A study of destination attractiveness through domestic visitors' perspectives: The case of Taiwan's hot springs tourism sector. *Asia Pacific Journal of Tourism Research*, *14*(1), 17–38.

Livina, A., & Atstaja, D. (2015) Understanding the philosophy and performance of tourism and leisure in protected areas for transition to a green economy. In Reddy, M.V. and Wilkes, K. (Eds), *Tourismin the Green Economy*, Routledge, ISBN 978-0-415-70921-7, pp. 71–86.

Livina, A., & Reddy, M. (2017). Nature park as a resource for nature based tourism. In *Environment. Technologies. Resources. Proceedings of the International Scientific and Practical Conference* (Vol. 1, pp. 179–183).

Margry, P. J. (2008). Secular pilgrimage: A contradiction in terms? In P. J. Margry (Ed.), *Shrines and Pilgrimage in the Modern World: New Itineraries into the Sacred* Amsterdam University Press: Amsterdam.

McKinney M. L. (2008). Effects of urbanization on species richness: A review of plants and animals. *Urban Ecosystem*, *11*, 161–176. DOI: 10.1007/s11252-007-0045-4

Nagaranjan, V. R. (1998). The earth as Goddess Bhu Devi: Towards a theory of embedded ecologies in folk Hinduism. In L. E. Nelson (Ed.), *Purifying the Earthly Body of God: Religion and Ecology in Hindu India* (pp. 269–296). Albany: State University of New York.

Nielson, K. G., Gill, K. M., Springer, A. E., Ledbetter, J. D., Stevens, L. E., & Rood, S. B. (2019). Springs ecosystems: Vulnerable ecological islands where environmental conditions, life history traits, and human disturbance facilitate non-native plant invasions. *Biological Invasions*, *21*(9), 2963–2981.

Pyšek, P., Jarošík, V., Hulme, P. E., et al. (2010) Disentangling the role of environmental and human pressures on biological invasions across Europe. *Proc Natl Acad Sci USA*, *107*, 12157–12162. DOI: 10.1073/pnas.1002314107

Rohman, F., Priambodo, B., & Akhsani, F. (2021). Evaluation of ecosystem services based on amphibian community of the seven water springs in Malang Indonesia. In *AIP Conference Proceedings* (Vol. 2353, No. 1, p. 030007). AIP Publishing LLC.

Rood, S. B., Samuelson, G. M., Weber, J. K., & Wywrot, K. A. (2005). Twentieth century decline in stream flows from the hydrographic apex of North America. *Journal of Hydrology*, *306*, 215–233.

Séraphin, H., & Gowreesunkar, V. (2019). What marketing strategy for destinations with a negative image?. *Worldwide Hospitality and Tourism Themes*, *11*(2).

Séraphin, H., Gowreesunkar, V., Zaman, M., & Lorey, T. (2019). Limitations of trexit (tourism exit) as a solution to overtourism. *Worldwide Hospitality and Tourism Themes*, *11*(5), 566–581. DOI: 10.1108/WHATT-06-2019-0037

Shrestha, R. B., Jayesh, D., Mukherji, A., Madhav, D., Kulkarni, H., Mahamuni, K., Bhuchar. S., & Bajrachary, S. (2018). *ICIMOD Manual. Protocol for Reviving Springs in the Hindu Kush Himalaya: A Practitioner's Manual*. p. 74.

Simanjuntak, B. H. (2005). Study of forest land use change to farming land use towards soil physical characteristic (case study of Kali Tundo Watershed, Malang). *AGRIC*, *18*(1), 85–101.

Smith, M. (2015). Baltic health tourism: Uniqueness and commonalities. *Scandinavian Journal of Hospitality and Tourism*, *15*(4), 357–379.

Smith, M. K., & Puczkó, L. (2016). Balneology and health tourism. In *The Routledge Handbook of Health Tourism* (pp. 299–310). Routledge.

Šnepste, E. (2014). Aglonas reliģisko svētku ietekmes novērtējums (Impact assessment of the Aglona religious festival). Master's Thesis, Turība.

Springer, A. E., & Stevens, L. E. (2009). Spheres of discharge of springs. *Hydrogeology Journal*, *17*(1), 83–93.

Springer, A. E., Stevens, L. E., Ledbetter, J. D., Schaller, E. M., Gill, K. M., & Rood, S. B. (2015). Ecohydrology and stewardship of Alberta springs ecosystems. *Ecohydrology*, *8*(5), 896–910.

Stevens, L. E., & Meretsky, V. J. (Eds.) (2008). *Aridland Springs in North America: Ecology and Conservation*. Tucson: University of Arizona Press.

Stevens, L. E., Schenk, E. R., & Springer, A. E. (2021). Springs ecosystem classification. *Ecological Applications*, *31*(1), e2218.

Stevens, L. E., Springer, A. E., Ledbetter, J. D. (2016) Springs ecosystem inventory protocols. Springs Stewardship Institute, Museum of Northern Arizona, Flagstaff, Arizona. http://docs.springstewardship.org/PDF/ProtocolsBook.pdf. Accessed 25 Nov 2018.

Tabacchi, M. H. (2010). Current research and events in the spa industry. *Cornell Hospitality Quarterly*, *51*(1), 102–117.

Tabayashi, A. (2010). Regional development owing to the commodification of rural spaces in Japan. *Geographical Review of Japan Series B*, *82*(2), 103–125.

Taff, B. D., Benfield, J., Miller, Z. D., D'antonio, A., & Schwartz, F. (2019). The role of tourism impacts on cultural ecosystem services. *Environments*, *6*(4), 43.

Tambovceva, T., Atstaja, D., Tereshina, M., Uvarova, I., & Livina, A. (2020). Sustainability challenges and drivers of cross-border greenway tourism in rural areas. *Sustainability*, *12*(15), 5927.

Tiwari, P. C. (2000). Land-use changes in Himalaya and their impact on the plains ecosystem: Need for sustainable land use. *Land Use Policy*, *17*(2), 101–111. DOI: 10.1016/S0264-8377(00)00002-8

Torraco, R. J. (2016). Writing integrative literature reviews: Using the past and present to explore the future. *Human Resource Development Review*, *15*(4), 404–428.

Urtāns Guntis Gūtmaņa ala. (2008). senākais tūrisma objekts Latvijā. Eiropas kultūras mantojuma dienas 2008. Neparastais mantojums. Rīga: Valsts kultūras pieminekļu aizsardzības inspekcija. 54.-55. Lpp.

Vaeliverronen, L., Kruzmetra, Z., Livina, A., Grinfilde, I., & Esaf, S. (2017). Engagement of local communities in conservation of cultural heritage in depopulated rural areas in Latvia. *International Journal of Cultural Heritage*, *2*, 13–21.

Willis, C. (2015). The contribution of cultural ecosystem services to understanding the tourism–nature–wellbeing nexus. *Journal of Outdoor Recreation and Tourism*, *10*, 38–43.

21 Tourism Ecosystem Services for MSMEs Post-COVID

The Case of Mauritius

Michael Pompeia

21.1 Introduction

With the advent of the COVID-19 pandemic, tourism players are facing unprecedented turmoil and have to reinvent their offerings, especially the Micro, Small and Medium Enterprises (MSMEs). Worldwide, this category of business has been acknowledged as the backbone of many economies. Those small businesses in the tourism sector had to face a plethora of challenges in recent times, even before the pandemic. As a result, some small actors had to temporarily or permanently stop their activities, affecting their turnover, profitability and reserve. These MSMEs in the hospitality sector have to review their offerings to continue finding in the coastal regions opportunities to start a small business (Gowreesunkar et al., 2014). By nature of their size, these economic players can readapt rapidly and work more inclusively to be resilient.

Similarly, other sectors have been able to cope and introduce new ways to operate, such as working from home or on a rotation. Yet, according to Smart et al. (2021) and Gandasari and Dwidienawati (2020), MSMEs in the tourism sector have to devise new ways of operation. In addition, several other endogenous factors such as economic recession and climate change are making the challenges for MSMEs in this sector fiercer as Small Island Developing States (SIDS) such as the Maldives, Seychelles and Mauritius are more exposed to the negative effect of climate change. As per (Gowreesunkar et al., 2014), these SIDS depend greatly on the tourism sector to generate opportunities for the local community. Gowreesunkar and Séraphin (2016) stressed that MSMEs are more reactive and work on a firefighting mode rather than being proactive and foresee potentially emerging storms on their business journey.

Therefore, the COVID-19 pandemic is an opportunity for the MSMEs in the tourism industry to reinvent their operating structure by implementing innovative, inclusive, sustainable, and socially responsible practices to enhance economic performance (Kim, Periyayya, & Li, 2013). Since it appears that travelling will continue even in financial worries as it needs self-fulfillment (Séraphin & Gowreesunkar, 2017), the small actors should be ready to tap other potential markets. MSMEs can instil a new mode of doing business based on putting the people and the environment first, especially for

DOI: 10.4324/b23145-27

SIDS. Small actors can focus on their competitive advantage within the value chain integration and sustainability through this strategy. Moreover, to cope with this paradigm shift, specific core aspects such as co-competition based on the respect of the ecosystem will need to be reinforced where necessary. In contrast, actors in the tourism sector depend on other MSMEs activities to satisfy the ever-changing tourist's needs. So, it is important to delight tourists by modifying the service delivery of MSMEs in the hospitality sector (Sirieix & Remaud 2010; Sarkar 2012).

Similarly, tourists want to know the impact their activity will have on the environment. Furthermore, it is also an indication of a strong commitment to Sustainable Development Goal (SDGs) 13, which is about climate change action. The model to be explored want to adopt sustainable practices by MSMEs. Therefore, this opportunity represents a potential benefit for the MSMEs and the local population (Bricker et al., 2012). Adhering to SDGs 13 proposed by the United Nations (UN) is more than a must, especially after the COVID-19 pandemic, as it helps MSMEs generate a healthy profit through the implementation of sustainable practices, which will reflect a potential competitive advantage within a small niche market and may help MSMEs to act as a '*good corporate citizen*' (George et al., 2016).

Nevertheless, MSMEs in this sector are working in a disconnected and haphazard way despite all the players being interdependent and working for the same objectives. A paucity of research exploring a practical operational model to bring all actors together has been noted. Due to the growing competition in the tourism sector, the destination must think out of the box and provide concepts that outweigh the basic offerings of destinations such as sea, sun and sand. This shall respond to the increasing debate about environmental issues focusing on sustainable living and environmental preservation. Due to their malleability, MSMEs are more appropriate to bring change and integrate further the concept of sustainability in their product offering. Therefore, the purpose of this chapter is to explore the suitability of an integrated business model, which promotes inclusiveness among MSMEs in the tourism sector. More than before, MSMEs in the same value chain should work together and support each other to provide tourists with a unique experience, and this at the same time is in line with SDGs 13 and 17.

21.2 Literature Review

21.2.1 Tourism Ecosystem Post Pandemic

According to Becker (2020), the tourism sector has witnessed the biggest slump ever experienced since its existence. The tourism players find ways to put their businesses back on track while some have completely phased out. Therefore, as per Lew et al. (2020), MSMEs in the tourism sector are reorganising their business structure by taking into account the latest standards and rules. MSMEs in the tourism sector are connected with MSMEs in the agricultural, handicraft, transportation and manufacturing sectors, among others. The tourism players are working on strategies to shape and define the tourism product. Government and financial institutions of different jurisdictions are assisting these players regarding grants and moratoriums on loans.

There has also been a change in the traveller's perception and demand. Travellers are becoming more responsible for their actions towards the environment. They want to create a positive footprint on the local community and the environment based on social, economic and environmental factors. Tourists are moving towards destinations that promote sustainable and equitable tourism as their actions significantly impact the environment. In addition, states are imposing new rules and regulations concerning travellers. Some jurisdictions ask for 14-day isolation, while others are asking for travellers to be fully vaccinated. Also, some countries have restricted access to travellers that come from high Covid-risk destinations.

21.2.2 Value Addition in the Tourism Sector

The travel experience consists of a series of activities experienced by tourists provided by multiple different entities (Rahmiati et al., 2018) in the value chain. The tourism value chain shows several sources of competitive advantage from the ability to create and effectively manage all players in the tourism industry, supported by the ability of local governance to guarantee to push the attractiveness of the area and differentiate it from competitors (Rahmiati et al., 2020). Having a proper design value chain will allow tourism sector actors to effectively respond to the actual COVID-19 period and boost their image vis-a-vis the society by contributing to the preservation of employment. It is important that hotels in Mauritius implement a sustainable value chain and ensure that each interconnected actor creates value in their action.

The hospitality sector in Mauritius has known a downward trend due to the COVID-19 pandemic, with the closure of several related businesses. However, some players of this sector have taken it as an opportunity to rethink and readapt their value chain to the actual COVID-19 situations. In Mauritius, there exist a divergence between the economic, socio-cultural and environmental dimensions. Therefore, MSMEs need to build their value chain so that the above-mentioned dimensions are met and that as many people as possible, particularly the disadvantaged ones, benefit from value-added activities. In addition, it is also vital for the MSMEs to make sure that the value-added activities do not damage the environment.

A properly designed value chain will positively impact on various indicators such as the employment level and poverty level (Goh & Kong, 2018). MSMEs are compelled to innovate and to differentiate themselves from their competitors while also being efficient, as they do not have the economies of scale possessed by larger organisations. Operations management in the hospitality sector changes with market conditions, prompting enterprises to adjust their business strategies. In facilitating a different travel experience, the tourism industry requires tourism service providers to collaborate, coordinate, and integrate into creating the best tourism products and services

(Yilmaz & Bititci, 2006). This will answer the changing demand of customers and adapt to their consumption patterns (Christian et al., 2011).

21.2.3 Inclusiveness in the Tourism Business Model

As mentioned above, MSMEs in the tourism sector contributes to the local economy through the linkage and inclusiveness with another economic operator. Moreover, due to the pandemic, these economic players must partner themselves to leverage strategic partnerships, thus creating a mutually beneficial output. This solid partnering will allow MSMEs to provide high-quality products to tourists while bringing a pool of innovative ideas into their business.

In addition, this will help boost the reputation of the business within the local community, among the guests and the government authorities. Being able to partner with the local producer for the raw materials will allow the tourism player to lower their cost of production while contributing to the reduction of carbon emissions in the air. In addition, the local community will benefit from employment creation since these MSMEs support the local actors of the economy instead of importing part of their resources. However, this is not always the case in SIDS. There is not a proper organisation between the different economic players. The COVID-19 pandemic has unveiled the importance of inclusiveness in the hospitality sector. The inclusiveness will help in the smooth running of a circular economy in which each actor in that particular model would benefit mutually.

21.2.4 Tourism Industry in Mauritius

Mauritius is part of the Mascarenes region and has a multi-ethnic population of approximately 1.3 million inhabitants and has an overall population density of 641 persons per square km based on 2016 and census carried out by the local Central Statistical Office (CSO) with a lineage of immigrants from India, Europe, China and Africa. The tourism sector contributes more than 20% (Approximately US$ 2.5 million) to Gross Domestic Product (GDP) and employs more than 25% of the country's formal workforce (WTTC, 2018). The majority of these figures are due to large operators' contribution and a percentage of it from actors, defined as hotels offering less than 50 rooms and employing fewer than ten staff (Nachmias & Walmsley, 2015) having no well-defined target market. The tourism industry in Mauritius is characterised by below-average company size, low growth rates, weak internationalisation, relatively low market entry barriers and relatively poor qualification levels – all of which have significant implications for the management and competitiveness of the hotels that dominate the industry.

As Mauritius Island mostly uses the coastal landscapes as a core touristic appeal, there is a reliance on the basic 3'S (sun, sea, and sand). Thus, it is of utmost importance for the protection of our coastal landscapes. Especially in recent years, climate change has negatively impacted the environmental, social and economic levels (Herrera-Cano & Gonzalez-Perez, 2019). As a SIDS, Mauritius also has a fragile ecosystem and limited natural resources, thus directly impacting climate change. In the past 20 years, Mauritius has started encountering:

- deforestation due to economic development,
- degradation of our coastal landscape, and
- water pollution due to high utilisation of pesticides and chemical.

Therefore, climate action has become a priority in the 2030 Development goals. In addition, MSMEs in this sector provide more personalised services to the customers despite a number of limitations and constraints that these businesses have to face (Haroon & Hoq, 2012; Rylance & Spenceley, 2016), especially due to their size.

21.2.4.1 Importance of Handicraft Sector in the Hospitality Sector

The Mauritius handicraft and locally made products sector for small- and medium-sized businesses have seen years of decline due to policies favouring cheap imported products over local products, a lack of support in developing entrepreneurship, and difficulty accessing the market on the island. The handicraft sector, characterised by authenticity and particular know-how, is facing several constraints at the level of production, design and marketing. This sector has suffered tremendously from the domino effect that the COVID-19 pandemic has had on the tourism industry. With the support of Mauritius's authorities, this sector has adapted its offerings by providing genuine local products made from locally available materials. Furthermore, many of the operators are adopting a greener approach in their production.

Though there have been several incentives from the authorities to promote the handicraft sector, it still has tremendous potential. A high dominance of women characterises this sector and various reasons can explain this. It represents an additional revenue to the family budget for most of them and is done mainly as a home-based activity. Several government-led initiatives have helped to empower those subsistence businesses through training and grant facilities. Furthermore, with the closure of many textile factories in the past decade, while the tourism industry was booming, many chose to go towards handicraft production. There are many reasons for this move, among which is that activity does not require heavy capital investment or infrastructure such as machinery, buildings and power (Dar & Parrey, 2013). The handicraft sector is also a good source of cultural and ecotourism, opening many job opportunities. In addition, it is considered as one of the main pillars of tourism development.

21.2.4.2 Constant Supply of Fresh Vegetables to Small Hotels

Throughout the years, the food habits of Mauritian consumers have shifted towards processed and convenience foods, with an exigency on quality, food safety and brands. This condition is highly related to the hospitality sector, where the local people want access to the same food quality as tourists. In this context, in 2018, the authorities came up with the Sustainable Diversified Agri-food Sector Strategy to ensure food security to feed both the locals and visitors. The main objectives of this programme are to improve food quality and safety through the adoption of good agricultural and management practices and certification by local vegetable producers. Feeding the local population of 1.26 million and 1.45 million visitors represents a true challenge.

The net food requirement is estimated to be at 690,000 tons annually, out of which 75% is made up of agricultural and food products imports. Some 8,000 small farmers and 375 hydroponic producers (often termed Agripreneurs) operate in the food crop sector cultivating food crops and fruits on a total area of 7,646 ha with a total annual production of 96,847 tons (Statistics, 2019). In this case, fruit production consists mainly of banana and pineapple, which are grown all year round in some established commercial orchards, and seasonal fruits, such as litchi and mango. On the other hand, it should be highlighted that fruit production relies heavily on backyard production. However, food waste is among the most prominent type of hospitality waste (Gössling et al., 2011; Wan et al., 2017; Filimonau & De Coteau, 2019).

Constant supply of fresh vegetables and fruits to MSMEs in the tourism sector is of significant importance in the daily operation of those businesses. Therefore, hotels should work in close collaboration with the suppliers in view of determining ways through which the waste collected can be recycled and reused. Again, the proposed model below explains how the hotels integrate the vegetables and fruits supplier into their value chain to be more sustainable and competitive.

21.3 Methodology

The method used for developing this business model was based on an extensive literature review, face-to-face interviews, and focus groups with different Mauritian's hospitality sector actors. The focus group method was used with 24 participants, among which 12 MSMEs were involved in the handicraft sector dealing with tourists and 12 others in different sectors of activities such as transportation, leisure activities and tour operators. This extensive panel was representative of the main actors in the tourism sector. Their opinions were gathered on the implications of having a sustainable business model composed mainly of MSMEs in the value addition chain. This qualitative instrument was favoured because it renders the interactions between participants possible. Furthermore, it allows a better understanding of the themes (Saunders, 2012) and possible changes needed to enhance the service provided.

Due to the unavailability of those in the agricultural sector for the focus group, this supplier category was interviewed via unstructured face-to-face interviews. This same research technique was adopted for other stakeholders, including one representative from the tourism sector association and government agencies supporting the tourist sector. The perception of the local population on having a sustainable business model importance within the tourism industry was captured through a survey carried out online via a google form. The focus group was mainly to understand the actors in the value chain and the dependency on each other. Different themes emerged from the focus group and these themes were further discussed during the face-to-face interview.

For both the focus group and online survey, a pilot study was carried out before administration. The purpose of the pilot study was to check the flow of the questions and if there was any misunderstanding regarding the preset questions. For the focus group, all questions were maintained, whereas questions 7 and 9 were modified for the online survey. The main limitations of the present study are that the research includes a modest sample size. Therefore, for the purposes of this research, there was a greater emphasis on targeting a modest number of key respondents who could provide expert opinions on the topic rather than a more extensive sample size of lower quality (Pompeia, 2021).

21.4 Findings and Discussion

Knowing the customers is key in the service sector as this help business to meet their demand better. All the participants for the Focus Group (FG) agreed that there had been a change in the number of tourists visiting the island during recent years. One of the main themes that emerge is the importance of offering an authentic experience to the tourist visiting the destination. Today tourists want to share the life of the locals and go deep into their culture. Furthermore, travellers are willing to pay more for certified offerings which often symbolise a higher quality of service (Karlsson & Dolnicar, 2016). So, to further differentiate the destination from other island destinations, more emphasis can be made to make shopping another important 'S' of tourism and this can be with the help of local artisans. Genuine local products can be proposed in-house of those MSMEs sharing a common brand. This represents an opportunity for MSMEs in this sector, especially in such a context of high competition.

Most of them concur that more tourists are interested in the green concept and want a higher degree of authenticity. Tourists are demanding more experience with the local people, not just the sun, sea and sand. All these are expected to be offered with a more environmentally friendly approach. From all over the globe, travellers are showing great concern about the impact of their activities on the environment, according to the majority of FG participants. Consumers are highly conscious that any activity affects three main pillars: economic, environmental, and social. These pillars are fundamental to business resilience and sustainability, especially in the tourism sector. This is in line with Lacy et al. (2014), Bocken et al. (2014), Reichel et al. (2016), and Korhonen et al. (2018), enforcing that sustainability should be applied in almost all sectors of the economy where natural resources are involved. It was highlighted that when the service providers operate within a closed cycle and use renewable energy for their operations, the impact will positively impact the environment and financial health. Thus, having MSMEs compete against each other will lead to a zero-sum game.

Therefore, connecting and making the tourism ecosystem work together is vital to reach a common social goal of providing a sustainable delivery to the customers. The participants emphasised that compared to large enterprises, the MSMEs struggled for capital requirements, research and development, human resource demands and information costs (Iordache et al., 2010; Thomas et al., 2011), and working together would help them tremendously. The majority of the participants (92%) highlighted that MSMEs would both benefit economies of scale and protect the environment by working together. With an integrated approach, MSMEs will also have a constant flow of work and cash, apart from the environmental impact. In such conditions, the economic viability of the small businesses engaged in the tourism sector will be higher. These MSMEs will better forecast their sales and plan material purchases and other economic activities relevant to the business. Therefore, the present subsistence state may evolve to a business model which is solid and enhance growth. Clustering will provide the MSMEs with a marketing tool since competition will be tougher post-COVID-19 as all destinations will develop new strategies to attract more tourists. This will encourage knowledge transfer, tourist education, communication, development of new cultural values, early development of small enterprises, cooperative activities, fostering a common purpose and focus, community support for a destination, enhanced product quality and visitor experience, and more repeat business. Within a conducive environment, MSMEs will become more competitive and may even think about targeting foreign markets. These changes give them new perspectives on their lives while giving MSMEs new resources to address their difficulties. Furthermore, customers today want a unique experience and to have the feeling of the local people.

The circular economy (CE) model could inspire businesses in the tourism industry as it ensures business practices are done in an eco-friendly way. The participants supported the urge to develop a model inspired by the CE for MSMEs in the tourism industry. This will allow the MSMEs to focus on the optimum use of resources by reusing, reducing, repairing and recycling (Reichel et al., 2016) while being profitable. Thus, the participants listed the areas where possible collaboration could be engaged and the following scored among the top five:

(i) Handicraft production

Handicraft is not merely souvenirs to take back from a vacation destination; it portraits the talents of the local producers and the know-how on specific manufacturing techniques. The participants stated that the local products are unique to the destination and the tourism industry must capitalise on this.

(ii) Fresh agricultural produce

The advent of hotels in Mauritius has been a game-changer in producing quality vegetables to meet the request for a premium product.

Two decades ago, actors in the vegetable sector shifted from open field vegetable production to sheltered vegetable production. Furthermore, it is mainly to produce fresh salad tomatoes, cherish tomatoes and English cucumber, and primarily for the tourism industry. The vegetable producers depends largely on the hospitality operators and is an important part of the value chain. It is the main reason why the vegetable producers have been considered chain members and must work together to improve the competitiveness of local MSMEs in hospitality.

(iii) Leisure activities

Leisure activities are an important part of the offering of the accommodation service providers. Proposing leisure programs, combining physical activity, nature and social components, have proved to be very effective in enhancing well-being, self-esteem and positive mood in individuals (Loureiro & Veloso, 2017).

(iv) Repairs and maintenance

It is important that the hotel management team have a good preventative maintenance plan. This is so because this plan will allow the maintenance team to run the hotel smoothly and at the same time contribute to the increase in client satisfaction. In addition, it allows the hotel to cut costs by avoiding large expenses and allows for the general safety of the hotel and the staff. Furthermore, hotels with good repairs and maintenance teams will always benefit from positive word-of-mouth advertising from travellers.

(v) Transportation

As per Page and Connell (2014), transportation is a crucial element for the success of the tourism industry. This is so since it is a medium for the movement of goods and services to satisfy travellers' desires. It also plays an essential role as it provides accessibility to travellers from different parts of the world. This is one of the most important. It has been claimed by Currie and Falconer (2014) that if actors from the tourism industry want to increase their success level, they should innovate their modes and routes of transportation.

These five (5) components are key in the hospitality offering and greatly influence the customer's overall experience. All participants highlighted that the tourism product is a complex one as it consists of both tangible and intangible aspects from diverse suppliers. The tourism sector can help reduce poverty, foster development, and create employment opportunities (Rahmiati et al., 2020), especially for youth and women. The customers consume the basic tourism products such as transportation, accommodation during the trip, and excursions (Rahmiati et al., 2020). Those activities could involve purchasing tangible products such as souvenirs, foods and drinks, or consuming the scenery of natural tourism. Those activities impact on the intention of revisiting. Therefore, positive referrals enhance loyalty (Mendes et al., 2010) to the destination by providing a sustainable tourism product. Especially today, where customers are extremely conscious of the triple bottom line concept of sustainability, businesses need to tap on this opportunity to readapt their business model to fit their expectations and enhance the consumer experience. This characteristic makes it difficult to regroup all suppliers within one model. However, all the respondents believed it would add substantial value to the ecosystem by clustering the services offered.

Most participants highlighted that a business model for MSMEs must mitigate the tourism industry's negative effect of pushing for economic growth. The model must consider all the environmental impacts on economic activity since it contemplates design, transportation, consumption, recycling, and disposal. The benefits of an integrated model will be manifolded as it:

(i) will enhance an integrated sustainable model,
(ii) will help actors in the hospitality sectors to use the sustainable aspect as a marketing tool, and
(iii) promote an ethical business environment.

21.4.1 Conceptual Model

Thus, the proposed model connects the hospitality sector's main elements with accommodation at the centre. It proposes that tourists experience the difference from their usual frame of reference and the touristic experience needs to be educationally appropriate. MSMEs should tell the tourist what the destination is and drive how customers perceive it (Valle et al., 2012). The impact of such a model will be manifold and, among others, counter competition from neighbouring countries.

No stone should be left unturned. Even relatively new hospitality concepts have been reviewed as a simple proposition that combines economic and ecological benefits for travellers and tourist areas. All players in the tourism chain need to work together and create value to deliver satisfied tourism products and services. So, the model aims to increase the economic and ecological well-being of actors in the hospitality sector. It abides by the findings of Bocken et al. (2014) that a a CE model should mainly involve the:

(i) maximisation of energy efficiency,
(ii) value creation from waste,
(iii) movement from non-renewable to renewable resources,
(iv) maximisation of ethical trading, and
(v) leverage of education to mitigate unsustainable consumption.

Therefore, real solutions can help actors in the hospitality sector, especially SIDS, achieve their major concerns. Furthermore, this model gives MSMEs a distinct competitive disadvantage, making forming close networks among MSMEs critically important to their survival and growth. The proposed model in this chapter is inspired by existing concepts such as the circular economy (CE). It contributes to interconnections between firms, suppliers and related institutions within a geographical boundary, resulting in knowledge spillover based on geographical proximity (Porter, 1990) and greater ability to innovate (García-Villaverde et al., 2017) and compete. Furthermore, MSMEs adopting this approach may use it as a communication cue showing during their advertising campaign.

21.4.1.1 Description of the Proposed Model

The application of the model shown in Figure 21.1 aims to have the MSMEs in the different sectors work together and be a real economic growth engine for SIDS. The model tends to incite small tourism operators to be socially responsible and, at the same time, profitable. As per Korhonen et al. (2018), the flow and connectivity between people can create additional value to the entrepreneurs' business. The above model encourages the procurement of all consumables such as soap and shampoo in rooms from local producers. It is expected that the producers use locally available raw materials, for instance, local fruits and other endemic herbs, for the above examples used. The accommodation will need to guide the producers about their customers' preferred products and profile to answer a specific need.

Moreover, all empty containers must be returned to the producers to reduce production costs and promote an environmentally friendly business ecosystem. Thus on this aspect, the model is adopting the basic principle of a circular economy with lesser waste. The proposed model will help to achieve relatively high differentiation and relatively low cost for business continuity. There is wide recognition of the need to go back to nature and manipulate the earth for all its benefits, namely those related to the body and mental health (Loureiro & Veloso, 2017).

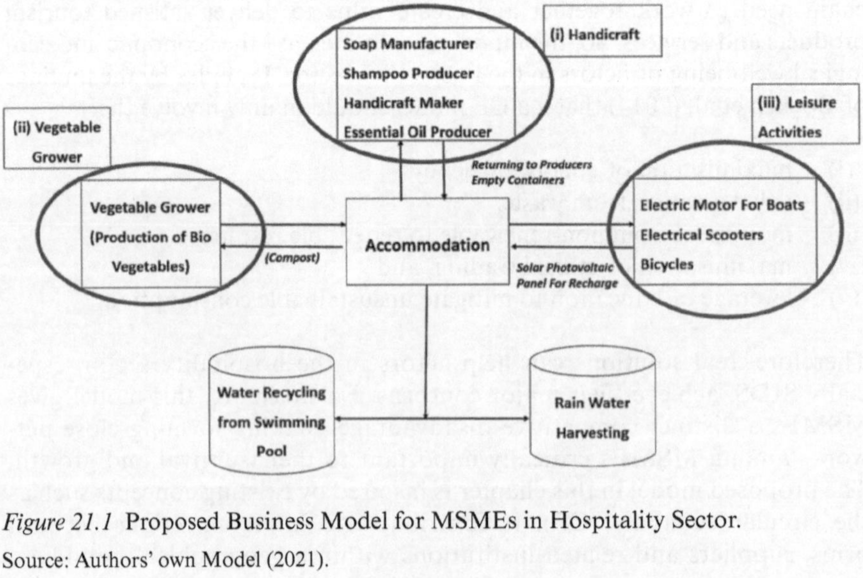

Figure 21.1 Proposed Business Model for MSMEs in Hospitality Sector.
Source: Authors' own Model (2021).

Vegetable growers within the business model will have the primary mission of producing organic vegetables. This is a growing concern for most people in the world of having chemical-free products to eat. The benefits of this offering are manifold, with growers protecting their immediate environment with fewer chemical products and the accommodations will be offering safe vegetables to the visiting tourist. The actual concept can be enhanced by allowing tourists to the field where the vegetables are grown. Tourists may also be given the opportunity to harvest vegetables, giving a unique experience, especially for these clients living in big cities with no access to the vegetable growing field. It is a good example of food traceability. The business knows exactly the route from the farm to fork and more importantly, the customers have even participated in the process. The vegetable growers can also organise other activities, for example, for kids, they may plan workshops on specific terms such as land preparation, seed sowing, irrigation and other related tasks. Such offering is becoming crucial since most youngsters are screen-addicted worldwide, and these on-field workshops will allow them to get away from their habit.

The present model includes vegetable producers to produce superior value vegetables respecting the eco-friendly, which will add value to the delivery to the end consumer (Koenig-Lewis et al., 2014). The accommodation will also promote sustainable vegetable growing through the supply of compost. It is commonly agreed that most food wastes arise from tourism activity and represents a major environmental and societal issue (Gretzel et al., 2019). This is related to the fact that there is generally a tendency for oversized food portions during the regular buffets with various fruits and vegetables with the pertaining holiday moods. This condition leads to food waste, and it is estimated that a third of food produced globally is lost or wasted (Blakeney, 2019). On this note, a growing number of consumers are becoming more ethically and environmentally conscious (Massa & Testa, 2012). Thus, MSMEs in the hospitality sector may set the pace have taken all their organic waste to be converted into compost, which will be given back to the vegetable growers. In this area again a workshop can be organised explaining to the visitors the process of compost making, which is of great interest nowadays.

The MSMEs part of this integrated model will become more resilient and contribute a higher proportion to the national economy (Hassan & Azman, 2014). Moreover, production will be demand-led and producers have to meet the required quality standard. In addition, dealing directly with the suppliers will reduce intermediaries meaning a higher profit margin for the entrepreneurs and a more competitive price for tourists. Being involved in a green business ecosystem will force local entrepreneurs to be more responsible in their production. This will encourage the use of environmentally friendly materials and equipment during the manufacturing process. The MSMEs skills will help to enrich the island's cultural heritage and characterise authenticity. An additional element is added in the actual model, namely, using eco-friendly equipment and green energy to address environmental issues. Being an island, Mauritius offers tourists a wide range of sea activities using fossil fuels. These fuels are used by boats to carry tourists to visit islets, parasailing, snorkelling and other nautical activities and causes pollution. To complete the green experience of the customers, the model encourages strict eco-friendly practices. For instance, most boats should be replaced by electrical engines. However, to adopt this measure and make it efficient, the accommodations should install a solar farm. It is then only that this measure will be complete as using electricity production from fossil fuel will defeat the purpose as pollution at source will be same or even higher. Therefore, having an individual solar farm per accommodation will help to increase energy consumption from a clean source and reduce running costs. Other activities proposed can make use of a clean source of energy. Another way to reduce wastage is to preserve water. For example, hotels can develop nautical activities using captured rainwater. In addition, rainwater harvesting can be used at places where non-drinkable water is needed. For example, to fill artificial lakes or ponds and allow marine life to grow. This will reduce water consumption associated with external use, thus saving water. In these artificial ponds, tourists may fish and kids may have water paddling activities safely without polluting water.

21.5 Recommendations for the Implementation of the Model

The government in many circumstances provides support schemes for the small actors to grow and sustain. However, MSMEs often claim that there are many laudable initiatives on paper, but it is quite difficult to have access and put all these into practice. In this context, a national certificate should be available for all those involved in such an integrated approach to obtain existing support rapidly. Furthermore, the role to be played by the relevant authorities should be clearly known to all small actors so that all can use these supports. Below are the guidelines to be taken on board to ensure a smooth integration of the MSMEs for the responsible tourism industry.

21.5.1 Quality Standards for All Stakeholders in the Model

The first element that all key actors should agree on is the quality of the product. Small producers quite regularly cannot meet quality standards or cannot supply products of consistent standards. This represents a potential pitfall for the proposed model and support to achieve a certain quality standard should be given. All those involved in the integration process must be aligned on the same standards and take all necessary precautions to reach the benchmark agreed. Furthermore, the quality standards to be imposed should allow the small operators to innovate within a conducive environment that is not too rigid.

21.5.2 Quota for Procurement

By nature of their activity, small players tend to employ mainly members of the family to help. But in the case of integration, all parties should agree on a quota to be supplied. This will help the handicraft supplier plan the production, which is essential for sourcing raw materials. The quota will also allow the small producer to make a forecast based on solid ground and manage their micro-enterprises properly. Nevertheless, regulatory bodies need to act as watchdogs to create a level playing field to avoid favouritism.

21.5.3 Green Loan at Low Interest

To meet demand and ensure a constant supply, a transformation of the small manufacturer will be required. MSMEs generally complain that in terms of procedures, it is very tedious to obtain a loan. In this context, special Loan incentives could be put at their disposal with the following characteristics:

(i) Low-interest rate around 1% to 2%
(ii) Smoother procedures for application
(iii) Rapid disbursement of the loan with

21.5.4 Other Schemes for Purchase of Green Equipment

Small actors are invited to propose authentic products produced using greener equipment and clean energy sources where applicable. Thus, the authorities should develop a series of schemes to encourage businesses to adopt greener sources such as energy solar. For example, rebate could be given to purchase solar panels, and tax-paying companies may allow tax holidays when implementing these projects.

21.5.5 Provision of Compost-Making Machines

Since the accommodation services provider will convert all their organic waste into compost for the vegetable growers, provision must be made for the supply of compost-making machines. These machines should be movable and free from odour within the hotel compound. In addition, these compost-making machines must be accessible for the tourist to visit if necessary.

21.5.6 Creation of an Association for the Implementation of the Model

A roadmap should be established for SMEs by an association in the hospitality sector, where the positioning of SMEs in this journey should be defined and improved accordingly. As SMEs move forward on the roadmap, additional demographic and behavioural data is expected to be added to the integration, even that private sectors predominantly control the tourism sector. However, the government regulates the industry (Baum et al., 1997) and acts as a watchdog to create a level playing field for all operators, such as creating the legal environment and supporting schemes for the small actors.

21.5.7 Get Everyone Involved

The management should inculcate a culture where the staff and the guest incorporate energy-saving practices in their day-to-day living. For example, housekeeping in the morning should turn off the lights and other hotel appliances that are not using electricity. Instead, they should use natural light and hotels should always look at ways to upgrade their premises so that more daylight is produced, thus making a huge difference in electricity consumption. On the other hand, the guests can embrace some practices that aim to reduce electricity consumption, such as the utilisation of towels more than once. This shall reduce not only the consumption of electricity but also a reduction in water consumption.

A genuine local corner can be established in all these accommodations and a Mauritian handicraft shop must be established by the national agency to promote entrepreneurship in the country. Nevertheless, the artisans should offer quality products at a competitive price with a clear Mauritian touch and more personalised service enterprises, including numerous MSMEs.

21.5.8 Creating Ambassadors

The key card control system activates electrical appliances once the occupant inserts the key card into the key pouch. The key pouch is installed at the door entrance and once the key card is inserted, the status of the room automatically switches on to occupied. Once the occupant leaves the room by removing the key card, all appliances will turn off, whereby some which are of great importance will come to normal mode.

21.5.9 Regular Maintenance

When addressing issues such as energy saving in the hotel sector, electricity consumption is regarded as one of the main elements that should be tackled with the appropriate solution. Lighting and other electrical appliances such as air conditioners and fridges are the appliances that consume the most electricity. Therefore, it is important to conduct regular maintenance to make the system last longer and perform efficiently. For example, having a well-maintained lighting system will make long-lasting lamp life that uses minimum energy and produces a quality lighting system. As for the AC, it is crucial to check the filter to minimise energy consumption regularly.

On the other hand, the occupancy sensor works on selected items and is based on the body movement of the occupant. Electrical appliances such as air conditioning, TV, and lighting systems are turned off because nobody is detected in the room. This system can work in parallel with the key card system. Those appliances will automatically switch off upon absence of movement or upon removing the card from the key pouch. Moreover, the maintenance team should ensure that pipes and mend leaky taps are well maintained. Very often, hotels lose water from tap leaking or pipes leaking. Therefore, the maintenance team should ensure that all the equipment that has been implemented are in good condition through periodic review.

21.5.10 Create an Energy and a Water Management Plan

The management should devote time and resources to design a water management plan that is realistic, relevant and measurable. The plan will analyse the current water consumption level in all key areas of the hotels and determine parameters against which the future water consumption level will be evaluated. This will allow the hotel to set realistic and achievable goals towards achieving water preservation. For example, the hotel should water the garden when the weather is cool to avoid evaporation and humidity to ensure that the garden is only water on demand.

21.6 Conclusion

This chapter explored the different aspects where MSMEs can benefit from an integrated model to enhance sustainable and responsible tourism. The model derived from interviews and focus groups with the main actors showed that most of the touching points with the customers have been considered, and the green touch is proposed wherever applicable. This model will help make the journey sustainable and make MSMEs in the tourism sector more competitive by offering a unique experience to the visitors. Regarding the success of its implementation, it is likely to be positive since the private operators have an important role in the economic development of Mauritius and other SIDS. It has been shown how clients play a major role in sustainable tourism. Hotels need to collaborate with clients through education and constant innovation to mitigate the ecological effects of their activities. Therefore, measures such as displaying information cards in the guest room depicting the importance and encouraging them to reuse their towel or linen are implemented. This will help in reducing wastage and other polluting activities. This model guides small and medium hotels to advise their guests on the importance of protecting the planet and acting responsibly in their actions. Together, tourists and hotels can play a major role in promoting sustainable and responsible tourism within a fragile ecosystem. In the same line, this model may guide future research by integrating other elements in the value chain to provide a greener experience to the tourist with the help of MSMEs.

References

Baum, F., Macdougall, C., & Putland, C. (1997). How can health bureaucracies consult effectively about their policies and practices?: Some lessons from an Australian study. *Health Promotion International, 12(4)*, 299–309.

Becker, E. (2020). How hard will the coronavirus hit the travel industry? Retrieved May 6, 2020, from https://www.nationalgeographic.com/travel/2020/04/how-coronavirus-is-impacting-the-travel-industry/

Blakeney, M. (2019). *Food Loss and Food Waste: Causes and Solutions.* UK Northampton: Edward Elgar Publishing.

Bocken, N., Short, S., Rana, P., & Evans, S. (2014). A literature and practice review to develop sustainable business model archetypes. *Journal of Cleaner Production, 65*, 42–56.

Bricker, S., Barkwith, A., MacDonald, A., Hughes, A., & Smith, M. (2012). Effects of CO_2 injection on shallow groundwater resources: A hypothetical case study in the Sherwood Sandstone aquifer, UK. *International Journal of Greenhouse Gas Control, 11*, 337–348.

Christian, M., Garza, A., & Slaughter, J. (2011). Work engagement: A quantitative review and test of its relations with task and contextual performance. *Personnel Psychology, 64(1)*, 89–136.

Currie, C., & Falconer, P. (2014). Maintaining sustainable island destinations in Scotland: The role of the transport–tourism relationship. *Journal of Destination Marketing & Management, 3(3)*, 162–172.

Dar, M., & Parrey, A. (2013). Socio-economic potential of handicraft industry in Jammu and Kashmir: Opportunities and challenges. *International Monthly Refereed Journal of Research In Management & Technology, 2*, 20–28.

Filimonau, V., & De Coteau, Delysia A. (2019). Food waste management in hospitality operations: A critical review. *Tourism Management, 71*, 234–245.

Gandasari, D., & Dwidienawati, D. (2020). Content analysis of social and economic issues in Indonesia during the COVID-19 pandemic. *Heliyon, 6*(11), e05599. https://doi.org/10.1016/j.heliyon.2020.e05599

García-Villaverde, P., Elche, D., Martinez-Perez, A., & Ruiz-Ortega, M. (2017). Determinants of radical innovation in clustered firms of the hospitality and tourism industry. *International Journal of Hospitality Management, 61*, 45–58.

George, G., Howard-Grenville, J., Joshi, A., & Tihanyi, L. (2016). Understanding and tackling societal grand challenges through management research. *Academy of Management Journal, 59(6)*, 1880–1895.

Goh, E., & Kong, S. (2018). Theft in the hotel workplace: Exploring frontline employees' perceptions towards hotel employee theft. *Tourism and Hospitality Research, 18(4)*, 442–455.

Gössling, S., Garrod, B., Aall, C., Hille, J., & Peeters, P. (2011). Food management in tourism: Reducing tourism's carbon 'foodprint'. *Tourism Management, 32(3)*, 534–543.

Gowreesunkar, V., & Séraphin, H. (2016). Entrepreneurship in Haiti: Toward an identification of the 'Blind Spots'. *Études caribéennes, (35)*.

Gowreesunkar, V., Van der Sterren J. & Séraphin H. (2014) Social Entrepreneurship as a tool for promoting Global Citizenship in Island Tourism Destination Management, pp. 7–23 /e-ISSN: 2014-4458

Gretzel, U., Murphy, J., Pesonen, J., & Blanton, C. (2019). Food waste in tourist households: a perspective article. *Tourism Review*, *75*(1), 235–238.

Haroon, A., & Hoq, M. (2012). *Support to Sustainable Management of the Bay of Bengal Large Marine Ecosystem (BOBLME) Project*. Bangladesh: Bangladesh Fisheries Research Institute.

Hassan, S., & Azman, A. (2014). Visible work, invisible workers: A study of women home based workers in Pakistan. *International Journal of Social Work and Human Services Practice*, *2(2)*, 48–55.

Herrera-Cano, C., & Gonzalez-Perez, M. A. (2019). *Representation of Women on Corporate Boards of Directors and Firm Financial Performance Diversity within Diversity Management: Types of Diversity in Organizations* (pp. 37–60). Bingley: Emerald Publishing Limited.

Iordache, C., Ciochină, I., & Asandei, M. (2010). Clusters-tourism activity increase competitiveness support. *Theoretical & Applied Economics*, *17*(5), 99–112.

Karlsson, L., & Dolnicar, S. (2016). Does eco certification sell tourism services? Evidence from a quasi-experimental observation study in Iceland. *Journal of Sustainable Tourism*, *24(5)*, 694–714.

Kim, V. W. E., Periyayya, T., & Li, K. T. A. (2013). How does logo design affect consumers' brand attitudes? *International Journal of Innovative Research in Management*, *2*(1), 43–57.

Koenig-Lewis, N., Palmer, A., Dermody, J., & Urbye, A. (2014). Consumers' evaluations of ecological packaging – Rational and emotional approaches. *Journal of Environmental Psychology*, *37*, 94–105.

Korhonen, J., Honkasalo, A., & Seppälä, J. (2018). Circular economy: The concept and its limitations. *Ecological Economics*, *143*, 37–46.

Lacy, P., Keeble, J., McNamara, R., Rutqvist, J., & Haglund, T. (2014). *Circular Advantage: Innovative Business Models and Technologies to Create Value in a World without Limits to Growth*. Chicago, IL, USA: Accenture.

Lew, A., Cheer, J. M., Haywood, M., Brouder, P., & Salazar, N. B. (2020). Visions of travel and tourism after the global COVID-19 transformation of 2020. *Tourism Geographies*, *22*(3), 455–466. https://doi.org/10.1080/14616688.2020.1770326

Loureiro, A., & Veloso, S. (2017). Green exercise, health and well-being. In *Handbook of Environmental Psychology and Quality of Life Research* (pp. 149–169). Cham: Springer.

Massa, S. & Testa, S. (2012) The role of ideology in brand creation: the case of a food retail company in Italy. *International Journal of Retail & Distribution Management* *40*(2), 109–127

Mendes, J., Silva, J., Rodrigues, P., & Pereira, L. (2010). A tourism research agenda for Portugal. *International Journal of Tourism Research*, *12(1)*, 90–101.

Nachmias, S., & Walmsley, A. (2015). Making career decisions in a changing graduate labour market: A hospitality perspective. *Journal of Hospitality, Leisure, Sport & Tourism Education*, *17*, 50–58.

Page, Stephen, & Joanne Connell (2014). Transport and Tourism. In Alan A. Lew, C. Michael Hall, & Allan M. Williams (Eds.), *The Wiley Blackwell Companion to Tourism* (pp. 155–167) John Wiley & Sons, Ltd. http://onlinelibrary.wiley.com/doi/10.1002/9781118474648.ch12/summar

Pompeia, M. L. F. (2021). Assessing the suitability of a single brand of MSMEs in the hospitality sector to boost sustainable development: The case of Mauritius. *Worldwide Hospitality Tourism Themes 13*, 109–123.

Porter, M. (1990). The competitive advantage of nations. *Competitive Intelligence Review*, *1*(*1*), 14–14.

Rahmiati, F., Othman, N., & Bonavisi, V. (2018). Travel motivation and domestic tourist satisfaction in Bali, Indonesia. *International Journal of Business Studies*, *2*(*2*), 105–110.

Rahmiati, F., Othman, N., Ismail, Y., Bakri, M., & Amin, G. (2020). The analysis of tourism value chain activities on competitive creation: Tourists perspective. *Talent Development & Excellence*, *12*(*1*), 4613–4628.

Rahmiati, F., Othman, N. A., & Tahir, M. N. H. (2020). Examining the Trip Experience on Competitive Advantage Creation in Tourism. *International Journal of Economics & Business Administration*, *8*(1), 15–30.

Reichel, A., De Schoenmakere, M., & Gillabel, J. (2016). *Circular Economy in Europe, Developing the Knowledge Base*. European Environmental Agency, Report No. 2/2016.

Rylance, A., & Spenceley, A. (2016). Applying inclusive business approaches to nature-based tourism in Namibia and South Africa. *Tourism: An International Interdisciplinary Journal*, *64*(*4*), 371–383.

Sarkar, A. (2012). Green branding and eco-innovations for evolving a sustainable green marketing strategy. *Asia-Pacific Journal of Management Research and Innovation*, *8*(*1*), 39–58.

Saunders, M. (2012). Organizational trust: A cultural perspective. *Development and Learning in Organizations: An International Journal*, 26(2) Available Online at: https://doi.org/10.1108/dlo.2012.08126baa.002.

Séraphin, H., & Gowreesunkar, V. (2017). Conclusion: What marketing strategy for destinations with a negative image? *Worldwide Hospitality and Tourism Themes* *9*(5), 570–576.

Sirieix, L., & Remaud, H. (2010). Consumer perceptions of eco-friendly vs. conventional wines in Australia. In *5 International Conference of the Academy of Wine Business Research*.

Smart, K., Ma, E., Qu, H., & Ding, L. (2021). COVID-19 impacts, coping strategies, and management reflection: A lodging industry case. International Journal of Hospitality Management, 94(June 2020), 102859. https://doi.org/10.1016/j.ijhm.2021.102859

Statistics, D. O. (2019). *Digests of Agricultural Statistics*. Retrieved from https://statsmauritius.govmu.org/Documents/Statistics/Digests/Agriculture/Digest_Agri_Yr18.pdf

Thomas, R., Shaw, G., & Page, S. (2011). Understanding small firms in tourism: A perspective on research trends and challenges. *Tourism Management*, *32*(*5*), 963–976.

Valle O, Pintassilgo P., Matias P., & Andre, F. (2012), Tourist attitudes towards an accommodation tax earmarked for environmental protection: a survey in the Algarve. *Tourism Management*, *33*(6), 1408–1416.

Wan, Y. K. P., Wan, Y. K. P., Chan, S. H. J., Chan, S. H. J., Huang, H. L. W., & Huang, H. L. W. (2017). Environmental awareness, initiatives and performance in the hotel industry of Macau. *Tourism Review*, *72*(1), 87–103.

WTTC (World Tourism and Trade Council). 2018. *Travel & Tourism Global Economic Impact & Issues 2018*. London: WTTC.

Yilmaz, Y., & Bititci, U. S. (2006). Performance measurement in tourism: A value chain model. *International Journal of Contemporary Hospitality Management 18*(4), 341–349.

22 Sustainable Cultural Ecosystem Services and Community-based Tourism (CBT) Models Post COVID-19 Pandemic within Aberdares Conservation Area and National Park, Kenya

Shem Wambugu Maingi, Felix Lamech Mogambi Ming'ate, and Vanessaa G.B. Gowreesunkar

22.1 Introduction

Ecosystems have inherent values in the societies that transcend beyond provisioning (food and water), regulating (disease and flood control), support (nutrient cycling) and cultural servicing (spiritual, cultural and recreational) (Kumar and Kumar, 2007). Historically, communities globally have had strong social linkages with their cultural ecosystem services; this connection with nature often was non-exploitative, aesthetic and spiritual in nature and often under the custody of the local and indigenous communities. Studies have shown that the increasing value of cultural ecosystem services within the knowledge economy has contributed significantly to the world economy and is the key driver for wealth creation in the developed economies (Borin and Donato, 2015; Edvinsson, 2013). The effects of globalization, colonization and mobility have significantly affected local milieu, identities and values. O'Dell and Billing (2005) note that the emergence of the experience economies has created a demand for cultural ecosystem services within the global tourism sector. Further, in seeking to commodify these experiences, the authors note that,

> Landscapes of experiences ... are not only organized by producers (from place marketers and city planners to local private enterprises), but are also actively sought by consumers. They are spaces of pleasure, enjoyment and entertainment as well as the meeting grounds in which diverse groups move about and come in contact with each other.
>
> O'Dell and Billing (2005), p. 16

The important nexus between cultural ecosystem services and niche tourism experiences are by no means an important aspect in tourism theory and practice. Some scholars have made arguments on the dialectical relationships

DOI: 10.4324/b23145-28

between local cultures and economy, thereby proposing a cultural economy that draws towards building local cultural economic identities, as well as providing cultural services to visitors and communities alike (Shin and Stevens, 2013; Simonsen, 2001). This relationship between biodiversity and communities has existed over the years. The emblematic places are ingrained within the history and local identities of the indigenous peoples. These historical linkages have been further enhanced by vital cultural ecosystem services provided by the local communities, therefore providing authentic visitor experiences for the tourists. Historically, biodiversity conservation in Kenyan context has relied on the critical connection between the communities and conservation agencies (Agrawal, 1997; Akama, Maingi and Camargo, 2011; Matheka, 2008). The Hanoi statement in 2016 recognizes the importance of engaging and integrating communities as partners in wildlife conservation and tourism (Hanoi, 2016). Globally, in numerous policy forums, the strategic roles of local community engagement have been emphasized to tackle conservation issues such as poaching and illegal wildlife trade (IUCN, 2021; Roe and Booker, 2019). However, in Kenyan context has been an exclusion of indigenous peoples in benefitting and sustainably using the conservation resources in certain areas.

Additionally, the COVID-19 pandemic has had devastating impacts on communities adjacent to protected areas in Kenya such as the Aberdares Conservation Area and National Park. For instance, a recent policy brief by the UNDP Kenya Country Office Strategic Policy Advisory Unit on Articulating the Pathways of the Socio-Economic impact of the Coronavirus (COVID-19) pandemic on Kenyan Economy reports that the COVID-19 pandemic has alienated local communities socially and economically by disrupting tourism supply chains globally, therefore increasing their socio-economic vulnerabilities in the face of adversity as well as reducing their level of socio-economic participation and control over tourism enterprise (UNDP, 2020a). In order to contain the impacts of COVID-19, the Government of Kenya and other governments throughout the world have introduced COVID-19 policy responses such as lockdowns, international travel restrictions, subsectoral closures, and adjustments in public transportation. The use of masks and social distancing have affected local communities that depend on the cultural ecosystem services to improve their socio-economic well-being through tourism in one way or the other (Blayac et al., 2021). For instance, Gössling (2020) reports that COVID-19 pandemic-related international, regional and local travel restrictions have affected the national economies, including tourism value and supply chains as well as destination systems.

Kenyan indigenous forests are underrated by park-adjacent communities as they do not reflect the socio-economic development priorities of communities adjacent to protected areas (Abukari and Mwalyosi, 2020). Well-managed small-scale and community-based tourism enterprises could allow the non-consumptive use of Kenya's increasingly scarce indigenous forest resources, provide alternative post COVID-19 livelihoods and thus make an important contribution to their management. The book chapter therefore uses the Aberdares Conservation Area and National Park as a case study to address two significant gaps in the current literature. First, the chapter investigates the linkages between COVID-19 policy responses, cultural ecosystem services, socio-economic well-being of communities and tourism. This question involves the assessment as to whether there is a relationship between COVID-19 and cultural ecosystem services, socio-economic well-being of communities and tourism. Second, the chapter examines the weakness of the current policy in addressing the challenges caused by COVID-19 pandemic in improving the socio-economic well-being of the communities and tourism. This gap is addressed by assessing the current policies in relationship to the challenges faced by COVID-19.

22.2 Literature Review

22.2.1 Theoretical Background

There have been various theoretical perspectives on the roles of cultural ecosystem services and biodiversity conservation in mitigating the effects of pandemics such as COVID-19. While environmental scholars have examined the concept of cultural ecosystem services as tool for addressing the ongoing attrition of the eco-cultural environment and as an approach for biodiversity conservation planning (Lennon and Scott, 2014), eco-cultural scholars have looked at the concept as a means of identifying the right balance between social progress and environmental conditions (Tallis et al., 2008). Research by the World Bank (2004) as well as the World Resource Institute (2007) vindicates Kaplan's Attention Restoration theory by indicating that biodiversity loss significantly affects human well-being and social progress. According to the Biophyllia Hypothesis theory, human beings since history continued to rely on nature for emotional, intellectual, spiritual, social and physical benefits (Gullone, 2000). Natural environments provide restorative benefits to the mind and soul.

There has been a school of thought noting that biodiversity loss has contributed to the majority of the challenges facing humanity today. Whitham (2015) advances a critical connection between human-wildlife conflicts and ecosystem services. There has been a movement towards environmental justice when it comes to finding the right balance between fair and meaningful engagement of communities as well as biodiversity conservation (Ortega, 2011; Schlosberg, 2007). The promotion of cultural ecosystem services perspective integrates ecocentrism values within society i.e. a moral consideration of nature and protection of the rights of the poor indigenous communities. Studies have shown that ecocentrism is key towards the promotion of positive conservation values (Taylor et al., 2020; Thompson and Barton, 1994). The rise of the eco-cultural industries has made tourism and tourism stakeholders acknowledge a greater responsibility towards the host environment and the host community (Jamal et al., 2010). However, Everard et al. (2020) posit that despite the promotion of ecocentrism in society, as well as the reduction of the regulations on ecosystem services, they are inextricably linked to degraded ecosystems as well as the spread of zoonotic diseases from wildlife to humans. According to Iroro (2020), research work on the conservation of the round-leaf bat *(Hipposideros curtus)* concluded that human action contributed to the spread of the COVID-19 pandemic. There is a strong evidence pointing that human actions (i.e. destruction of the bat habitats as well as wildlife trade) enabled the coronavirus SARS-CoV-2 to jump from the wildlife to the humans (Tointon, 2020). Therefore, this study makes the proposition that to enhance post-COVID response, sustainable ecosystem services, conservation and community livelihoods, there needs to have the right balance between conservation action, social progress and environmental justice.

22.3 The Concept of Cultural Ecosystems

Ecosystems as defined by the National Geographic are geographical areas where the animals, habitats as well as landscapes and weather conditions work together to form the bubble of life (National Geographic, 2021). These human ecosystems are essential in sustaining and fulfilling human life. These positive ecosystem benefits to humans are referred to as ecosystem services. These ecosystem services are classified by MEA (2005) as provisioning services (e.g., nutrition, shelter, fresh water), regulating services (e.g., climate regulation, water purification), cultural services (e.g., aesthetic, spiritual, recreational experiences), and supporting services (e.g., nutrient cycling, soil formation). Cultural ecosystem services have been further as the non-material benefits that communities obtain from nature (Daniel et al., 2012; IUCN, 2015a; MEA, 2005). Tourism to national parks and conservation areas provide tourism cultural ecosystem services vital for human well-being and the quality of life, which can be categorized into physical (rest, relaxation, sport, tourism, health, adventure and leisure); psychological (restoration, peace, solitude, and education), spiritual (reflection, a sense of place and connection with nature), social (reconnection with families and friends), cultural (aesthetic appreciation of the cultural values of a place) and environmental (recreation, preservation and conservation of the ecosystems) (Husk et al., 2016; IUCN, 2015b; Kim et al., 2015).

Cultural ecosystem services are important for communities, and various research studies have shown the critical roles of cultural ecosystem services in improving the quality of life of both urban and rural people (Andersson et al., 2015; Dou et al., 2017; Sen and Guchhait, 2021). However, the Millennium Ecosystems Assessment Reports called by the United Nations to assess the effects of ecosystem changes on human well-being evidenced a declining status of ecosystem services (Hassan, Scholes and Ash, 2005). There have been increasing concerns on the decline in cultural ecosystem services as well as the aesthetic value of the local cultural ecosystem services as the degradation of ecosystems continues (Muller et al., 2019). A degradation of the environment translates to fewer human benefits and lower spiritual, aesthetic and cultural experiences to tourism visitors and local communities.

22.4 The Concept of Community-based Tourism (CBT)

The community-based tourism (CBT) model centres around the involvement of the host community in tourism activities and tourism development in an area. According to Pawson, Archy and Richardson (2017), the concept gained momentum within the research world, with the need towards entrenching the benefits of globalization to the local communities and economies and making the sector much more sustainable. This community-based approach has also raised proponents and opponents in equal measure. Research by Kibicho (2008) and Akama, Maingi and Camargo (2011), in Kenyan context, see CBT as a means of addressing the post-colonial legacies, shedding the colonial image and empowering the communities to participate in the management of their tourism resources. This school of thought has looked at CBT as the ideal anti-oppressive 'communitarianism' that seeks to promote social justice and sustainable cultural industries. Further, Mitchell and Ashley (2009) and Maingi (2021) argue that CBT is an ideal poverty alleviation strategy for developing countries. CBT ensures local control over tourism development as well as equitable benefits for the locals (Hall, 1996; Pearce, 1992). According to Blackstock (2005), it provides an alternative model for communities that guarantee consensus-based decision making, social justice and community ownership of the resources within their locality. Researches on CBT approach have identified unique benefits of the CBT model or approach. These include direct economic impact on families, social well-being improvements, and sustainable diversification of lifestyles (López-Guzmán et al., 2011; Manyara and Jones, 2007; Rastegar, 2010). Ashley and Garland (1994) have further classified the benefits of CBT as follows: (1) to communities – boosting welfare, economic growth, and empowerment in the communal areas, (2) to conservation – encouraging community commitment to wildlife conservation and sustainable management of the natural resource base and to the country, it helps in diversifying and developing the tourism product, particularly eco-tourism, and ensuring the long-term sustainability of its resource base. CBT is certainly an effective way of implementing policy coordination, avoiding conflicts between different actors in tourism, and obtaining synergies based on the exchange of knowledge, analysis, and ability among all members of the community (Kibicho, 2008; López-Guzmán et al., 2011).

22.5 Current Policy Frameworks on Cultural Ecosystems Services and CBT in Kenyan Context

The Aberdares ecosystem in Kenya is of outstanding exceptional resource value (ERVs), and it has been the priority of the Kenyan Government through the Kenya Wildlife Services to protect and conserve the principal water catchment for Kenya's major rivers, wilderness character and cultural resources and threatened and endemic species for present and future generations (KWS, 2010). Kenya's National Environmental Policy, 2013, as well as Kenya's Forest Policy No. 9 of 2005, recognizes the state of decline in the national forest biomes such as the Aberdares ranges and ecosystem services within the country (GoK, 2013; KFS, 2010). About 80% of Kenyan population depends on these forest biomes and related resources as sources of their livelihoods (GoK, 2013; Mulinge et al., 2016). There are several attempts in the current literature addressing policy frameworks on cultural ecosystem services in practice and geographical contexts (Daniel et al., 2012; MEA, 2005), but none of these has attempted to feature on a policy framework that links cultural ecosystem services and CBT in Kenyan context (Abunge, Coulthard and Daw, 2013; Chaigneau et al., 2019; Mutoko, Hein and Shisanya, 2015). Further policy frameworks for sustainable cultural ecosystem services post COVID-19 pandemics remain completely unavailable in the current literature (Chaigneau et al., 2019; Mahajan and Daw 2016; Maithya et al., 2020). Thus in this chapter, we seek to make a theoretical contribution in proposing policy options for sustainable cultural ecosystems services post COVID-19 pandemics.

22.6 Methodology

The Aberdare ranges indicated in Figure 22.1 being on the tentative list of the UNESCO World Heritage sites has significant historical, eco-cultural and touristic value to the tourism industry, visitors and the indigenous

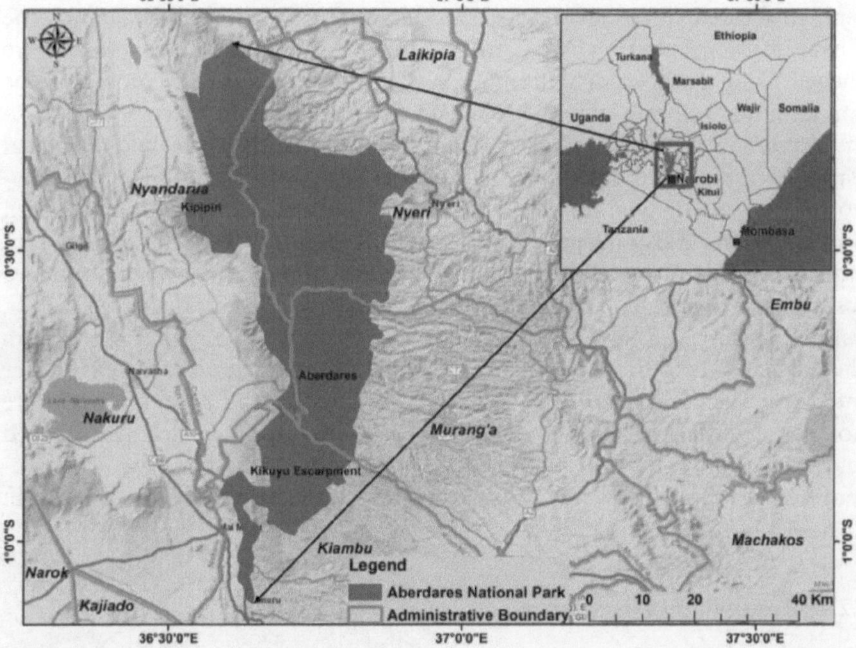

Figure 22.1 Map of Aberdares National Park.

Source: Researchers.

communities (i.e. the maasai and the kikuyu communities). The land of the majestic peaks, moorlands and falls has a significant colonial heritage as being home to the Mau Mau freedom struggle and the forest was used by the freedom fighter, Dedan Kimathi, as a post office. The Aberdare ranges are the main water catchments for Sasumua and Ndakaini Dams, which provide most of the water for Nairobi City, the capital city of Kenya. It is also one of the important bird areas (IBAs) and one of the five water towers and an important catchment area for the Tana River, Ewaso Nyiro River and Lake Naivasha. The sustainability of the Aberdare forests is critically important to the viability of park-adjacent communities and intervention measures are needed to protect these indigenous forests and cultural ecosystem services and enhance future sustainability as this supports the adjacent forest communities with their livelihood. While the remaining patches of Kenya's indigenous forests are currently threatened by human activity, previous attempts to protect them through exclusion of host communities has resulted in hostility and open human-wildlife conflict.

The research methods used were designed to elicit information based on the local inhabitant's experiences during the period of COVID-19 and before. The reason for doing this was the fact that the research wanted to compare the effects of COVID-19 on the cultural ecosystem services and community-based tourism within Aberdares Conservation Area and National Park. Data collection combined overtly observing the study communities and participating in their activities, document analysis and administering 74 semi-structured interviews to the communities living adjacent to the Aberdares Conservation Area and National Park. These interviewees were chosen using purposeful sampling by carefully identifying members of the communities who are versed with the cultural ecosystem services and community-based tourism within Aberdares Conservation Area and National Park. All the participants' information was kept confidential by using pseudo-names and coding to protect them and enhance confidentiality when doing data analysis and presenting results. The researchers also sought for ethical approval from Kenyatta University Ethical Review Committee and a research permit from NACOSTI before undertaking the research. Further, before any interviews were conducted, the researchers sought consent from the participants to first accept to participate in the research before and after interview was conducted. Those who did not accept to be interviewed were not victimized in any way. All the interviews were voice recorded, transcribed and transcripts generated and coded manually by the researchers. The content from the trancripts was then grouped into themes that were relevant in answering the study objectives.

22.7 Results and Discussions

The results in this chapter examine impacts of COVID-19 on tourism, cultural ecosystem services and the socio-economic well-being of communities within Aberdares Conservation Area and National Park in Kenya and then provide policy options, conclusions and recommendations that can reduce the impacts associated with COVID-19.

The results reveal that the local communities adjacent to the National park are majorly farming communities and small business communities. From further discussions, majority of the locals were highly aware of the COVID-19 pandemic and to a particular extent exhibited anxiety related to the COVID pandemic. These anxieties were intensified by the fears of mandatory quarantine regulations imposed on 28 March 2021 on people who were in contact with identified COVID cases through contact tracing at designated Government facilities at the cost of the patient. The *'coronaphobia'* also resulted from the negative socio-economic ramifications associated with the virus. The resultant economic impacts of the COVID-19 pandemic have also had significant effects on the community's well-being. These emotions were expressed by some of the respondents who noted that:

> We have lost livelihoods as a result of the pandemic. We have lost our business as well as jobs. Supply and demand for goods and services have been significantly hampered. Our small businesses have been severely affected.

This state of uncertainty affected them immensely and in particular when meeting and communicating with foreigners from the community. Currently, the social and economic shocks to local communities arising from the COVID pandemic have been widely documented in Kenyan context (NRT, 2020; UNDP, 2020b). These shocks existed beyond the Kenya Government, which introduced the KES 8.5 billion National Cash Transfer Program to the elderly and vulnerable communities (Bowmans, 2020). The findings of these studies are comparably consistent with other regions in terms of the social impacts of the global health crisis on the local communities. To a particular extent, the local communities felt that they were not supported adequately during the COVID-19 pandemic outbreak. The community further noted that the tourism facilities and industry did not provide any support during the pandemic.

The respondents were asked on how COVID-19 has affected tourism and conservation within the Aberdares National Park. The locals noted that tourism and travel were significantly affected by the intercountry cessation of movement in and out of Nairobi Metropolitan area, Kilifi, Kwale and Mombasa counties. This had a severe impact on the tourism supply chains as noted by one of the respondents who noted that,

> Supply and Demand of goods and services has been significantly hampered with the transportation sector owing to the COVID-19 related lockdowns and curfews.

The state of supply chains in the communities has been adversely affected by the pandemic. The pandemic has developed into an unprecedented socio-economic and human crisis, where the communities experienced supply disruptions as a result of the closures. This struggle for communities to manage disruptions to supply chain has been widely documented after the COVID-19 pandemic (Butt, 2021). Vulnerabilities of the local communities adjacent to the Aberdares conservation area have an important effect on the state of conservation of forests and nature within the conservation area.

Loss of jobs by mostly youth and women working in the hotel industry was also found as a major impact of COVID-19 on the inhabitants of the Aberdares National Park. This was also reported in the Caribbean states that women and children have been adversely affected by the pandemic (Bryon et al., 2021; OECS, 2020). Further, tourists no longer visited the Aberdares National Park as they used to before the pandemic. This followed a national ban on international travel and international passenger flights, effective 25 March 2020 (Bowmans, 2020). This situation has had an impact on the local economy as well as the employees' welfare.

22.8 Policy Options, Conclusions and Recommendations

The study provides a policy analysis of the existing conservation model in Kenya as well as underpinning a global, regional and national context. Given the fact that the Government is focusing on reviving tourism in the country, there is a need to create an emergency fund to caution the communities depending on cultural ecosystem services. This was reported by the community that most of the tourism proceeds are taken by the government, and communities do not get a share of the revenue yet Aberdares forest and National Park are critical to the resources to the community. Findings of this study propose a review of national tourism policies and development priorities in conservation and tourism sectors and factors in the social impacts of pandemics such as COVID-19 in Kenyans tourism development policies and priorities. The diversity of ecosystem benefits and conservation policies in Kenya needs to address indigenous cultural ecosystem services. Following the above policy options, it is concluded that there are no clear policies that link impacts of pandemics such as COVID-19 with conservation and community livelihoods. The study therefore recommends the following:

- In accordance to the Hanoi declaration of 2016, there is a need to engage and integrate local communities as partners in conservation of cultural ecosystem services and tourism;
- During pandemics such as the COVID-19, there is need to review forest management policies to address the socio-economic challenges and communities livelihoods;
- There is a need for government to allow consensus-based conservation and tourism planning, thus guaranteeing social justice and community ownership of the resources within their locality.

Acknowledgment

It is with gratitude that we acknowledge a grant provided by the British Institute in East Africa in September 2020 which led us to undertake the research and publication of this chapter.

References

Abukari, H. and Mwalyosi, R.B. (2020) Local communities' perceptions about the impact of protected areas on livelihoods and community development. *Global Ecology and Conservation*, 22, p. e00909.

Abunge, C., Coulthard, S. and Daw, T. (2013) Connecting marine ecosystem services to human well-being: Insights from participatory well-being assessment in Kenya. *Ambio*, 42(8), pp. 1010–1021.

Agrawal, A.A. (1997) *Community in Conservation: Beyond Enchantment and Disenchantment. Conservation and Development Forum*. Retrieved on 2 February 2021. Retrieved through: http://dlc.dlib.indiana.edu/dlc/bitstream/handle/10535/3963/Community_in_Conservation.pdf

Akama, J.S., Maingi, S.W. and Camargo, B.A. (2011) Wildlife conservation, safari tourism and the role of tourism certification in Kenya: A post colonial critique. *Tourism Recreation Research*, 36(3), pp. 281–291. DOI: 10.1080/02508281.2011.11081673

Andersson, E., Tengo, M., McPhearson, T. and Kremer, P. (2015) Cultural ecosystems services as a gateway for improving urban sustainability. *Ecosystems Services*, 12, pp. 165–168. DOI: 10.1016/j.ecoser.2014.08.002

Ashley, C. and Garland, E.B. (1994) *Promoting Community-based Tourism Development: Why, What, and How?* (Vol. 4). Directorate of Environmental Affairs, Ministry of Environment and Tourism.

Blackstock, K. (2005). A critical look at community based tourism. *Community Development Journal*, 40(1), pp. 39–49.

Blayac, T., Dubois, D., Duchêne, S., Nguyen-Van, P., Ventelou, B. and Willinger, M. (2021) Population preferences for inclusive COVID-19 policy responses. *The Lancet Public Health*, 6(1), p. e9. Retrieved from: https://www.thelancet.com/journals/lanpub/article/PIIS2468-2667(20)30285-1/fulltext

Borin, E. and Donato, F. (2015) Unlocking the potential of IC in Italian cultural ecosystems. *Journal of Intellectual Capital*, 16(2), pp. 285–304. DOI: 10.1108/JIC-12-2014-0131

Bowmans (2020) *COVID-19: Tracking Government Response in Kenya Update as at 30th November 2020*. Bowmans Law. Retrieved from: https://www.bowmanslaw.com/wp-content/uploads/2020/12/Bowmans-Kenya-Government-Response-Tracker.pdf

Butt, A.S. (2021) Strategies to mitigate the impact of COVID-19 on supply chain disruptions: A multiple case analysis of buyers and distributors. *International Journal of Logistics Management*. Vol. ahead-of-print No. ahead-of-print. DOI: 10.1108/IJLM-11-2020-0455

Byron, J., Martinez, J. L., Montoute, A., & Niles, K. (2021). Impacts of COVID-19 in the Commonwealth Caribbean: key lessons. *The Round Table*, 110(1),pp. 99–119.

Chaigneau, T., Brown, K., Coulthard, S., Daw, T.M. and Szaboova, L. (2019) Money, use and experience: Identifying the mechanisms through which ecosystem services contribute to wellbeing in coastal Kenya and Mozambique. *Ecosystem Services*, 38, p. 100957.

Daniel, T.C., Muhar, A., Arnberger, A., Aznar, O., Boyd, J.W., Chan, K.M. and von der Dunk, A. (2012) Contributions of cultural services to the ecosystem services agenda. *Proceedings of the National Academy of Sciences*, 109(23), pp. 8812–8819.

Dou, Y., Zhen, L., De Groot, R., Du, B. and Yu, X. (2017) Assessing the importance of cultural ecosystem services in urban areas of Beijing municipality. *Ecosystems Services*, 24, pp. 79–90.

Edvinsson, L. (2013) IC 21: Reflections from 21 years of IC practice and theory. *Journal of Intellectual Capital*, 14(1). DOI: 10.1108/14691931311289075

Everard, M., Johnson, P., Santinno, D, and Staddon, C. (2020) The role of ecosystems in mitigation and management of COVID-19 and other zoonoses. *Environmental Sciences and Policy*, 111(1), pp. 7–17. DOI: 10.1016/j.envsci.2020.05.017

Government of Kenya (2013) *National Environmental Policy 2013*. Ministry of Environment and Mineral Resources. Government of Kenya. Retrieved through: http://www.environment.go.ke/wp-content/uploads/2013/06/13-NEP-No-trackch-3.pdf

Gössling, S. (2020). Risks, resilience, and pathways to sustainable aviation: A COVID-19 perspective. *Journal of Air Transport Management*, 89(1), pp. 1–4, 101933.

Gullone, E. (2000) The biophilia hypothesis and life in the 21st century: Increasing mental health or increasing pathology? *Journal of Happiness Studies*, 1, pp. 293–322. Retrieved from: https://link.springer.com/article/10.1023/A:1010043827986

Hall, C.M. (1996) *Introduction to Tourism in Australia: Impacts, Planning and Development*. Addison, Wesley and Longman, Melbourne, Australia.

Hanoi (2016) *Hanoi Statement on Illegal Wildlife Trade, Hanoi Conference on Illegal Wildlife Trade*, Vietnam held on 17–18 November 2016. Retrieved from: https://www.traffic.org/site/assets/files/2808/hanoi-statement-on-illegal-wildlife-trade.pdf

Hassan, R., Scholes, R. and Ash, N. (2005) *Ecosystems and Human Well-being: Current State and Trends, Volume 1. Findings of the Conditions and Trends Working Group of the Millenium Eosystem Assessment Board*. Island Press, Washington. Retrieved through: https://www.millenniumassessment.org/documents/document.766.aspx.pdf

Husk, K., Lovell, R., Cooper, C., Stahl-Timmins, W. and Garside, R. (2016) Participation in environmental enhancement and conservation activities for health and well-being in adults: A review of quantitative and qualitative evidence. *Cochrane Database of Systematic Reviews*, Issue 5. Art. No.: CD010351. DOI: 10.1002/14651858.CD010351.pub2

Iroro, T. (2020) Roost and habitat protection for short-tailed roundleaf bats in Nigeria, Conservation Leadership Programme. Available at: https://www.conservationleadershipprogramme.org/project/roost-and-habitat-protection-for-short-tailed-roundleaf-bats-in-nigeria/

IUCN (2015a) *Cultural Ecosystems Services: A Gateway to Raising Awareness for the Importance of Nature for Urban Life*. The Urban Biodiversity and Ecosystems Services Project, IUCN. Retrieved from: https://www.iucn.org/downloads/urbes_factsheet_08_web_1.pdf

IUCN (2015b) *Healthy Parks, Healthy People: The State of the Evidence*. Parks Victoria. Retrieved through: https://www.iucn.org/sites/dev/files/content/documents/hphpstate-evidence2015.pdf

IUCN (2021) *Communities and Illegal Wildlife Trade*. IUCN Commission on Environmental, Economic and Social Policy. Retrieved from: https://www.iucn.org/

commissions/commission-environmental-economic-and-social-policy/our-work/specialist-group-sustainable-use-and-livelihoods-suli/communities-and-illegal-wildlife-trade

Jamal, T., Camargo, B., Sandlin, J. and Segrado, R. (2010) Tourism and cultural sustainability: Towards an eco-cultural justice for place and people. *Tourism Recreation Research*, 35(3), pp. 269–279.

KFS (2010). *Aberdares Forest Reserve Management Plan (2010-2019)*. Kenya Forest Service. Retrieved through: http://www.kenyaforestservice.org/documents/Aberdare.pdf

Kibicho, W. (2008) Community-based tourism: A factor-cluster segmentation approach. *Journal of Sustainable Tourism*, 16(2), pp. 211–231.

Kim, H., Lee, S., Uysal, M., Kim, J. and Ahn, K. (2015) Nature-based tourism: Motivation and subjective well-being. *Journal of Travel and Tourism Marketing*, 32(1), pp. S76–S96. DOI: 10.1080/10548408.2014.997958

Kumar, M. and Kumar, P. (2007) Valuation of the ecosystems services: A psycho-cultural perspective. *Ecological Economics*, pp. 1–12. DOI: 10.1016/j.ecolecon.2007.05.008

KWS (2010) *Aberdares Ecosystem Management Plan, 2010-2020*. Kenya Wildlife Services. Retrieved from: https://rris.biopama.org/sites/default/files/2019-03/Aberdare_Ecosystem_Final_plan_2010-2020.pdf

Lennon, M. and Scott, M. (2014) Delivering ecosystems services via spatial planning: reviewing the possibilities and implications of a green infrastructure approach. *The Town Planning Review*, 85(5), pp. 563–587. Retrieved from: https://www.jstor.org/stable/pdf/24579269.pdf?casa_token=9j2ivHfdaL0AAAAA:THx0qiPCW5aEpIGEpnr2OJH_PlxPToqrXDNTIVs39vvnH7j1QhaLOAovL2PT0-yO4HZQAjwjGFlhZeBVuFc-Fj-mIkeQh-gt5SPop3Fkkt0qup-j2OG2LQ

López-Guzmán, T., Sánchez-Cañizares, S. and Pavón, V. (2011). Community-based tourism in developing countries: A case study. *Tourismos*, 6(1), pp. 69–84.

Mahajan, S.L. and Daw, T. (2016) Perceptions of ecosystem services and benefits to human well-being from community-based marine protected areas in Kenya. *Marine Policy*, *74*, pp. 108–119.

Maingi, S.W. (2021) Safari tourism and its role in sustainable poverty eradication in East Africa: The case of Kenya. *Worldwide Hospitality and Tourism Themes*, 13(1), Vol. ahead-of-print No. ahead-of-print.

Maithya, J., Ming'ate, F. and Letema, S. (2020) A review on ecosystem services and their threats in the Conservation of Nyando Wetland, Kisumu County, Kenya. *Tanzania Journal of Science*, 46(3), pp. 711–722.

Manyara, G. and Jones, E. (2007) Community-based tourism enterprises development in Kenya: An exploration of their potential as avenues of poverty reduction. *Journal of Sustainable Tourism*, 15(6), pp. 628–644.

Matheka, R.M. (2008) Decolonisation and wildlife conservation in Kenya, 1958-68. *The Journal of Imperial and Commonwealth History*, 36(4), pp. 615–639. DOI: 10.1080/03086530802561016

Millennium Ecosystem Assessment (2005) Ecosystems and Human Well-Being Synthesis. Island Press, Washington, DC.

Mitchell & Ashley (2009) *Tourism and Poverty Reduction: Pathways to Prosperity*. Taylor & Francis, Abingdon, OX.

Mulinge, W., Gicheru, P., Muriithi, F., Kihiu, E., Kirui, O.K. and Mirzabaev, A. (2016) Economics of land degradation and improvement in Kenya. In Nkonya, E., Mirzabaev, A. and von Braun, J. (Eds), *Economics of Land Degradation and*

Improvement – A Global Assessment for Sustainable Development. Springer, Cham. DOI: 10.1007/978-3-319-19168-3_16

Muller, S.M., Peisker, J., Bieling, C., Linnemann, K., Reidl, K and Schmieder, K. (2019) The importance of cultural ecosystems services and biodiversity for landscape visitors in the biosphere reserve Swabian Alb (Germany). *Sustainability*, 11, pp. 1–23.

Mutoko, M.C., Hein, L. and Shisanya, C.A. (2015) Tropical forest conservation versus conversion trade-offs: Insights from analysis of ecosystem services provided by Kakamega rainforest in Kenya. *Ecosystem Services*, 14, pp. 1–11.

National Geographic (2021) *Ecosystem*. Encyclopedia Resource Library, National Geographic. Retrieved from: https://www.nationalgeographic.org/encyclopedia/ecosystem/

NRT (2020) The community-level impacts of COVID-19 in Northern and Coastal Kenya – Insights from NRT. Northern Rangelands Trust. Retrieved from: https://www.nrt-kenya.org/covid19-impact

O'Dell, T. and Billing, P. (2005) "Experiencescapes: Tourism, Culture and Economy", Copenhagen Business School Press, Copenhagen, 2005. 196pp. ISBN 87-630-0150-0.

OECS Commission (2020) COVID-19 and beyond: Impact assessments and responses. Retrieved from: https://oecs.org

Ortega, B.A.C. (2011) *Justice and fairness in tourism: A Grounded theory study of cultural justice in Quintana Roo, Mexico*. Unpublished PhD Thesis, Texas A & M University. Retrieved from: https://oaktrust.library.tamu.edu/handle/1969.1/ETD-TAMU-2011-05-9102

Pawson, S., Archy, P. and Richardson, S. (2017) The value of community-based tourism in Banteay Chhmar, Cambodia. *Tourism Geographies*, 19(3), pp. 378–397.

Pearce, D. (1992) Alternative tourism: Concepts, classifications and questions. In Smith, V.L. and Eadington, W.R. (Eds.), *Tourism Alternatives: Potentials and Problems in the Development of Tourism*. John Wiley and Sons, New York, pp. 18–30.

Rastegar, H. (2010) Tourism development and residents' attitude: A case study of Yazd, Iran. *Tourismos*, 5(2), pp. 203–211.

Roe, D. and Booker, F., (2019) Engaging local communities in tackling the illegal wildlife trade: A synthesis of approaches and lessons for best practice. *Conservation Science and Practice*, 1(5), pp. 1–26. DOI: 10.1111/csp2.26

Schlosberg, D. (2007) *Defining Environmental Justice: Theories, Movements and Nature*. Oxford University Press, New York.

Sen, S. and Guchhait, S.K. (2021) Urban green space in India: Perception of cultural ecosystem services and psychology of situatedness and connectedness. *Ecological Indicators*, 123, pp. 2–16.

Shin, H. and Stevens, Q. (2013) How culture and economy meet in South Korea: The politics of cultural economy in culture-led urban regeneration. *International Journal of Urban and Regional Research*, 37(3), pp. 1707–1723.

Simonsen, K. (2001) Space, culture and economy – A question of practice. *GeografiskaAnnaler: Series B, Human Geography*, 83(1), pp. 41–52. DOI: 10.1111/j.0435-3684.2001.00089.x

Tallis, H., Kareiva, P., Marvier, M. and Chang, A. (2008) An ecosystems services framework to support both practical conservation and economic development. *PNAS*, 105(28), pp. 9457–9464. Retrieved from: https://www.pnas.org/content/pnas/105/28/9457.full.pdf

Taylor, B., Chapron, G., Kopnina, H., Orlokowska, E., Gray, J. and Piccolo, J.J. (2020) The need for ecocentricism in biodiversity conservation. *Conservation Biology*, 34(5), pp. 1089–1096.

Thompson, S.C.G. and Barton, M.A. (1994) Ecocentric and anthropocentric attitudes towards the environment. *Journal of Environmental Psychology*, 14(2), pp. 149–157.

Tointon, K. (2020) *How Conservation Can Prevent Future Pandemics*. Birdlife International. Retrieved from: https://www.birdlife.org/worldwide/news/how-conservation-can-prevent-future-pandemics

UNDP (2020a) *Policy brief on articulating the pathways of the socio-economic impact of the coronavirus (COVID-19) pandemic on the Kenyan economy*, Issue No. 4/2020. United Nations Development Programme Kenya Country Office Strategic Policy Advisory Unit. Assessed through: https://www.undp.org/content/undp/en/home/coronavirus/socio-economic-impact-of-covid-19.html

UNDP (2020b) *COVID-19 Pandemic: Humanity Needs Leadership and Solidarity to Defeat COVID-19*. United Nations Development Programme, Kenya. Retrieved from: https://www.ke.undp.org/content/kenya/en/home/coronavirus.html

Whitham, C. (2015). Combining patterns of human-wildlife conflicts & ecosystem services for more efficient conservation management. Beijing: Beijing Forestry University (PhD Thesis). Available at https://www.researchgate.net/profile/Charlotte-Whitham/publication/294089440_Combining_patterns_of_human-wildlife_conflicts_and_ecosystem_services_for_more_efficient_conservation_management/links/56be087108ae44da37f88984/Combining-patterns-of-human-wildlife-conflicts-and-ecosystem-services-for-more-efficient-conservation-management.pdf

World Bank (2004) *How Much Is an Ecosystem Worth? Assessing the Economic Value of Conservation*. The World Bank. Assessed through: http://documents1.worldbank.org/curated/en/376691468780627185/pdf/308930PAPER0Ecosytem0worth01public1.pdf

World Resource Institute (2007) *Restoring Nature's Capital; an Action Agenda to Sustainable Ecosystem Services*. World Resources Institute, Washington, DC.

Part VII

Conclusion

23 Reflections and Conclusion

Management of Tourism Ecosystem Services Post Pandemic – Looking Ahead

Vanessaa G.B. Gowreesunkar, Shem Wambugu Maingi, and Felix Lamech Mogambi Ming'ate

As the world continues to navigate between the "*new normal*" and "next *normal*" of an ongoing pandemic, recovery plans for the global tourism ecosystem services are still not bringing desired results. This is reflected in the World Travel and Tourism Council (WTTC) Economic Impact Report, which shows that in 2020, the global Travel & Tourism sector lost almost USD 4.5 trillion dollars, with over 62 million jobs lost (WTTC, 2021). The sector's contribution to the global economy decreased from 10.4% to just 5.5% last year and leisure spending decreased by 49.4% and business spending by 61%. According to the Report, the pandemic crisis was 18 times bigger than the global financial crisis of 2008, and it has exposed long-standing structural weaknesses as well as gaps in tourism ecosystem management, policy decision, and restoration (Gowreesunkar et al., 2021). As travel restarts in some parts of the world, poorly managed tourism ecosystem services, misinformation about the pandemic and its unknown evolution, and weaker consumer confidence continue to plague the global tourism industry. The COVID-19 pandemic and its allies (Delta and Omicron) have presented an unprecedented opportunity to reconsider the management of tourism ecosystem services and the following actions are required:

- A re-examination of how the evolution of the pandemic is further affecting the tourism ecosystem services in the near and long term;
- A re-assessment of its impacts on the natural and cultural ecosystem services;
- A review of the collaboration strategies among tourism stakeholders;
- A reconsideration of how the tourism ecosystem services may build on existing framework of sustainable tourism;
- and ultimately, application of lessons learnt from global case studies.

With this as background, the book *Management of Tourism Ecosystem Services in a Post Pandemic Context: Global Perspectives* proposes a synthesis of views and case studies that foster understanding of the above points. Published at a time when the world is still struggling with the caprices of the pandemic, the book proposes a revised agenda for the management and restoration of tourism ecosystem services post pandemic. Drawing from 20

DOI: 10.4324/b23145-30

global case studies written by 46 authors from varied backgrounds and interdisciplinary interests, the book provides some practical solutions regarding the management and restoration of tourism ecosystem services in countries such as India, Kenya, Poland, Argentina, United Kingdom, Lamu island, Mauritius island, Spain, Bali, Indonesia, Ethiopia, Uganda, Latvia, Malaysia, Nigeria, Sweden, Pakistan, Sri Lanka, China, and South Africa. The book centers efforts in serving the global tourism industry, and the general conclusion of each case study summarizes the need for collaboration among stakeholders and explores new opportunities offered by the "*next normal*" of the pandemic situation. While COVID-19, Delta, and Omicron are collectively still ravaging the global tourism business, this book offers 23 chapters on the theme "management and restoration of tourism ecosystem services in the context of the pandemic". The five thematic areas of the book cover contemporary issues related to the management and restoration of ecosystem services in the global tourism industry.In so doing, the book addresses the following key objectives: analysis of ecosystem services serving the global tourism industry, assessment of their vulnerabilities in the face of the ongoing pandemic, management and restoration of natural and cultural ecosystem services, sustainability of the tourism ecosystem services, and future trends affecting tourism ecosystem services.

23.1 Looking Ahead

Looking ahead, the overall impression captured from the book shows that unlike most reviews that highlight the dark sides of the pandemic, the case studies point toward the possibility of transforming the tourism system in ways that contribute to a more hopeful future for tourism practitioners, tourists, and host communities. For instance, case studies from Uganda, Poland, Kenya, Pakistan, and Sweden show positive changes for the environment with the most noticeable effect, at the global level, being the reduction in greenhouse gas and air pollution. Some of the tourism resources are restored and the scene is greener and ready to welcome visitors. In addition to this, overtourism brought to destinations such as Bali, Sri Lanka, China, India, the UK, and Spain has also eased since international bans on travel have been applied. Raising this point to another level, it would seem that with the changing trend of the pandemic, most Governments' ability to manage the sustainability issues is diminished, undermining efforts to tackle the pandemic and its environmental, economic, and social impacts. Moreover, efforts are being hampered to attain the sustainable development goals (SDGs). This is particularly reflected in the case studies from India, Nigeria, Latvia, Mauritius, and Spain. It is therefore crucial to guarantee that the tourism ecosystem services are sustainably managed and restored, so that progress can continue to be made toward achieving the SDGs related to natural resources management, in particular: SDG 6 (ensure availability and sustainable management of water and sanitation for all); SDG 7 (ensure access to affordable, reliable, sustainable, and modern energy for all); SDG 2 (end hunger, achieve food security and improved nutrition, and promote sustainable agriculture); SDG 14 (conserve and sustainably use the oceans, seas, and marine resources for sustainable development); SDG 15 (protect, restore and promote sustainable use of terrestrial ecosystems, sustainably manage forests, combat desertification, halt and reverse land degradation, and halt biodiversity loss) (UNDP, 2021). The management techniques proposed in the book are adaptive in nature, and they are meant to protect and sustain natural and cultural ecosystem services utilized by the tourism industry. To sail through the COVID-19 pandemic and its impacts on tourism ecosystem services, the management of natural and cultural ecosystem services is therefore very important. One assumption is that different destinations may arrive at a different baseline that reflects degraded conditions of the tourism natural and cultural ecosystem services. If in a post-COVID-19 world a new awareness of the vulnerability of the tourism ecosystem is built, this could open the door for improved management and restoration of the natural and cultural tourism ecosystem. Finally, tourism destinations have a part to play in this interconnected world and need to be more responsible and accountable toward the ecosystem services.

23.2 Concluding Insight

This book enables an appreciation of the implications of not reflecting on the way that tourism ecosystem services are currently and/or recently managed. The book shares a positive note on the pandemic as well as a vision of global transformation for the natural and cultural tourism ecosystem services. This book will therefore remind future generations of this pandemic and of the tough moments encountered by the tourism industry. Nonetheless, perceived differently, a good crisis is never to be wasted (Winston Churchill). Contributors, editors, and well-wishers of this book believe that the industry will recover if appropriate management solutions are applied. Successful management and restoration of tourism ecosystem services will continue to depend largely on a coordinated effort among countries, policy decisions on natural and cultural tourism ecosystem services, new management guidelines, harmonized safety, hygiene protocols, and effective communication to help restore consumer confidence. The speech of the WTTC President and CEO during the opening of the Global Summit of the WTTC in Mexico is hopeful: "While 2020 was not the year we expected it to be, it was a year during which people really came together. We saw our sector join forces to support our workforce and the local communities that host us" (WTTC Report, 2021). Likewise, according to the edited volume of Gowreesunkar et al. (2021) on *Tourism Destination Management in a Post-Pandemic Context*, tourism will rebound as it has from previous crises. This hope is also echoed in the statement of Ezekiel Emanuel, Member of the Biden-Harris COVID-19 Advisory Board:

> Travel will explode after the pandemic. People like novelty and changes of scenery. We have all been locked down with the monotony of the same rooms, same walking routine, and inability to see new things. When it is safe to travel, people will go, go, go.
>
> (BBC World News, 2021)

As goes the old adage, every cloud comes with a silver lining, we also found a silver lining inside the pandemic, and altogether as a terrific team, we are glad to offer readers another memorable gift produced during the reign of COVID-19, Delta, and Omicron. Editors of the book wish you all a pleasant reading.

References

BBC World News (2021) What will we be craving in a post-pandemic world?. https://www.bbc.com/worklife/article/20201109

Gowreesunkar, V.G., Maingi, S.W., Roy, H. and Micera, R. (2021), "Conclusion: Rebuilding Tourism Post Pandemic: Some Reflections on Destination Management Solutions!", Gowreesunkar, V.G., Maingi, S.W., Roy, H. and Micera, R. (Ed.), *Tourism Destination Management in a Post-Pandemic Context (Tourism Security-Safety and Post Conflict Destinations)*, Emerald Publishing Limited, Bingley, pp. 329–334. https://doi.org/10.1108/978-1-80071-511-020211023

UNDP (2021) Sustainable Development Goals. United Nations Development Programme. Retrieved from: https://www.undp.org/content/dam/undp/library/corporate/brochure/SDGs_Booklet_Web_En.pdf

WTTC (2021). Opening of WTTC's Global Summit hears praise for private sector's united approach to building recovery. https://wttc.org/News-Article

Index

Note: Page numbers in **bold** refer to tables and page numbers in *italics* refer to figures.

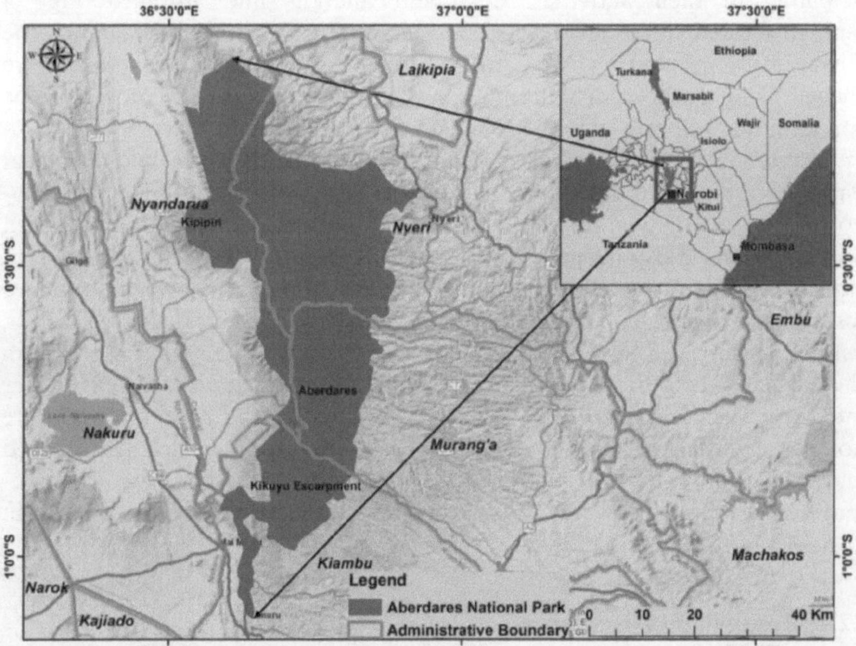

